ETHICAL ISSUES IN HEALTH CARE
ON THE FRONTIERS OF THE TWENTY-FIRST CENTURY

Philosophy and Medicine

VOLUME 65

The titles published in this series are listed at the end of this volume.

ETHICAL ISSUES IN HEALTH CARE ON THE FRONTIERS OF THE TWENTY-FIRST CENTURY

Edited by

STEPHEN WEAR
JAMES J. BONO
GERALD LOGUE
ADRIANNE McEVOY
University at Buffalo, New York, USA

KLUWER ACADEMIC PUBLISHERS
DORDRECHT / BOSTON / LONDON

Library of Congress Cataloging-in-Publication Data

Ethical issues in health care on the frontiers of the twenty-first century / edited by
Stephen Wear ... [et al.].
    p. cm. -- (Philosophy and medicine ; v. 65)
    Includes index.
    ISBN 0-7923-6277-2 (hardbound : alk. paper)
    1. Medical ethics. I. Wear, Stephen. II. Series.

    R724.E7879 2000
    174•.2--dc21

                                                                    00-035406

ISBN 0-7923-6277-2

Published by Kluwer Academic Publishers,
P.O. Box 17, 3300 AA Dordrecht, The Netherlands

Sold and distributed in North, Central and South America
by Kluwer Academic Publishers,
101 Philip Drive, Norwell, MA 02061, U.S.A.

In all other countries, sold and distributed
by Kluwer Academic Publishers, Distribution Center,
P.O. Box 322, 3300 AH Dordrecht, The Netherlands

Printed on acid-free paper

Printed and bound in Great Britain by MPG Books Ltd., Bodmin, Cornwall.

JOHN NAUGHTON, M.D.

Professor of Medicine, Physiology, Rehabilitation Medicine
and Social and Preventive Medicine

Former Dean, School of Medicine and Biomedical Sciences,
and Former Vice President for Clinical Affairs
at the State University of New York at Buffalo

CLINICIAN
    SCHOLAR
        LEADER
            MENTOR
                COLLEAGUE

TABLE OF CONTENTS

PART III: THE PHYSICIAN/PATIENT RELATIONSHIP

JOHN NAUGHTON

PREFACE:
THE CONTINUED ROLE OF BIOMEDICAL ETHICS IN
THE NEXT MILLENNIUM*

The School of Medicine and Biomedical Sciences of the State University
at Buffalo (UB) was honored to sponsor and host the Sesquicentennial
Symposium "Ethics and Values in Health Care and Medicine on the
Frontiers of the Twenty-First Century" on November 15 and 16, 1996. It
represented the closing phase of a celebration that extended from May 11,
1994 through November, 1996. This symposium was unique in that it
highlighted the important changes and adaptations that will continue to
develop in biomedical ethics and it melded characteristics and values
intrinsic to the essence of Buffalo, its medical school and its university.
The Sesquicentennial Planning Committee co-directed by Harold Brody,
M.D., Ph.D., Distinguished Teaching Professor of Anatomical Sciences,
Ronald Batt, M.D., Professor of Gynecology and Obstetrics, and Joyce
Vana, Ph.D., Assistant Professor of Social and Preventive Medicine,
envisioned this symposium as one of its most significant undertakings.
Accordingly, Jerry Logue, M.D., Professor of Medicine and Co-Director
of the Center for Clinical Ethics and Humanities in Health Care together
with Stephen Wear, Ph.D., Associate Professor of Medicine and Co-
Director of the Center, and James Bono, Ph.D., Associate Professor of
History and Medicine, accepted the responsibility for organizing and
implementing the symposium – UB and the Center are forever grateful
for the hard, dedicated work provided by these leaders and their many
associates who made the dream come true. We thank the many speakers
who agreed to participate in the symposium and to recast their important
contributions for this volume.

It is conceivable that not every medical school would choose to
highlight a symposium on biomedical ethics as a part of a
sesquicentennial celebration in the twentieth century. After all, since
Abraham Flexner's Report to the Carnegie Commission in 1910, medical
education and health care have expanded to influence every aspect of
American life. As one reviews the array of activities that UB sponsored
throughout the past two and one-half years, it spanned that wide range of
involvement to include special emphasis and celebration of its leadership

Stephen Wear, James J. Bono, Gerald Logue and Adrianne McEvoy (eds.), Ethical Issues in
Health Care on the Frontiers of the Twenty-First Century, ix–xii.
© 2000 *Kluwer Academic Publishers. Printed in Great Britain.*

and involvement in medical student education, the conduct of biomedical research, its involvement in innovation and change, and the important, rather unique bond between this medical school and its community.

Perhaps more important than any other function, it is the latter that serves to promote a continued commitment to the importance of biomedical ethics to UB. In contrast to most American medical schools, the UB medical school has always existed as a major component of a university. Its origins began in a small community with leadership that foresaw the need for a university. Among the leaders were two important stalwarts of Buffalo who had a significant impact on the formation of UB, and on the nation, Millard Fillmore and James Platt White, M.D. Fillmore was a member of the New York State Legislature and it was he who secured the Charter necessary to form a university; he was UB's first Chancellor and remained so until 1881. Dr. White was a nationally recognized gynecologist and obstetrician who was a founding faculty member, an innovator, and an excellent educator. These two gentlemen were the pacesetters, but it was Fillmore whose dream UB represented, and it was Fillmore who died with the knowledge that only a small part of the dream had been realized. I don't think he despaired, but from his writings, one can detect his disappointment that UB had not fulfilled the more noble social purposes he thought so necessary for Buffalo citizens and for its nourishment, enrichment and survival in the years ahead. What seems ironic is that he couldn't appreciate that he, White and their collaborators had sponsored an institution of higher education rather singularly unique in the developing and expanding American enterprise not only for the nineteenth century, but also for the twentieth century and surprisingly for the approaching new millennium.

What was this unique institution of higher education? While Fillmore had envisioned UB as a university, his personal intellectual appreciation was that of a college designed to prepare a broad range of young people with sufficient knowledge and skills to be able to earn a level of income sufficient to be individually independent and to nourish the community through lifelong service and other contributions. For him, UB had to be "imminently useful." However, the reality was that UB began with one academic unit, the medical school, and throughout his tenure as Chancellor, UB and the medical school remained as a single, undistinguishable unit. Even when it expanded, UB added other professional schools that included Dental Medicine, Nursing, Pharmacy and Law. It was not until 1915, in large part due to Flexner's evaluation of UB's medical school, that UB developed its School of Arts and Sciences, and thus, assumed its place among the other institutions of higher education.

Had Fillmore lived throughout UB's first seventy years, he would probably have been elated by the success of his university, and he should have been satisfied and pleased that UB remained intrinsically bonded to its community while at the same time engrafting the values and standards important to higher education's mission in the region.

UB and its medical school have undergone many challenging transitions since 1846. Included among them were: (1) the completion of an academic campus in the far northeast corner of the City of Buffalo while leaving its medical, dental and law schools firmly situated in the core of downtown Buffalo; (2) the eventual relocation, after the second world war, of the law school to the newer campus in Amherst, and the medical and dental school to the original academic campus; and (3) the merger with the State University of New York System in 1962. Despite these significant transitions, any one of which could have changed the intrinsic integrity of UB and disrupted the bonding between community and university, that did not happen. To this day, the ties between community and academe persist. Fillmore and White should celebrate their success and important contribution to Buffalo and Western New York.

This symposium dedicated to the importance, relevance and need of biomedical ethics is indeed a dedication to the memory and spirit of Fillmore and White and to those who followed them. UB and its community represent an environment in which the changing attitudes and role of biomedical ethics are tested every day. As the content of this program demonstrates, the issues to be faced by health professionals, patients, lawyers, clergy, and payors will continue to expand and to become ever more complex. UB is fortunate to have a Center for Clinical Ethics that provides the necessary leadership and guidance to ensure that open dialogue and critical decision can be attained in a society that will continue to become more diverse with increasing conflict between the Judeo-Christian traditions of biomedical ethics and the ever-evolving and growing need for more populist forms of biomedical ethics.

Once again, Happy Birthday UB, and thanks to all who participated in this event.

* Supported in part by The Robert Wood Johnson Foundation Grant: The Generalist Physician Initiative, No. 024254.

REFERENCES

Flexner, A.: 1910, *The Flexner Report*, Carnegie Foundation for the Advancement of Teaching, Washington, D.C.

Sentz, L.: 1996, 'Medical history in Buffalo: Collected essays, 1846-1996, Introduction,' History of Medicine Collection, State University of New York at Buffalo, xiii.

STEPHEN WEAR, JAMES BONO,
GERALD LOGUE AND ADRIANNE McEVOY

ETHICAL ISSUES IN HEALTH CARE ON THE
FRONTIERS OF THE TWENTY FIRST CENTURY

This volume is in part comprised of papers evolved from presentations
given at a conference held in Buffalo, N.Y. on November 15-16, 1996.
This conference was itself part of a much larger series of conferences and
other activities that marked the sesquicentennial celebration of the
University at Buffalo during that year.

In pulling this volume together, we, the editors, elected to provide
introductory detail to the three specific sections themselves, rather than
provide an extensive introduction here. A few brief notes in this place,
however, are called for to assist the reader to appreciate the overall unity
of focus and concern that guided both the conference organizers, and
those who contributed to it. Some basic historical and futuristic concerns
are at play throughout the volume.

It was, first of all, desired that the conference proceed in light of the
history of the University at Buffalo. This history pivots on the fact that
the university began as a free-standing medical school in 1846, and only
after much development, inspired in part by the desires and concerns of
the medical school faculty themselves, did the school become a full-
fledged university, complete with numerous professional schools, as well
as graduate departments in the arts and letters, and the social sciences.
The private University of Buffalo was then eventually incorporated into
the State University of New York system, of which it is the flagship
campus. In part, then, the hope for this conference was that it would arise
out of an appreciation for the 150 years of community service and
concern that produced the medical school, and made the broader
university its logical completion. We have, in this regard, elected to
include, as a Preface to this volume, the remarks of Dr. John Naughton,
the then current Dean of the School of Medicine, regarding the history of
the school, the broader concerns and contributions of its students and
faculty over this period, and how a conference on ethics and values in
medicine was seen as a necessary, core component of its sesquicentennial
celebrations.

As any such anniversary celebration should be both backward and
forward looking, so was the development of this conference, the selection

*Stephen Wear, James J. Bono, Gerald Logue and Adrianne McEvoy (eds.), Ethical Issues in
Health Care on the Frontiers of the Twenty-First Century, xiii–xvi.*
© 2000 *Kluwer Academic Publishers. Printed in Great Britain.*

of its topics and speakers, and the charge to them regarding their presentations. Shored up by the further fact that we are entering a new millenium, the organizing committee for this conference decided early on that its topics should be chosen on the basis of the pivotal character they will have as we move into the new millenium, and that the contributions should have a decidedly futuristic character, i.e. should attempt to identify what sort of choices or dilemmas these issues will present us in the future, and what sorts of resolutions might be made to them.

Though one might argue about the selections, the three main topic areas in this edition, viz. the dilemma of funding health care, the human genome project, and the physician/patient relationship, were actually chosen early on, and for reasons that remained cogent and contributory to the conference, and the development of this volume.

The issue of the dilemma of funding health care is clearly a natural choice for such a futuristic conference, not just intrinsically, but in the fact that whatever resolutions are made to this dilemma, they will clearly and profoundly effect most other bioethical issues, such as the level of funding and access that citizens will have to the findings of the human genome project, or the time available for, and resultant character of, the physician/patient relationship. No matter what else we focused on, it seemed, from the beginning, that the clear movement of health care into some sort of managed practice had to be canvassed for its ethical issues and options. H. Tristram Engelhardt, Jr., M.D., Ph.D., Lawrence McCullough, Ph.D., and E. Haavi Morreim, Ph.D. were recruited to address this area of concern.

The human genome project, for its part, commended itself to the conference organizers as that area of medical science and research that was likely to be the most innovative and profound in its effects on health care in the future. As we move from a medicine focused on humours or systems, to one focused on pathogens or cells, and now, with the genome project, to one focused on the role of our genetic structures in our understanding and approach to wellness and disease, it seems that we are in the midst of a fundamental paradigm shift, and are obliged to get as clear as we can about it upfront. Eric Juengst, Ph.D., Dorothy Nelkin, Ph.D., and Diane Paul, Ph.D. were our major presenters in this area, and the section is completed by an integrating commentary supplied by Jonathan Moreno, Ph.D.

Paradigm shifts, and societal issues of health care access and provision aside, however, it was finally felt that we should also focus on what will remain the basic unit of health care provision, viz. the physician/patient relationship. Particularly with the advent of managed care, enormous concern has arisen regarding how to avoid the conflicts of interest that any "managed" system may fall prey to, and how to protect and foster the fiduciary character that this relationship must, according to most commentators, somehow retain. Kathryn Montgomery, Ph.D., Julie Rothstein, M.D., and Howard Brody, M.D., Ph.D. supplied the main contributions in this area, with a general commentary added by Scott DeVito, Ph.D.

In sum, our three chosen topics seem, at this point, somehow inevitable to us, as if we could have chosen no others. Correctly or not, this edition thus ends up focusing on (1) the most fundamental unit of care, i.e. the physician/patient relationship, (2) what may well be the area of medical science and research that most radically effects the nature of that care in the foreseeable future, i.e. the human genome project, and (3) the societal options for financing and distributing that care, i.e. the dilemma of funding health care.

There is one other basic factor that needs to be identified as a basic contributor to the nature of the individual contributions in this volume, as well as the further significance that we take it to have. In a time when the concerns of multiculturalism, and the recognition of human diversity, are basic touchstones of any attempt to understand and evaluate our world, any inquiry into ethics and values must grapple with the fact that there seems to be little agreement amongst us about many of the basic dilemmas we face, or even, more generally, regarding how we should approach them. For all of our scientific and technological progress, particularly in the last half century, less and less do we seem to have any real agreement as to the ethics and values that should guide their usage. It rather seems, in fact, that we enter the new millenium encased within a sort of philosophic Tower of Babel.

This more philosophic problem repeatedly arose as we attempted to specify which topics we would focus on, and how they might be treated. To meet it, we asked Professor H. Tristram Engelhardt, Jr., a holder of doctoral degrees in both philosophy and medicine, to deliver the keynote address to the conference on precisely this issue. The reader is directed especially to this contribution, following immediately after this introduction, for a sense of the more basic philosophical dilemma that we all face, and that each contributor was charged to be mindful of. Variously expressed, Professor Engelhardt's charge to us all was to be honest about and mindful of the profound diversity of views and approaches that people take to any issue within the realm of ethics and values. What was thus requested in all of these contributions was more than just one more scholar's "spin" on what a given issue amounts to, or how it might be coherently resolved, but rather the provision of a deeper sense of the nature of these issues that we might all agree to, and resolutions that leave ample space for different views of what human flourishing amounts to, and the place of medicine within such views. We believe that the reader will find all of our contributors were mindful of this further burden in their specific discussions, and suggest that the reader first also embrace the challenge that Professor Engelhardt set before us all.

Beyond these generic introductory remarks, the reader is directed to the specific editors' introductions to each of the three sections.

University at Buffalo
Buffalo, New York
August, 1999

H. TRISTRAM ENGELHARDT, JR.

KEYNOTE ADDRESS

BIOETHICS AT THE END OF THE MILLENNIUM: FASHIONING HEALTH-CARE POLICY IN THE ABSENCE OF A MORAL CONSENSUS

I. REVISITING THE ENLIGHTENMENT DREAM AND WAKING TO REALITY

In the United States and later in Europe and then across the world, contemporary secular bioethics developed over against a cluster of diverse religious examinations of moral probity, as well as professional statements regarding appropriate medical practice. In the 1950s and 1960s, religious reflections on health care policy produced a considerable literature that went far beyond the explorations of medical ethics and medical deontology that had taken shape over the previous two centuries (Gregory, 1770; Percival, 1803; Saundby, 1902). There was a recognition of the challenges posed by new knowledge and new technology. Much of the growth in religious reflections was achieved by Protestant thinkers (Fletcher, 1954; Hauerwas, 1974 and 1977; Ramsey, 1970a and 1970b; Smith, 1970; Vaux, 1977). However, even the Roman Catholics, who had a considerable history in such undertakings (Bonnar, 1944; Coppens, 1897; Cronin, 1958; Finney, 1922; La Rochelle and Fink, 1944; McFadden, 1946a and 1946b), came to address health care and the biomedical technologies with new vigor (Ficcara, 1951; Healy, 1956; Kelly, 1979; Kelly, 1958; McCormick, 1973). It was during this period that one of the classic studies of Jewish medical ethics was published in English (Jakobowits, 1959). Medical ethics both professional and religious acquired a new identity within the emerging rubric of bioethics.[1]

The new field of bioethics did not disclose a uniform theoretical foundation or understanding of moral probity and appropriate moral conduct. This could at first be attributed to the diverse religious roots from which contemporary bioethics had drawn its first energies. It was taken for granted that religious views would be diverse, ecumenical aspirations to the contrary notwithstanding (Engelhardt, 1995b). As a consequence, the remedy for the diversity of voices characterizing

Stephen Wear, James J. Bono, Gerald Logue and Adrianne McEvoy (eds.), Ethical Issues in Health Care on the Frontiers of the Twenty-First Century, 1–16.
© 2000 Kluwer Academic Publishers. Printed in Great Britain.

bioethics was an appeal to secular reason. Even if religious bioethics drew from incompatible theoretical foundations and gave voice to a diversity of views regarding abortion, third-party-assisted reproduction, organ transplantation, suffering, and death, it was assumed that philosophy could bring unity.

Bioethics thus sought the unity that had been promised by the modern philosophical project. The modern philosophical project, which had taken shape in reaction to the moral diversity engendered by the Reformation, was further fortified by the Enlightenment hope to establish a universal secular society. Even if religion was not one, reason and humanity were presumed to be one: a rational examination of human sympathies, sensibilities, and reason should lead to a secular culture that could unite all.

The Enlightenment search for the unity promised by a philosophically grounded morality was undertaken anew by bioethics in the 1970s and 1980s. This search for a rationally unified bioethics was rooted in more than theoretical interests. If health care law and policy were not to have only the authority of force, and if society could not agree regarding the policies authorized by God, then an appeal to reason might suffice. In the face of a diversity of religious views and a culture that was significantly disunited on important issues such as the morality of abortion and euthanasia, an appeal to reason offered the promise of securing a deeper unity behind the apparent plurality of moral visions. If one could secure a rationally grounded bioethics, then one could distinguish justifiable from unjustifiable health care policies in terms of that which could be shown to be rational or irrational. Moreover, one would have the authority of reason to impose rational health care policy. Those who resisted could be dismissed or constrained as irrational. The rationally justified policy to which they objected could be shown to be congenial with the rational commitments that should bind all. This in turn meant that at least in principle one could show that, since all are united by this reason, all are members of the community of rational persons bound by reason, a community transcending the apparent diversity of religious commitments. The bottom line was that, if such a project were successful, bioethical law and health care policy could find a justification that should in principle appeal to all.

This search for a unifying justification for bioethical pronouncements became especially important as governments turned in response to new technological developments to create policy bearing on human reproduction, birth, suffering, medical treatment, and death. In the United States, the first major endeavor in this vein was undertaken by the National Commission for the Protection of Human Subjects of Biomedical and Behavioral Research (National Commission, 1975, 1976, 1977a, 1977b, 1977c, 1978a). A significant body of administrative law and biopolitics emerged in areas where there seemed to be less than moral unanimity. A justification was needed to establish and enforce such policy. In the case of the National Commission, this was sought in a set of middle-level principles designed to reach across theoretical divisions (National Commission, 1978b). These principles did not resolve controversies regarding the use of fetuses in research, nor did they prove able to bring closure to debates about health care allocations. Over the last two decades, matters have not become easier. The debates regarding abortion and third-party-assisted reproduction, which commanded the attention of many in the discussions of the 1970's, have not gone away. If anything, the debates are nested in opposing moral communities which are ever more divided one from another. If anything, the positions taken have hardened along lines that show themselves to be frontiers in the cultural or moral wars. To these disputes, debates regarding health care reform and euthanasia have emerged as both divisive and further reflective of the fundamentally different perspectives from which bioethics, biomedical law, and health care policy can be regarded.

The plurality of religious moral views has been recapitulated by a plurality of secular moral views. Secular bioethics has been shown to be as much of a plural noun as religious bioethics. It is fractured by a diversity of views about how to balance concerns with freedom and equality, liberty and security, not to mention a plurality of understandings of the morality of abortion and physician-assisted suicide. More fundamentally, this plurality of visions is in principle irreconcilable. One cannot choose among policies on the basis of which has the best consequences without already knowing how to compare consequences. One cannot establish that set of laws that best maximizes preferences unless one already knows how to compare rational preferences with irrational preferences, as well as how to discount preferences over time. So, too, one cannot appeal to intuitions without already knowing which intuitions should govern. Nor will an invocation of a reflective equilibrium between moral intuition and moral rules, or among moral intuitions, aid unless one already knows how to guide one's reflection in discounting certain intuitions and giving greater weight to others, or unless one begs the question by presupposing that there will be only one equilibrium point. Appeals to disinterested observers, hypothetical choosers, and hypothetical contractors require that the observers, choosers, or contractors already be fitted out with a particular thin theory of the good or a particular set of moral sentiments in order to make one choice rather than another. Decision-theoretic appeals to rational decision-making also require a weighting of outcomes and a ruling out of absolute or transcendent goals. With any attempt to deliver a secular morally content-full moral vision as canonical, one must either beg the question or engage in an infinite regress (Engelhardt, 1996).

In sum, the attempt to secure in reason a foundation for a content-full bioethics fails. Nor will it help to appeal to sympathies, caring, commitment, or community. Once again, one will need to be able to make a canonical choice as to which sympathy, sense of caring, commitment, or community should have priority. Nor will appeals to humanism and medical humanism, such as they arose in the 1960s and 1970s, bridge the gulf (Pellegrino and McElhinney, 1979, 1982). Here, again, one encounters diversity. In the case of an unwanted pregnancy, for example, should one have sympathy for the woman or the fetus? In the case of intractable pain, should one show care by providing aid in suicide or protect against self-murder? There is also not one but many senses of humanism, each structured around a different understanding of the normatively human (Engelhardt, 1991). To offer a recasting of the title of MacIntyre's volume (MacIntyre, 1988), just as one must ask 'Whose justice? Which rationality?' one must as well ask: Whose caring? Whose sympathy? Whose moral community? Whose medical humanism? Rather than finding a canonical ethics to apply in bioethics, one finds instead numerous competing ethics grounded in diverse understandings of justice, fairness, moral rationality, caring, and sympathy. This confrontation with foundational moral diversity embeds bioethics not simply in a sociological post-modernity in which there is not one canonical or universal moral narrative (Lyotard, 1984), but one in which the plurality of narratives is unresolvable.

At the end of the millennium, we appear also to be at the end of the modern philosophical and Enlightenment hope unambiguously to discover in or through reason the canonical human sympathies and sentiments that should guide us. There is no canonical Ariadne's thread of sound rational argument to lead us to the morally canonical policy to which all should accede. One must note that the differences that separate us as moral strangers[2] need not be those that characterize us as alien to each other. It is enough that we give a different ranking to important goods, such as liberty, equality, prosperity, and security. Depending on how one will rank such goods, one will endorse either the moral assumptions of a democratic Cambridge, Massachusetts, or an authoritarian Singapore. Moral diversity will emerge from even a different ordering of important priorities. To take this predicament seriously is to face the project of rethinking secular ethics and as a consequence secular bioethics and health care policy. Such a project will at the very least require acknowledging moral diversity and seeking grounds for common policy to establish moral authority without undercutting moral diversity.

II. PURSUING THE MIRAGE OF CONSENSUS

Faced with these difficulties, contemporary bioethics did not turn to the project of developing a bioethics that acknowledges moral diversity. Instead, it endeavored to set it aside by seeking sources of authority that could bind despite diversity. When it became clear that neither ethics nor bioethics could be grounded in an uncontroversial account of moral probity, accounts were developed to reinterpret and explain away this diversity. In the field of bioethics two prominent strategies have been invoked to discount the significance of apparent moral diversity: appeals to middle-level principles and to casuistry.

The appeal to middle-level principles that lie between foundational theoretical principles and rules for particular cases has had its best articulation in the work of Beauchamp and Childress (1979). The proposal is that such principles can guide in the face of theoretical, moral disagreements. Although one author is a confessed rule utilitarian and the other a deontologist, they both found agreement in the application of the principles of autonomy, beneficence, non-maleficence, and justice to the analysis and resolution of controversial bioethical cases. To demonstrate that there is in fact a common morality, they both elaborated the principles and then applied them to cases regarding which they and like-minded bioethicists found agreement in the practical decisions to be made. They could then explain this communality of agreement while acknowledging a diversity of theoretical accounts through claiming that there is indeed a common morality. It is this which allowed the principles to be applied in the resolution of particular cases.

While theoretical diversity was accepted, moral diversity was denied. This denial was supported by the success of the principles in analyzing actual cases. However, the success presupposed that those who came to apply the principles already shared a common morality. Difficulties arise if a Nozickian libertarian and a Rawlsian social democrat make appeals to the principle of justice to resolve a controversy regarding health care allocation; all they discover is how much they disagree (Nozick, 1974; Rawls, 1971). They are not able by an appeal to the principle of justice to disclose a harmony in their analysis of cases. The same difficulty can be encountered with the other principles as well. A libertarian's application of the principle of autonomy to practical cases will in many circumstances lead to different conclusions than those of a rule utilitarian of Tom Beauchamp's persuasion. Even a Kantian's invocation of the principle of autonomy will produce different resolutions for truth-telling cases than would be embraced by James Childress, in that Kant would not allow lying, even to save the life of the innocent.

How could this state of affairs have come about? Why does the invocation of middle-level principles work for some, but not others? One would expect the middle-level principles to function as they do for Beauchamp and Childress if the two authors had begun the project of writing the principles of bioethics sharing one particular common morality (Beauchamp and Childress, 1979). This morality need not have been common to all or most members of society. It would be sufficient if Beauchamp and Childress shared the same general political views, ideology, sentiments, and moral inclinations. From such a community of moral sentiments, they could then proceed to reconstruct the general contours of this morality around the four middle-level principles. One of the pair could then deploy their four principles to display implications of their common moral views by engaging consequentialist concerns placed within these four principles. The other could proceed to use the same principles, all the while implicitly invoking deontological concerns. That the two would be in agreement with regard to the use of their middle-level principles in the analysis of particular cases would not be amazing if they had both succeeded in giving apt theoretical reconstructions of their common point of departure. Though there would be a diversity of theoretical languages and theoretical foundations, one would expect a concord regarding middle-level principles and cases if the theoretical reconstruction had been successfully directed to the morality common to Beauchamp and Childress.

Matters are quite different when individuals with different moral visions theoretically reconstruct their diverse moral understandings and then turn to the application of middle-level principles. In such circumstances, middle-level principles divide rather than unite, as in the case of the Nozickian and the socialist both invoking the principle of justice. Faced with the persistent challenge of dispelling moral diversity, one might then take the step of not appealing to principles at all, but rather instead turn directly to the analysis of cases (Jonsen and Toulmin, 1988). One might claim that the discipline of examining cases will by itself lead to a common casuistry which can unite those sundered by differences in theoretical commitments or divergent moral sentiments, intuitions, or sensibilities. The problems of post-modernity recapitulate themselves here as well, appeals to cases to the contrary notwithstanding. One must already know which cases are paradigmatic, who is an authority in their analysis and explication, as well as who is in authority to resolve controversies. The casuistry of the Roman Catholic church functioned precisely because there was a common understanding of exemplar cases, as well as an agreement regarding foundational metaphysical assumptions.[3] In addition, one knew who were authorities as casuists, as well as who was in authority to adjudicate cases. The crisis of contemporary post-traditional, post-modern morality is that we are not in agreement about who is in authority or even regarding who is an authority with respect to settling moral controversies.

There are many other variations on the theme of appealing to middle-level principles or casuistry. For example, one can attempt to find one's way through the maze of contrary moral intuitions, sentiments, principles, and claims by attempting to balance and judge the application of different moral appeals to the analysis of particular cases. Such refined and nuanced appeals to intuitions regarding the applicability of different moral appeals will either beg the question or reveal that a particular moral perspective has been used in the establishment of a particular moral equilibrium or a particular application of competing moral appeals. Then, of course, one will need to justify that moral perspective.

Others may attempt to define themselves out of these problems by stipulating in advance what will count as a canonical cluster of agreements. Thus, one finds Rawls appealing to that "overlapping consensus" likely to persist and gain adherents in a "more or less just constitutional democratic society" (Rawls, 1985). In this fashion, one can then subtly reimport content-full Enlightenment moral presuppositions in identifying a particular cluster of opinions constituting an over-lapping consensus in terms of what will count as a "just constitutional democratic society," thus begging the question (Rawls, 1993).

An even bolder strategy is simply to affirm the contingent character of one's own morality without expecting it to find any grounding in reason or human nature. Thus, one finds Richard Rorty claiming that

> We can keep the notion of "morality" just insofar as we can cease to think of morality as the voice of the divine part of ourselves and instead think of it as the voice of ourselves as members of a community, speakers of a common language. We can keep the morality–prudence distinction if we think of it not as the difference between an appeal to the unconditioned and an appeal to the conditioned but as the difference between an appeal to the interests of our community and the appeal to our own, possibly conflicting, private interests. The importance of this shift is that it makes it impossible to ask the question: "Is ours a moral society?" (Rorty, 1989, p. 59).

The difficulty is that in a multi-cultural and morally diverse society in which there are disagreements regarding the nature of justice and fairness, not to mention the significance of birth, reproduction, and death, it will not suffice to endorse the contingently held moral commitments of those who are "'twentieth-century liberals' or 'we heirs to the historical contingencies which have created more and more cosmopolitan, more and more democratic political institutions'" (Rorty, 1989, p. 196).

In the face of moral diversity and the difficulties of spanning this diversity by invoking middle-level principles, casuistic analysis, the balancing of moral appeals, the achievement of moral equilibria, the endorsement of a particular overlapping consensus, or the canonization by fiat of a particular contingent morality, many have simply proceeded to invoke the concept of moral consensus as if it existed when all of the evidence is to the contrary. In such appeals there is often a quiet desperation born of the recognition of the need to provide moral authority for the coercive power of law and public policy. The concept of moral consensus appears to function in a quasi-secular theological vein to legitimate particular policies and laws (Bayertz, 1994).

Serious issues in public policy disclose that controversies regarding the moral issues at stake in bioethics are substantive. One might consider, for example, the recent debate regarding health care reform (White House, 1993). Actual bioethical investigations of the moral issues reveal significant areas of disagreement (Engelhardt, 1994b; Strosberg, 1992). Where consensus is obtained, this is often secured by the force of circumstance and political pressure in which agreement is, for example, encouraged by excluding certain key moral considerations, setting deadlines, and providing uncomfortable circumstances for deliberation (Dubler, 1993; Secundy, 1994). Or more adroitly, consensus is manufactured by impaneling a commission and carefully choosing the members, the agenda, and appointing a powerful chair to keep discussions on track for the goals one wishes to achieve. One does not impanel persons with truly diverse moral understandings. One need only imagine how a bioethics commission would function that incorporated real moral diversity and included libertarians, socialists, conservatives, feminists, atheists, and the devoutly believing. For instance, imagine an ethics commission whose members include Jesse Jackson, Jesse Helms, Bella Abzug, and Mother Theresa.

III. LIVING WITH MORAL STRANGERS: CONSENT, MARKETS, AND LIMITED DEMOCRACIES

If individuals do not share a common, taken-for-granted moral culture, and in addition fail to convert to the understanding of God's moral purposes for humans (Engelhardt, 1995a), and if sound rational argument does not disclose a canonical, content-full secular morality, individuals can still draw moral authority from common agreement. Under such circumstances, the authority of collaboration, including that collaboration found in law and public policy, is not the authority of God or reason, but the consent of the participants. Given that hypothetical consent accounts cannot proceed without begging the question, consent must be actual consent among actual persons, or at least it must be that consent implicit in the practice of resolving issues involving innocent persons only with their agreement. Authority must be drawn from the collaboration of those who join together in a common venture. The idea is that of a practice into which one can enter so as to discover and understand a web of moral authority as long as one uses others only with their consent and with respect to which one cannot protest in general secular terms if visited by punitive or defensive force in response to one's having used others without their consent. The practice of appealing to the authorization of those who participate in a practice invokes the transcendental possibility for an intersubjective moral framework that can bind individuals in the face of moral diversity.

Just as metaphysical strangers can enter into the transcendental practice of resolving controversies regarding the empirical world by appeal to mutually agreed-upon ceteris paribus conditions and ways of testing appearances, while abandoning the hope of objectivity as correspondence with the object of knowledge as it is in itself and settling instead for objectivity as intersubjectivity, so, too, moral strangers can fashion a common secular world of intersubjective moral authority. A common endeavor becomes a *res publica* whose authority is the authority of those who participate. Moreover, the authority of such common endeavors, since it is limited by the consent of those who enter into them, will be characterized by circumscriptions that can best be characterized as secular moral rights to privacy.

The point is that one can resolve controversies by the authority of force, taken-for-granted customs, the will of God, sound rational argument, or consent. A resolution other than by unanimous consent requires a true knowing of moral probity, and this is what is blocked by moral diversity. In the face of moral diversity, there will not be agreement regarding taken-for-granted customs and the will of God. Nor will appeals to sound rational argument succeed in the absence of commonly acknowledged moral premises, as well as rules of evidence and inference (Engelhardt and Caplan, 1987). As a consequence, those practices in morality that can be most easily justified in the face of the diversity of our post-traditional, post-modern circumstances are those that one in fact already finds functioning quite well under just those conditions, namely, free and informed consent, the market, contracts, and limited democracies.

None of these practices requires an agreement requiring a common, content-full understanding of the good and the right. It is enough that individuals by themselves or in different communities approach each other from their own wellsprings of concerns and interests, however diverse, so as to collaborate in a web of agreements, however sparse. As with all secular morality, there will remain a disjunction between justification and motivation. The general justification for this secular moral authority will lie in the common agreement of those who participate. The motivation for entering into such practices will need to be found in the particular lives and concerns of particular individuals. The motivations will be diverse.

With recognizing moral diversity and the remaining secularly justifiable possibilities for common moral authority there is a confrontation with the secular moral obligation to acquiesce in a diversity of approaches to morally significant issues. Given real moral diversity, there will be a need for a robust toleration, which does not require accepting other viewpoints but forgoing force in constraining them, insofar as they are peaceable. For all its difficulties, *Roe vs. Wade* provided in its notion of rights to privacy the recognition of the limits of state authority, which leave people at liberty to do that which many will recognize as highly immoral. The recognition of the failure of sound rational argument to provide secular moral authority for health care policy and the reliance instead on the consent of the governed unavoidably supports rights to privacy as a moral limit on public authority.

The result of this is that rights to privacy should encompass not simply issues of abortion, reproductive freedom, physician-assisted suicide, and euthanasia, but they must extend as well to economic issues bearing on the organization of the use of resources for health care. It is in this second area that there has been a failure to take seriously rights to privacy or liberty interests, if one were to rebaptize rights to privacy under a new name but maintain their full force (*Cruzan*, 1990). In part, this may be the case because those influential in framing public policy are more at peace with acknowledging liberty in areas of reproduction and death and less willing to accept such liberty when it bears on economic matters. Yet, nearly all over the world one finds multi-tiered health care systems. If moral diversity is to have its proper recognition, one will need to devise strategies that acquiesce in the expression of moral and economic diversity in the fashioning of policies that bear most intimately on our lives and regarding which we have deep and important disagreements.

The recognition of moral diversity involves acknowledging as well that community, society, and state are not one. Large-scale, morally diverse societies encompass diverse moral communities. This moral diversity can be encompassed with secular moral authority only if one does not confuse society and community and then recognizes the state as a neutral guarantor of the peaceable collaboration of its citizens (Engelhardt, 1994a). In such circumstances, the recognition of the difference between community and society should lead to strategies for moral strangers to collaborate peaceably. In part, this may be achievable by invoking the moral equivalent of constitutional rights to privacy. These, for example, may be expressed in voucher systems for health care welfare rights that allow individuals to join in communities that respect their moral commitments.

IV. BIOETHICS IN THE 21ST CENTURY: FROM CULTURAL WARS TO LIVING WITH MORAL DIVERSITY

Moral diversity will not go away. There are substantive, secular and religious disagreements regarding nearly every area upon which health care touches. From reproduction and abortion to physician-assisted suicide and euthanasia, the disagreements are substantial. Moreover, they are likely only to become more emphatically stated as more issues come to divide moral communities. Though *Roe vs. Wade* opened an era of deep and profound controversy regarding matters at the beginning of life, even further divisions will become apparent as debates focus on physician-assisted suicide and euthanasia. The concerns that brought the 1996 *Compassion in Dying* and *Quill* cases and the debates regarding appropriate health care reform are also unlikely to disappear, even if these cases or similar cases are not in the near future upheld by future holdings of the Supreme Court. The moral lines of differences are also likely to become even more stark as one breaks through what can at best be described as the American false consciousness or ideology regarding health care (Marx, 1960), which has presumed that one could give to all Americans equal care, the best of care, allow physician-patient choice, and still have cost containment. The future will inevitably be one in which a health care system will emerge which will explicitly treat citizens unequally, guarantee adequate but not the best of care, in most cases restrict choices, and thus contain costs.

Nor will there be the possibility to establish one canonical vision of the physician-patient relationship. In the absence of a single normative ranking of values, patients may be invited to enter into (and thereby implicitly to consent to) arrangements that vary in their paternalistic character and the salience they give to traditional lines of family authority in medical decision-making. The diversity of moral visions will be even more manifest in discussions of germline genetic engineering, the genetic enhancement of human abilities, and the engagement of reproductive techniques that depart radically from traditional forms of reproduction such as cloning or the transfer of the nucleus of the ovum of one woman to the enucleated ovum of a second. In principle, moral judgments in these matters (i.e., concerns for more than minimizing any untoward side effects) depend on very content-rich views of proper human self-dominion, human evolution, and human reproduction. Definitive guidance in these matters will likely require religious guidance, about which there will not be secular moral agreement (Engelhardt, 2000).

At stake in the controversies we will face are deep and fundamentally different understandings of the sanctity of life, equality, and fairness. These differences show no promise in fact or in principle of disappearing. We should prepare to meet a future in which moral diversity can peaceably persist. This will require acquiescing in differences without suppressing them by force. Secular bioethics can help contribute by devising strategies for diverse moral communities peaceably to use health care resources without imposing on each other conflicting, content-full views of moral probity. At the very least, this will require something approximating rights to privacy and social devices such as voucher systems to support health care costs, through which persons can with their resources and collaborating others pursue their own moral vision of proper health care. The last will also require accepting the circumstance that there will always be multiple tiers of health care. This will at least require understanding that matters of toleration must be accepted not only by those who have strong views regarding right-to-life issues, but also by those who have strong views regarding equality, justice, and fairness. Since there will not be a resolution of our moral disagreements regarding sanctity of life nor substantive understandings of justice and fairness, we are best advised to proceed in general secular public policy not by invoking justice, fairness, and rights to privacy, but rather by engaging models of basic insurance against losses at the natural and social lotteries. Such strategies will require citizens to acknowledge that, when they meet as moral strangers in a secular society, they do not meet with the authority of God, nor even with an authority of reason that can be a surrogate for God.

Baylor College of Medicine and Rice University
Houston, Texas

NOTES

[1]   The term bioethics may very well have first been coined by Van Rensselaer Potter in 1971
      (Potter, 1971). Its contemporary currency was established by the first edition of *The
      Encyclopedia of Bioethics* (Reich, 1978).
[2]   "Moral stranger" is used to identify the circumstance that persons meet without the necessary
      common value premises, rules of evidence, and rules of inference, so as to allow the
      resolution of the controversies by sound rational argument, and without the recognition of a
      common moral authority. See Engelhardt, 1996.
[3]   I am significantly in debt to Kevin Wm. Wildes, S.J. for his discussions and for his *Moral
      Acquaintances* (Wildes, 2000).

BIBLIOGRAPHY

Bayertz, K. (ed): 1994, *The Concept of Moral Consensus*, Kluwer, Dordrecht.
Beauchamp, T.L. and Childress, J.F.: 1979, *Principles of Biomedical Ethics*, Oxford University
    Press, New York.
Bonnar, A.: 1944, *The Catholic Doctor*, Burns Oates & Washbourne, London.
Compassion in Dying v. Washington, 49 F.3d 586 (9th Cir. 1995).
Coppens, C.: 1897, *Moral Principles and Medical Practice*, 3rd ed, Benziger Brothers, New
    York.
Cronin, D.A.: 1958, *The Moral Law in Regard to the Ordinary and Extraordinary Means of
    Conserving Life*, Typis Pontificiae Universitatis Gregorianiae, Rome.
Cruzan vs. Director, Missouri Department of Health, 497 U.S. 261 (1990).
Dubler, N.N.: 1993, 'Working on the Clinton Administration's Health Care Reform Task Force,'
    Kennedy Institute of Ethics Journal 3 (December), pp. 421-431.
Engelhardt H.T., Jr. and Caplan, A.L. (eds.): 1987, *Scientific Controversies*, Cambridge
    University Press, New York.
Engelhardt, H.T., Jr.: 1991, *Bioethics and Secular Humanism: The Search for a Common
    Morality*, Trinity Press International, Philadelphia.
Engelhardt, H.T., Jr.: 1994a, 'Sittlichkeit and post-modernity: An Hegelian reconsideration of
    the state,' in H.T. Engelhardt, Jr. and T. Pinkard (eds.), *Hegel Reconsidered: Beyond
    Metaphysics and the Authoritarian State*, Kluwer, Dordrecht, pp. 211-224.
Engelhardt, H.T., Jr.: 1994b, 'Health care reform: A study in moral malfeasance,' *Journal of
    Medicine and Philosophy* 19 (October), 501-516.
Engelhardt, H.T., Jr.: 1995a, 'Moral content, tradition, and grace: Rethinking the possibility of a
    Christian bioethics,' *Christian Bioethics* 1 (March), 29-47.
Engelhardt, H.T., Jr.: 1995b, 'Christian bioethics as non-ecumenical,' *Christian Bioethics* 1
    (September), 182-199.
Engelhardt, H.T., Jr: 1996, *The Foundations of Bioethics*, 2nd ed., Oxford University Press, New
    York.
Engelhardt, H.T., Jr.: 2000, *The Foundations of Christian Bioethics*, Swets and Zeitlinger, Lisse.
Ficarra, B.J.: 1951, *Newer Ethical Problems in Medicine and Surgery*, Newman Press,
    Westminster, Md.
Finney, P.A.: 1922, *Moral Problems in Hospital Practice*, 2nd ed., Herder, St. Louis.

Fletcher, J.: 1954, *Morals and Medicine*, Princeton University Press, Princeton, New Jersey.

Gregory, D.: 1770, *Observations on the Duties and Offices of a Physician*, Strahan, London.

Hauerwas, S.: 1974, *Vision and Virtue*, Fides Publishers, Notre Dame, Indiana.

Hauerwas, S.: 1977, *Truthfulness and Tragedy*, University of Notre Dame Press, Notre Dame, Indiana.

Healy, E.F.: 1956, *Medical Ethics*, Loyola University Press, Chicago.

Jakobowits, I.: 1959, *Jewish Medical Ethics*, Bloch, New York.

Jonsen, A. and Toulmin, S.: 1988, *The Abuse of Casuistry*, University of California Press, Berkeley.

Kelly, D.F.: 1979, *The Emergence of Roman Catholic Medical Ethics in North America*, Edwin Mellen Press, New York.

Kelly, G.: 1958, *Medico-Moral Problems*, Catholic Hospital Association, St. Louis.

La Rochelle, S.A. and Fink, C.T.: 1944, *Handbook of Medical Ethics*, M.E. Poupore (trans.), Newman Book Shop, Westminster, Md.

Lyotard, J.F.: 1984, *The Postmodern Condition*, G. Bennington and B. Massumi (trans.), Manchester University Press, Manchester, p. 37.

Marx, K. and Engels, F: 1960, *The German Ideology*, International Publishers, New York.

McCormick, R.: 1973, *Ambiguity in Moral Choice*, Marquette University Press, Milwaukee.

McFadden, C.J.: 1946a , *Medical Ethics*, Davis, Philadelphia.

McFadden, C.J: 1946b, *Medical Ethics for Nurses*, Davis, Philadelphia.

MacIntyre, A.: 1988, *Whose Justice? Which Rationality?* University of Notre Dame Press, Notre Dame, Indiana.

National Commission for the Protection of Human Subjects of Biomedical and Behavioral Research: 1975, *Research on the Fetus*, HEW, Washington, DC.

National Commission for the Protection of Human Subjects of Biomedical and Behavioral Research: 1976, *Research Involving Prisoners*, HEW, Washington, DC.

National Commission for the Protection of Human Subjects of Biomedical and Behavioral Research: 1977a, *Report and Recommendations on Psychosurgery*, HEW, Washington, DC.

National Commission for the Protection of Human Subjects of Biomedical and Behavioral Research: 1977b, *Psychosurgery (Appendix)*, HEW, Washington, DC.

National Commission for the Protection of Human Subjects of Biomedical and Behavioral Research: 1977c, *Research Involving Children*, HEW, Washington, DC.

National Commission for the Protection of Human Subjects of Biomedical and Behavioral Research: 1978a, *Research Involving Those Institutionalized as Mentally Infirm*, HEW, Washington, DC.

National Commission for the Protection of Human Subjects of Biomedical and Behavioral Research: 1978b, *The Belmont Report: Ethical Principles and Guidelines for the Protection of Human Subjects of Research*, HEW, Washington, DC.

Nozick, R.: 1974, *Anarchy, State and Utopia*, Basic Books, New York.

Pellegrino, E.D.: 1979, *Humanism and the Physician*, University of Tennessee Press, Knoxville, p. 17.

Pellegrino, E.D. and McElhinney, T.: 1982, *Teaching Ethics, the Humanities, and Human Values in Medical Schools*, Society for Health and Human Values, Washington, D.C.

Percival, T.: 1803, *Medical Ethics*, Russell, Manchester.

Potter, V.R.: 1971, *Bioethics, Bridge to the Future*, Prentice-Hall, Englewood Cliffs, N.J.

Quill vs. Vacco, 80 F.3d 716 (2d Cir. 1996).

Ramsey, P.: 1970a, *Fabricated Man*, Yale University Press, New Haven.

Ramsey, P.: 1970b, *The Patient as Person*, Yale University Press, New Haven.

Rawls, J.: 1971, *A Theory of Justice*, Harvard University Press, Cambridge, Mass.

Rawls, J.: 1985, 'Justice as fairness: Political not metaphysical,' *Philosophy and Public Affairs* 14 (Summer), 223-251.

Rawls, J.: 1993, *Political Liberalism*, Columbia University Press, New York.

Reich, W.T. (ed.): 1978, *Encyclopedia of Bioethics*, Free Press, New York.

Roe vs. Wade, 410 U.S. 113 (1973).

Rorty, R.: 1989, *Contingency, Irony, and Solidarity*, Cambridge University Press, Cambridge.

Saundby, R.: 1902, *Medical Ethics: A Guide to Professional Conduct*, Wright, Bristol.

Secundy, M.G.: 1994, 'Strategic compromise: Real world ethics,' *Journal of Medicine and Philosophy* 19 (October), 407-417.

Smith, H.: 1970, *Ethics and the New Medicine*, Abingdon Press, Nashville.

Strosberg, M.A., Wiener, J.M., Baker, R. (eds.): 1992, *Rationing America's Medical Care: The Oregon Plan and Beyond*, Brookings Institution, Washington, D.C.

Vaux, K.: 1977, *This Mortal Coil*, Harper & Row, New York.

White House Domestic Policy Council: 1993, *The President's Health Security Plan*, Times Books, New York.

Wildes, K.W., S.J.: 2000, *Moral Acquaintances*, University of Notre Dame Press, Notre Dame.

PART I

THE DILEMMA OF FUNDING HEALTH CARE

STEPHEN WEAR

THE DILEMMA OF FUNDING HEALTH CARE

In his Keynote address, H. Tristram Engelhardt, Jr. provides us all with a
vision that those who are already familiar with his work will recognize as
vintage Engelhardt. Spelled out most fully in his seminal work *The
Foundations of Bioethics*, this vision is especially concerned to be
realistic about both the human condition in general, and the scholarship
that is possible within the field of bioethics in particular. In the latter case,
it is the exuberance, if not hubris, of the typical bioethics commentator
that especially draws his fire. Offered initially in the form of a history of
bioethics scholarship over the past few decades, that history is portrayed
as if it was a parallel, disanalogous universe to the history of
philosophical thought over the last millenium. From initial, primarily
religion-based offerings in the 1950s and 60s, Engelhardt illustrates how
a plurality of religious views on basic bioethical themes led to a plurality
of secular views on these themes in the 1970s and beyond. His point,
however, is not simply to illustrate the richness and diversity of viewpoint
and recommendation within bioethics, but more basically to identify a
more basic result of all this discussion, a result that many still do not want
to recognize or accept, viz., that the diversity of viewpoints and results
are indicative of a much more fundamental problem, that is, our inability
to generate any truly content-rich, systematic account that will be
satisfying to and binding upon all persons regarding the bioethical issues
that we face.

Engelhardt's *The Foundations of Bioethics* provides the fullest account
of his argument here, but his keynote efficiently portrays its essence: in
sum, the Enlightenment project that aimed to secure a detailed and
systematic view of the nature of right and wrong and the proper character
of human flourishing, *on the basis of reason alone*, has failed, utterly and
irreversibly. What we have instead, and bioethics perhaps most fully
illustrates this, is a broad diversity of viewpoints and experience that will
admit of no summary consensus statement that well-intentioned, rational
persons are obliged to accept. The student of philosophy will surely
recognize much of this critique, and result, whether he or she recalls the
readings of MacIntyre, or Rorty, or many other thinkers in main-line
philosophy. There is, in fact, very little that is agreed upon in ethics,

*Stephen Wear, James J. Bono, Gerald Logue and Adrianne McEvoy (eds.), Ethical Issues in
Health Care on the Frontiers of the Twenty-First Century, 19–24.*

among persons, or schools of thought, or cultures, and the reasons for this, according to Engelhardt and many other philosophers, are systematic and intrinsic to the inquiry itself. And rather than offer some new synoptic vision, Engelhardt's basic recommendation is that we accept this diversity of viewpoint and result as inherent in the human condition, not just as some temporary failure to reach consensus that brighter minds will somehow, shortly, resolve for us.

As a philosopher, and his keynote is the most "philosophical" of the offerings contained in this volume, Engelhardt's view includes the following basic recommendations: first, that we must accept moral diversity as a basic fact of the inquiry into ethical issues, not attempt to explain it away, or ignore its reality as we offer our pet solution to whatever problem we choose to reflect upon; second, that this diversity entails that, ethically, we inhabit a world composed of "moral strangers," i.e., a world where moral diversity should be recognized as so robust and persuasive that we should assume that well-intentioned, rational persons may well have fundamentally different views on the nature of human flourishing and excellence, as well as on any of the myriad issues that arise within our experience; and third, that much of what passes for bioethics scholarship over the last few decades involves an attempt to ignore or explain away this diversity. Again, there is nothing new to the student of philosophy here; rather, there are only what seem to be the well-agreed-upon facts of the matter.

Not so for many bioethicists, as Engelhardt advises us. He, in fact, proceeds to describe recent bioethics scholarship as, in large part, an attempt to explain away such diversity. Whether it be via casuistry, where paradigm cases somehow are taken to guide us toward consensus, or the work of Beauchamp and Childress, where a utilitarian and a deontologist report that they mainly come up with the same conclusions regarding various specific bioethical issues, even though their philosophical views radically differ, Engelhardt's suggestion is that any such consensus only occurs among those who already share specifically similar "political views, ideology, sentiments, and moral inclinations" (Engelhardt Keynote, p. 6). More mundanely, Engelhardt suggests, "many have simply proceeded to invoke the concept of moral consensus as if it existed when all evidence is to the contrary. In such appeals there is often a quiet desperation born of the recognition of the need to provide moral authority for the coercive power of law and public policy" (Engelhardt Keynote, p. 8). More specifically, and to the focus of this section, he goes on to note the recent, abject failure of the Clintons' attempt at health care reform, a failure he would in large part attribute to a strategy where "consensus is manufactured by impaneling a commission and carefully choosing its members, the agenda, and appointing a powerful chair to keep discussions on track for the goals one wishes to achieve." (Engelhardt Keynote, p. 9) The failure *and* inappropriateness of this strategy are apparent, he suggests:

> One does not impanel persons with truly diverse moral understandings. One need only imagine how a bioethics commission would function that incorporated real moral diversity and included libertarians, socialists, conservatives, feminists, atheists, and the devoutly believing (for instance, imagine an ethics commission whose members include Jesse Jackson, Jesse Helms, Bella Abzug, and Mother Theresa (Engelhardt Keynote, p. 9).

Those familiar with Engelhardt's writings will recognize that his recommendations for a full-blown libertarian approach to bioethics is just around the corner. For the purposes of this introduction, and this work, however, it is not Engelhardt's own libertarian views, embraced by relatively few bioethics scholars, which should concern us. The overall philosophical review and critique of bioethics that his keynote offers us is perhaps most fruitfully borne in mind as a basic cautionary recommendation that all the contributors to this volume were charged with at the outset. The editors of this volume also presumed to suggest to all the other contributors that they review this keynote preparatory to re-writing and polishing the contributions that are contained herein. In sum, however prudential their own recommendations, all the authors in this volume were charged to reflect further on how generically telling their arguments and positions are, as well as consider, à la Engelhardt, what possibilities for alternative views exist, however persuasive they see their own views to be.

For the purpose of introducing this specific section on the dilemma of funding health care, however, a further feature of Engelhardt's writings allows us to place his second offering in this volume on a par with the contributions of Laurence McCullough, and Haavi Morreim. That is, beyond his generic sorts of reflections on the unavoidable diversity of views on bioethics issues in general, Engelhardt usually goes on to make a specific case within the issue at hand for the wider philosophical view he commends to us. In this instance, Engelhardt's second contribution to this volume, entitled "Toward Multiple Standards of Health Delivery: Taking Moral and Economic Diversity Seriously", begins by critiquing what might be seen as the contemporary ethos regarding health care funding. Specifically focusing on what he refers to as the "American Health Care Ideology," he points out that its basic precepts are simply incoherent and unrealistic. These basic precepts are: (1) that "Americans have the right to the very best of health care;" (2) that "all Americans have the right to equal access to health care;" (3) that "patient and provider choice" must be provided for; and (4) that the first three precepts can be "pursued without costs becoming inordinate" (Engelhardt, this volume, pp. 24-25). He goes on to reflect on the incoherence (and intransigence) of the preceding ethos by reviewing the highly contentious and negative reception that the Oregon plan received, a plan that, as he notes, "proposed to do what nearly every other health care system in the world has done" (Engelhardt, this volume, p. 25). That is, the Oregon plan:

1. does not provide the best of health care to all, but rather defines a basic package stipulated as adequate;
2. does not provide equal care for all but rather adequate care for all, allowing those who wish to purchase better basic care as well as luxury care;
3. does not guarantee physician and patient choice; but
4. does contain costs (Engelhardt, this volume, p. 25).

Having retailed the sorts of considerations that gave rise to the Oregon plan, Engelhardt then proceeds, as might be expected, to parlay this into an argument for there being multiple forms of ethically acceptable modes of health care delivery, the different modes of delivery arising from the specific moral views, as well as economic resources, of the people that might be served by them. He thus ends up talking about "Cheapcare," which is a sort of minimally adequate care that all would have access to, and other forms of more luxury care that some people might wish to purchase at a higher premium. In effect, Engelhardt offers us a vision of a multiple-tier system of health care delivery that mimics the Oregon plan, and in doing so, argues for its ethical appropriateness, in contrast to the unrealistic and incoherent presumptions of the "American health care ideology." As a start in addressing the dilemma of funding health care, then, Engelhardt provides us, as expected, with the grounds for ethically considering multiple forms of delivery, keyed to people's differing moral and economic preferences and resources, and all this is shored up by an extended argument against the "one size fits all" views that are so common in the literature.

The other two contributions to this section, by Professors McCullough and Morreim, may perhaps be most usefully appreciated as taking on a similar task. That is, neither of them offers a "one size fits all" view of health care funding and delivery, but concern themselves instead with what they see as basic considerations that any health care delivery system must be mindful of.

Laurence McCullough's contribution is especially concerned with the fiduciary obligations of providers within a managed practice context, and his strategy is to attempt to identify ethically troublesome features of managed practice so that a preventive ethics approach may be taken to them, i.e., as he puts it, to deal with "ethical challenges" before they become "ethical crises." (McCullough, this volume, p. 33). McCullough proceeds by offering us a review in which he retails what he sees as the emergence of managed practice from a "cottage industry" sort of model. This is accomplished first by describing basic features that managed practice models of health care delivery seem to have in common, and the resultant qualms about the erosion of the fiduciary commitments of providers that we all have, i.e., that they will remain wholly focused on the best interests of their individual patients. Having detailed some of these problems, McCullough then goes on to provide us with an extraordinary historical review that presumes to make the case that these issues have remarkable continuity from the 18th century onward. Relying on the writings of John Gregory and Thomas Percival especially, McCullough does us all the unique and crucial service of advising us that there is, in fact, nothing really new in the ethical issues that managed care presents us; rather, prior generations have wrestled with quite similar issues in some very specific ways that should inform our own reflections. Having provided this seminal backdrop, McCullough then proceeds, in the last section of his paper, to cash all this in by means of a preventive ethics approach to certain current ethical issues in managed care, viz. gag orders, conflicts of interest in how physicians are paid, and the rationing of demand and supply.

Haavi Morreim, for her part, is ultimately concerned to specify the various competing notions of justice that she believes any health care delivery system must be guided by. Perhaps mindful of Engelhardt's cautionary remarks, she does not argue for one single form of justice as binding on all, but rather gives character and force to various competing forms of justice, leaving open which form(s) a given system might legitimately choose to be guided by. She proceeds with all this, however, by first reviewing the various sorts of costs and considerations that arise from the use and development of health care technology, arguably the most basic force in health care by which system budgets can be quickly broken, and where all sorts of ethical and fiduciary issues arise. Again, arguably in line with Engelhardt, she recognizes that various differing choices might well be made regarding the availability of such technology, but she goes on to emphasize and detail the complex choices that would be involved for any specific plan.

Morreim then proceeds to end this section of the volume by raising the further issue of how all such considerations and choices might be adequately retailed in contracts between managed care organizations and their clients, particularly the issue of gaining some sort of legally forceful agreements as to the scope and limits of a given policy. Again, we have a portrayal of diverse, complex and legitimate choices that currently face us as we attempt to grasp the specific lineaments of the dilemma of funding health care, both at present and, accepting the basic vision of our three authors, well into the millennium to come.

University at Buffalo
Buffalo, New York

H. TRISTRAM ENGELHARDT, JR.

TOWARD MULTIPLE STANDARDS OF HEALTH DELIVERY: TAKING MORAL AND ECONOMIC DIVERSITY SERIOUSLY

I. INTRODUCTION: FACING HUMAN FINITUDE

Framing American health care policy has been encumbered not simply by a limitation of resources, but by the failure to be honest about the human condition. Health care policy in the United States has been undertaken as if resources were unlimited, secular moral vision united in a canonical, content-full understanding of proper health care policy, and secular authority sufficiently robust so as to authorize the imposition of one view on all others. Policy has been fashioned with quasi-divine aspirations in the face of all too unavoidable human finitude. Or to rephrase matters, American health care policy has been encumbered by the failure to face finitude, to endorse an altruistic egalitarian, and to gamble prudently with lives. In order to take account of the human condition, health care policy must acknowledge the inevitability of death, the limits of resources, the diversity of moral visions, and the limits of secular moral authority. In particular, the limits of secular moral authority, and the presence of diverse moral views regarding the proper allocation of health care resources, would make the secular establishment of one view of justice, fairness, or equality analogous to the imposition on all of the Roman Catholic proscription of contraception. The presence of a moral diversity that cannot in principle, in secular moral terms, be set aside provides strong moral grounds against imposing a uniform health care policy whose enforcement would only be justified by denying the legitimacy of this diversity.

One begins with a moral pluralism as the background against which one must frame health care policy. Since moral pluralism is grounded in an epistemic disability (i.e., the inability to choose in a rationally principled fashion among a range of alternative, diverse moral visions – Engelhardt, 1996), a metaphysical skepticism is not engaged (i.e., the existence of a canonical moral understanding is not denied, only the possibility of identifying it in general secular moral terms – Engelhardt,

Stephen Wear, James J. Bono, Gerald Logue and Adrianne McEvoy (eds.), Ethical Issues in Health Care on the Frontiers of the Twenty-First Century, 25–33.
© 2000 *Kluwer Academic Publishers. Printed in Great Britain.*

1995) and there is no commitment to a moral relativism. As a consequence of our moral epistemological predicament, there is no commitment to a moral relativism.

II. CONFRONTING THE AMERICAN HEALTH CARE IDEOLOGY

Why has it been so difficult for Americans to come to terms with our obvious limitations? The best explanation can be found in the account given by Karl Marx of an ideology, a false consciousness. According to Marx (Marx, 1960), an ideology functions somewhat like well-entrenched canons of political correctness. Though everyone may in some sense recognize that we have deceived ourselves, few frankly acknowledge it. Most will tend to act as if we did not have to face our finitude. This false vision or false consciousness which constitutes the American health care ideology has four cardinal components.

First, there is the assertion that Americans have a right to the very best of health care. Indeed, there has been the presumption that most, if not all, in fact have access to such a level of care. Moreover, there has been the often-repeated platitude that all should have access, even if all in fact do not.

Second is the claim that all Americans have a right to equal access to health care, often acting as if the care to which most have equal access will in fact be the best. In particular, health care is given a special, if not unique, place in moral considerations regarding equality. This is the case even though, once one reaches a modest level of per capita health care expenditures, the most reliable predictor of who will live the longest is the sex of the person, not the per capita health care investment within a country. Women in the 24 OECD countries on average live 6.3 years longer than the males in their cohort (Schieber, 1994). Though generally inequality in housing, education, and even police protection is accepted, there is a near fascination, if not obsession, with inequalities in health care. For example, with respect to education, those who so wish are free to buy better basic private education, not simply luxury add-ons.

Third, the American health care ideology has been robustly committed to allowing provider and patient choice. The view has been that patients should be able to choose their health-care provider and physicians should be able to choose whom they will see as patients. As everyone knows, managed care has already significantly altered this ideological expectation.

Finally, the ideology's force as a false consciousness is especially demonstrated by the announced expectation that the first three goals could be pursued without costs becoming inordinate. Even a modicum of an appreciation of the lure of health care technology should have led American health policy analysts to have clearly understood this point in 1965, as Medicare and Medicaid were being enacted. They should have openly acknowledged that a significant escalation of American health care costs would occur, given the first three expectations.

The so-called health care reform proposals of the Clinton administration were offered fully within the embrace of this American health care ideology (White House, 1993). The Clinton administration promised health care reform that would have offered everyone equal care, the best of care, as well as physician and patient choice if one paid the additional premium for fee-for-choice services. There would be managed care, no rationing, and cost containment. Many who participated in shaping the plan understood that this was not possible (O'Connell, 1994).

The difficulty in piercing the American health care ideology has also been demonstrated by the disputes that have surrounded the Oregon proposal (Strosberg, 1992). The Oregon proposal was controversial first and foremost because it proposed to do what nearly every other health care system in the world has done, namely, to develop a health care system that:

1. does not provide the best of health care to all, but rather defines a basic package stipulated as adequate;
2. does not provide equal care for all but rather adequate care for all, allowing those who so wish to purchase better basic care as well as luxury care;
3. does not guarantee physician and patient choice; but
4. does contain costs.

The so-called Oregon proposal was radical in that it announced to Oregonians that they would receive less than the best and be treated unequally. The proposal was mainly shocking to non-Oregonians who had not yet examined their ideological blindness. Had it not been for the American ideology, who would have been perturbed by the proposal? The American health care system has always been unequal and has provided many with less than the best, ideological commitments to the contrary notwithstanding. Moreover, if one accepted the inescapable character of the human condition, being treated unequally and receiving less than the best would have also been acknowledged as unavoidable.

Indeed, the commitment to equality has shown itself to carry significant health care risks. It has distracted health care policy from attempting to provide all with a basic package of adequate care and has instead led to many having only episodic basic care. Rather than rationing health care for the poor, the poor have been rationed in the sense that some have been promised the best of care and others nothing, putting the latter class at risk. On the other hand, a commitment to providing unequal and less than the best care can in fact avoid the needless death of those in need of health care. First, if one gives priority to equality, one will not accept a health care system that will cover all, but not all equally. As a result, one will ration persons, not resources, for those under the poverty line. As currently administered, some of the poor are covered by Medicaid and others excluded. This approach will leave many without adequate care, thus putting their health and life in danger. The Oregon proposal was precisely to stop rationing people and instead to provide all under the poverty line with health care, albeit not all the health care they would have wanted or received under ordinary Medicaid. Second, if one focuses on providing equal care for all, one will not as easily ask the more answerable scientific question: how one can as cheaply as possible provide adequate care? Since it is impossible to provide all with the best of care without shifts in resources to health care that most will not accept, it is not possible to realize the American health care ideal. However, it is possible to explore how most cheaply to provide a basic adequate package of health care. The second offers by far the more attractive policy if one is more interested in alleviating suffering and preventing death than in pursuing equality.

In the 21st century, we will likely step away from the constraint of the current American health care ideology and its commitment to the egalitarianism of envy – most countries have done so under fiscal pressures. Until then, significant changes are impossible without first undermining the American health care ideology. One will need to abandon an egalitarianism of envy in which one is concerned that some may have better basic health care or more luxury health care than others. Instead, one will have to direct egalitarian passions within an egalitarianism of altruism through which one seeks as cheaply as possible to provide for the impecunious the morbidity and mortality protection generally available to the rich. Rather than worrying about those who have more, one will need to direct energies to determining how to give more health care cheaply to those who have less.

III. TOWARDS MULTIPLE STANDARDS OF HEALTH CARE

Once one has abandoned the ideological commitment of providing all equally with the best of care, one can then openly explore the morbidity and mortality protection given by different standards of health care. One should also be able to acknowledge that there are alternative plausible competitors for what might be accepted as a basic adequate package. One could then envisage moving towards an informed health care culture in which choices are made among different approaches to morbidity and mortality protection. Policy of this sort will require acknowledging that we as humans with our finite lives, resources, and probabilistic knowledge are forced to gamble with morbidity and mortality risks, that we are forced to gamble with our own lives and our own suffering.

The notion of a prudent gamble might be captured by imagining the licensing of an airline that would begin to sell flights to Paris at ten times the chance of crashing but with tickets at one-tenth the usual price. Since the chance of crashing is still quite low, one might very well decide, considering the higher risks per mile traveled one assumes by driving to the airport, that such tickets would be quite a bargain. If on the way to Paris one notices that the plane is about to crash, all that one can say with justification is that one has lost the gamble. Nothing has happened that is wrong or unfair. One has freely assumed a reasonable gamble and is about to lose. Life is finite and marked by unfortunate limits.

So, too, when addressing the issue of health care plans, one could imagine a proposal to license a new and very cheap managed health care plan: Cheapcare. Cheapcare would offer general access to the cheap and effective morbidity and mortality risk protection available in the British National Health Service, a standard below that promised by Oregon's basic package. In general, these benefits will compare well with current American outcomes with respect to mortality protection (Schieber, 1994). However, admissions to critical care units will be curtailed if one's likelihood of surviving were not good. There will in general be limits on high-cost, low-yield screening programs. In general, one will do well, but every now and then people will have the experience of those who have bought the 90% off tickets to Paris. They will lose.

The standard of care offered by Cheapcare would be one stipulated by agreement (Hall, 1993). One could bundle into the original subscription contract an agreement to all the consequences of participating. There would be advantages. By purchasing Cheapcare, one would be able to expend resources in areas other than health care, perhaps for better education and housing. One might even imagine the indigent being able to use a welfare voucher which they could partially redeem for Cheapcare while using the rest of the voucher to defray the costs of private education for their children. Cheapcare would begin from a candid acknowledgment of the diversity of human concerns and the limits of human knowledge, while also respecting the free choice of persons to fashion their own destinies. If a subscriber to Cheapcare wanted more care than Cheapcare offered, it would be an act of fraud for a Cheapcare health professional to use the resources of the Cheapcare system to provide that care. If such were done, the cost for Cheapcare would escalate and the Cheapcare bargain would no longer be as good. Subscribers would be defrauded. Would-be subscribers would be denied an opportunity which Cheapcare directors wish to offer. Of course, Cheapcare health professionals could mortgage their own houses and give their own private funds to those who wish to purchase better than Cheapcare.

These fantasies are offered to bring into question commitments to a single uniform standard of health care defended by some (Morreim, 1995). The more we examine any ideological commitment to a single uniform standard of health care, the less plausible it is likely to appear, especially as we confront our moral diversity and the limits of secular moral authority. The future promises more possibilities and more choices. As we become a culture more at home in making technological and scientific decisions, different persons and different communities will likely wish to come to terms in different ways with the unavoidable morbidity and mortality risks we all confront.

IV. ADDING MORAL DIVERSITY TO ECONOMIC DIVERSITY

It is not just that people have different views about how to use resources, not to mention different amounts of resources at their disposal. It is also the case that people have starkly different understandings of how appropriately to use health care. The sustained and often acrimonious debates that have characterized bioethics and health care policy over the last 30 years with regard to abortion and third-party-assisted reproduction reveal the depths of our disagreements. These disagreements will be underscored even further as physician-assisted suicide and euthanasia inevitably become a part of American health care. Individuals and communities will be committed to different forms and standards of health care, not simply out of economic reasons, but out of both secular and religious moral reasons.

As we move to acknowledge economic diversity, we should as well move to acknowledge moral diversity. The two together have in the future the promise of leading to the fashioning of health care packages with both special economic and moral commitments. One might for example imagine Roman Catholics in the United States joining their interests with other Roman Catholics throughout the world to create an international Vaticare health care system. Vaticare would have special moral commitments. Obviously, Vaticare would not pay for or allow abortion, sterilization, third-party-assisted reproduction, physician-assisted suicide, or euthanasia on its premises. In addition, because of its spiritual focus, Vaticare could articulate a special moral commitment to providing ordinary but not extraordinary care, unless one had purchased a special Fanaticare rider. Health care savings in these areas could be passed on to impecunious Roman Catholic hospital systems in the developing world. Since Roman Catholicism has traditionally been committed to providing health care *proportionem status*, Vaticare could offer the options of different levels of access both to amenities and expensive interventions. One might imagine a diversity of tiers ranging from Mother Theresa Strawmatcare to Cardinalcare and Papal Platinum Tiaracare.

Because deeply held religious and secular moral commitments in the area of health care have force and substantively divide, something like the foregoing scenario is probably unavoidable. When the bioethical issues that set people against standard health care involve issues other than those restricted to a few relatively small groups such as Jehovah's Witnesses and Christian Scientists, and instead cut across religions to include traditional Christians, Jews, and Moslems, the divisions will be significant and make the reemergence of traditional religious health care systems very likely. As one looks to the 21st century, we will need to explore the ways in which health care can reflect and come to terms with significant and enduring moral differences.

V. LOOKING TO THE FUTURE: LET A THOUSAND BIOETHICS
BLOOM AND A HUNDRED HEALTH CARE POLICIES COMPETE

The examples in this paper regarding high-risk discount airlines, Cheapcare, Vaticare, and Fanaticare may appear tongue-in-cheek and advanced solely as a humorous aside. They are offered in complete seriousness. The moral differences that separate individuals and communities regarding the proper use of health care will not evanesce. We have numerous incompatible views about the sanctity of life, proper human reproduction, equality in health care, and justice and fairness in health care, as well as with respect to ways in which death should be faced or expedited. Given this diversity, the imposition of uniformity is arbitrary and without moral authority.

The seriousness of these suggestions can be appreciated by recognizing that health care entitlements are best regarded as nested within insurance policies against being desirous of health care without the funds to purchase that care. Since there are not only different views about how much should be set aside to face future risks, but also moral disputes regarding which health care interventions are morally allowable, there should be a range of economically and morally diverse health care packages available. On the one hand, these observations bring us to acknowledging the implications of moral diversity. On the other hand, they take seriously the force of human freedom. The choice to engage a partially economic and religiously conditioned package of health care leaves one with no grounds for complaint when one finds in facing illness that one wished that one had purchased a more ample package of entitlements. Nor can one complain after a recent conversion that the moral texture of the package one has bought no longer fits one's moral commitments. One will need to turn to the charity of others, having no moral claim on services to which one did not engage an entitlement.

If the 21st century we are entering is accepting of human freedom and able to acknowledge human moral diversity, then we will need to rethink our commitment to uniform approaches to health care policy. We have up until now acted as if our society were one moral community, although the evidence is stronger every day that we are a multi-cultural, morally diverse society compassing moral communities with quite different understandings of the importance of birth, equality, suffering, and death. If we take moral diversity seriously as well as the limits of secular moral authority, we will need to acquiesce in the development of health care plans built around different understandings of risk-taking, as well as committed to different moral appreciations of the significance of reproduction, birth, suffering, and death. Taking moral diversity seriously will lead us to allowing diverse moral and economic approaches to health care, along with the fashioning of different acceptable standards for its provision.

Baylor College of Medicine and Rice University
Houston, Texas

REFERENCES

Engelhardt, H.T., Jr.: 1995, 'Christian bioethics as non-ecumenical,' *Christian Bioethics* 1, 182-199.

Engelhardt, H.T., Jr.: 1996, *The Foundations of Bioethics*, 2nd ed., Oxford University Press, New York.

Engelhardt, H.T., Jr.: 2000, *The Foundations of Christian Bioethics*, Swets and Zeitlinger, Lisse.

Hall, M.A.: 1993, 'Informed consent to rationing decisions,' *The Milbank Quarterly* 71, 645-668.

Marx, K. and Engels, F.: 1960, *The German Ideology*, International Publishers, New York.

Morreim, E.H.: 1995, *Balancing Act*, Georgetown University Press, Washington, D.C.

O'Connell, L.: 1994, 'Ethicists and health care reform: An indecent proposal?' *Journal of Medicine and Philosophy* 19, 419-424.

Scheiber, G.J., Poullier, J.P., Greenwald, L.M.: 1994, 'Health system performance in OCED countries, 1980-1992,' *Health Affairs* 1, 100-112.

Strosberg, M.A., Wiener, J.M., Baker, R., (eds): 1992, *Rationing America's Medical Care: The Oregon Plan and Beyond*, Brookings Institution, Washington, D.C.

White House Domestic Policy Council: 1993, *The President's Health Security Plan*, Times Books, New York.

LAURENCE B. McCULLOUGH

A PREVENTIVE ETHICS APPROACH TO THE
MANAGED PRACTICE OF MEDICINE:
PUTTING THE HISTORY OF MEDICAL ETHICS TO
WORK

We know already what the twenty-first century practice of medicine will
involve, namely, the managed practice of medicine. How far into the next
century managed practice will persist, we do not know. My purpose in
this chapter is to explain this concept and then present and defend a
preventive ethics approach to it. The managed practice of medicine
confronts physicians and other health care professionals, institutional
managers, private and public payers, patients, and society with interesting
ethical challenges. We have enough experience with the managed
practice of medicine to know what those ethical challenges are. In
response to them, we have two choices. The first, a bad choice in my
judgment, is to wait for these ethical challenges to turn into ethical
conflicts and then respond to them. To be sure, putting out fires is more
engaging than preventing them in the first place. Unfortunately, the
former is far more exhausting than the latter. Therefore, we should
embrace the second choice, preventing these ethical challenges from
turning into ethical conflicts for which physicians, medical institutions,
payers, patients, and society all pay an unnecessary price. If I am right,
that price may well include the loss of the medical profession as a
fiduciary profession, a price, I will argue, that we should not want to pay.
Seen in this light, preventive ethics, while it may not be as exciting as
reactive ethics, serves us all better in the long run.

In this chapter I begin by explaining briefly how managed practice has
come into existence and defining its two basic business tools. I also
identify two tools of ethics that we should use in a preventive ethics
approach to the managed practice of medicine: the physician as moral
fiduciary of the patient and informed consent. The first comes from
eighteenth-century medical ethics and the latter from our century – hence,
the importance of the history of medical ethics for twenty-first-century
medicine. I put the history of medical ethics to work in a consideration of
some of the strategies of managed practice, including gag orders, creating

*Stephen Wear, James J. Bono, Gerald Logue and Adrianne McEvoy (eds.), Ethical Issues in
Health Care on the Frontiers of the Twenty-First Century,* 35–63.
© 2000 *Kluwer Academic Publishers. Printed in Great Britain.*

conflicts of interest in how physicians are paid, rationing demand and supply, and population-based concepts of quality.

I. THE EMERGENCE OF THE MANAGED PRACTICE OF MEDICINE: FROM COTTAGE INDUSTRY TO A LARGE-SCALE INSTITUTIONAL MODEL OF AMERICAN MEDICINE

In the past two decades the United States has experienced far-reaching change in the organization and delivery of the private practice of medicine, from a cottage industry, fee-for-service approach, to a large-scale, prepaid or discounted institutional enterprise (Starr, 1982). This change has important implications for medical ethics, which become clear when the two models of medical care are contrasted.

There are four main features of the cottage industry model of medical care. First, medical care is delivered in a face-to-face fashion, in the context of a physician-patient relationship. An individual physician or a team takes care of patients, one at a time. Second, under fee-for-service, conflicts of interest exist, for physicians in particular. That is, reimbursement that is based on procedures and interventions creates a conflict between the physician's self-interest in remuneration and the ethical obligation to provide patient care that meets an acceptable standard of care. In addition, conflicts of interest also exist for physicians regarding job security and advancement, both of which depend ultimately on steady and growing revenue streams. It is simply a mistake to assume that fee-for-service medicine lacks such conflicts of interest and is therefore in this respect ethically problem-free (Rodwin, 1993).

Third, there tends to be little – or even, at times, no – accountability for scientific rigor in clinical judgment, decision making, and practice. This lack of accountability harkens back to eighteenth-century debates within medicine about whether medicine should be "laid open" to scientifically qualified lay persons to monitor its scientific rigor and quality. Those who argued that medicine should be "closed" to such outside monitoring won the day, with the result that physicians gained a great deal of autonomy over their enterprise (Lawrence, 1975). Whether that autonomy was in all cases restrained in its exercise by scientific discipline remains an open question. Indeed, numerous studies documenting regional differences in medical practice, and showing that this variation lacks scientific rationale, suggest strongly that the closing of medicine resulted in just what the proponents of laying medicine open predicted: the distortion of the standard of care by a heavy dose of physician self-interest.

Fourth, economic variables shape clinical practice. In the roughly twenty-five year period after World War II an economics of relative (to American economic history) abundance shaped clinical practice. This contrasted sharply with the economic condition of American medicine before World War II.

The result of these four features of the cottage industry of health care was a variable standard of care shaped by four factors (at least): an uncertain and often poorly understood mix of scientific discipline and rigor, clinical experience, consensus and individual clinical judgment; habits and customs acquired during training; tradition; and an economics of relative abundance that could afford the cost of sometimes scientifically undisciplined clinical judgment, decision making, and practice, and the enormous economic inefficiencies of scientifically poorly disciplined medicine. In short, the practice of medicine was, at best, poorly managed and, at worst, simply unmanaged from both business and scientific points of view.

The large-scale, institutional model of medical care also has four features. First, the delivery of medical care occurs in the context of an institution-patient relationship. The institution bears responsibility for the medical care of patients and discharges this responsibility through its employers and contractors. Military medicine, the VA system, and other public systems have been organized on this model for decades, providing often-overlooked models of what is mistakenly thought by some to be something brand new in the practice of medicine. What is new in the private practice of medicine – really, new on a broad, national scale – is not, therefore, new to medicine generally.

Second, conflicts of interest exist, but are imposed on physicians and subject them to economic penalties in the form of incomes put at risk (e.g., through withhold payment, discounted fee-for-service, and capitation), compressed incomes (simply paying doctors lower salaries), and incentives that can positively or adversely affect job security and job advancement. Third, in the institutional setting there exists the opportunity for increased accountability for the scientific rigor and quality of clinical judgment, decision making, and practice. Institutions are in the position to develop data on patient care and to seek reduction of the sometimes broad – and scientifically unexplainable – variability in patterns of patient management. In effect, the large-scale institutional model of medicine involves the "laying open" of medicine to scrutiny and management by qualified lay persons, two centuries after John Gregory (1772) and William Buchan (1769) made the arguments for doing so.

Fourth, economic variables shape clinical practice. For the past twenty-five years the economics of relative scarcity (to the quarter century after World War II) have shaped health care. In the larger history of the American economy the last quarter century is closer to the norm: very modest economic growth, punctuated by recessions, occurring in what is still the world's richest national economy.

Payers and providers must respond rationally to these constrained economics or else risk extinction, which they are neither prepared nor obliged to do. In particular, payers have brought to medical care a management philosophy that is jarring, because unfamiliar, to physicians, but commonplace in business: the relentless drive to identify, control, and reduce the cost of medical care. As a result, for many private payers (i.e., private employers) medical care no longer enjoys a privileged status; it is one service among others being purchased by the business concern and will be held to the same, relentless standards of quality and cost reduction as any other service purchased by the business concern. Finally, from the payer's perspective, there is an oversupply of physicians and medical institutions, hospitals especially. The carrying costs of this excess supply artificially inflate prices and so, to reduce costs, this excess carrying capacity must be eliminated.

The upshot of this transition of medical care from a cottage industry model to a large-scale, institutional model is the following. First, payers will not accept or tolerate (for very much longer) the unmanaged practice of medical care. We have, as a consequence, entered the era of *managed practice* (Chervenak and McCullough, 1995; McCullough, 1996). Second, physicians have lost two "freedoms:" (1) to have only favorable economic and other forms of conflict of interest and (2) to practice with variable, little, or even – in some cases – no scientific discipline and rigor. Third, the standard of care must be carefully and rigorously identified, in response to the shift to an economics of relative scarcity and the management philosophy of relentless control and reduction of costs.

Managed practice can now be defined; it is a creature of the large-scale, institutional organization and delivery of medical services. Institutions use two common, well-known, and reliable business strategies with physicians:

1. Creating conflicts of interest in how physicians are paid that can positively or adversely affect remuneration, job security, or job advancement (or any combination thereof); and
2. Regulation of clinical judgment, decision making, and practice.

Both of these business strategies aim at maximal economic efficiency. Creating conflicts of interest aims at such efficiency directly, by incentivizing physicians to reduce costs, especially in the short-term. This makes imposing stringent conflicts of interest very attractive to start-up managed care organizations that aim only to capture and hold a small market share until the MCO can be sold to a larger entity. Their commitment to quality of care is fragile, if it exists at all. By contrast, regulation of clinical judgment and practice aims at such efficiency indirectly and over the long term, roughly following the principles of total quality management and subsequent economic benefit as set out by W. Edwards Deming (1986). Well-capitalized and well managed MCOs in the market for the long term will, I suspect, increasingly turn to this business tool to achieve ever increasing economic efficiency.

This shift to a large-scale, institutional model of medical care raises significant ethical issues. First, there is an increasing dominance of economic considerations – principally cost-benefit and cost-effectiveness – as the shift to an economics of relative scarcity affects private and public medical care alike. These economic values tend to trump or automatically override all other considerations. Second, threats to remuneration, job security, and job advancement constitute powerful sources of conflicts of interest. Third, given their growing capacity to generate and analyze patient data, medical institutions now can effectively regulate clinical judgment, decision making, and practice through the use of such tools as practice guidelines, utilization review, etc.

In response to these changes in medical care, the ethical challenge to physicians and institutional managers can now be stated: to respond effectively to the management tools of managed practice – creating conflicts of interest and regulating clinical judgment, decision making, and practice – with two powerful tools of ethics:

1. The physician as the moral fiduciary of the patient; and
2. Informed consent from patients to be cared for under the management strategies of managed practice.

In the next section of this chapter I provide a more detailed account of the origin of the ethical concept of the physician as fiduciary. This concept has its origins in eighteenth-century British medical ethics, specifically the medical ethics of John Gregory (1724-1773) and Thomas Percival (1740-1803).

Why should the world's first universal culture and its health care professions look to works of medical ethics by an eighteenth-century Scotsman and an eighteenth-century Englishman for its ethical concept of a medical professional?, one might ask, especially for the twenty-first century. One might, indeed.

The answer has to do with the fact that American medicine of the present grew from an American medicine of the seventeenth through nineteenth centuries and that during this period American medicine was deeply influenced – indeed formed – by British institutions and thought, just as were our institutions of law and self-government. This is especially true for American medicine during the period from the latter decades of the eighteenth century through the middle of the nineteenth century (Baker, 1995). This period reflects a profound influence of British medical ethics on American medical ethics, culminating in the 1847 Code of Ethics of the American Medical Association. Major figures of late eighteenth-century American medical ethics, notably Samuel Bard and Benjamin Rush, were students at Edinburgh and read Gregory's medical ethics. The codes of state medical associations, sometimes known as "medical police," also show Gregory's strong influence on American medicine. Moreover, Gregory had a profound influence on the medical ethics of Thomas Percival, which provided the major source for the AMA 1847 Code.

I read the history of American medicine since the late eighteenth century, in one of its many dimensions, as the attempt to develop the actual social practice of the physician as the moral fiduciary of the patient on the basis of Gregory's concept of the physician as the moral fiduciary of the patient. Indeed, the history of American medical ethics and therefore of American medicine would, simply, not be what it has been had Gregory never given his medical ethics lectures in the University of Edinburgh nor published them, nor Percival published his little book.

II. THE ETHICAL CONCEPT OF THE PHYSICIAN AS FIDUCIARY: THE INVENTION OF MEDICINE AS A PROFESSION IN THE EIGHTEENTH CENTURY

A fiduciary is understood to be both "a person holding the character of a trustee, in respect to the trust and confidence involved in it and the scrupulous good faith and candor which it requires" and also to be "a person having duty, created by his undertaking, to act primarily for another's benefit in matters connected with such an undertaking" (Black, 1979, p. 563). Put more generally, the health care professional as a fiduciary in its ethically substantive sense "(1) must be in a position to know reliably the patient's interests, (2) should be concerned primarily with protecting and promoting the interests of the patient, and (3) should be concerned only secondarily with protecting and promoting the physician's own interests" (McCullough and Chervenak, 1994, p. 12).

John Gregory set out to write his medical ethics at a time when, as we shall see below, neither the concept nor the practice of being a fiduciary existed in medicine, although the word 'profession' was then in increasing currency. I read Gregory's medical ethics as a response to the then current state of medicine in Britain. We now follow Gregory in our writing of bioethics and clinical ethics in response to the problems of our time. A great deal that we now take for granted did not then exist. I rely in the following account on the work of Dorothy and Roy Porter (1989) on the experience of patients and physicians in the care of patients in the outpatient setting, and on the work of Gunther Risse (1986) on the Royal Infirmary of Edinburgh, the inpatient setting. The Royal Infirmary was a new medical institution, set up to provide free medical care to the deserving working poor.

There wasn't, as there is now, any uniform pathway into medicine. One could apprentice or go to a university and at the university there was no standardized curriculum. One could obtain a medical degree or stop short of doing so in one's studies; one could also solicit an unearned medical degree. Medical education, therefore, played little or no role in professional formation. Nor was there universal licensure. Although the Royal Colleges did attempt to assert monopoly control within particular geographic areas, they were not uniformly successful in achieving monopoly control of medical practice. Obviously, none of the formal means for certifying specialty knowledge with which we are quite

familiar existed. Thus, uniform education for becoming a professional did
not yet exist.

In such a context the concepts of health and disease were themselves
contested. Various schools and individuals put forth their concepts and
competed for their success in the market place. Indeed, there was a great
oversupply of practitioners who competed fiercely – very fiercely indeed
– in the medical market place for their concepts of health and disease,
their treatments, and therefore their livelihoods, as the Porters document
thoroughly (Porter and Porter, 1989). Patients had their own concepts of
health and disease, engaged routinely in "self-physicking" or self-care,
and traded physician's prescriptions – concoctions to be made up by the
housewife. Preparing prescriptions became an essential homemaking skill
(Lochhead, 1948). Thus, physicians competed with other physicians, with
apothecaries, with surgeons, with midwives, with other unorthodox or
"quack" practitioners, and with patients for market share, success, money,
power, and prestige. The alternative was economic extinction in the form
of unemployment as a medical practitioner. Such intellectual and marketplace competition and its possible dire economic consequences make self-
interest, not service to others, paramount; further evidence that a
profession of medicine did not yet exist.

Medicine exhibited little scientific discipline in its accounts of disease
and in determining the efficacy of treatments. Gregory (1743) lamented
the absence of adequate scientific method in medicine as a medical
student; there was no marked improvement when he began to give his
medical ethics lectures nearly a quarter of a century later. Nor was there
uniform agreement on how to work up the patient and so diagnosis was
often guesswork or, worse, *a priori* dogma. In addition, treatments failed
as often, perhaps more often, than they succeeded in benefiting patients.
Thus, there was no stable knowledge base for a profession of medicine.

Physicians, Gregory taught, could not hope to *control* human biology,
though they could aim to manage its processes well. As a consequence,
again from his medical student days, Gregory (1743) emphasizes again
and again the limited capacities of medicine. His concept of disease
involved observed abnormal functions, which nature would try to correct.
This self-correcting process, he notes, works better in non-human animals
than it does in humans, so the task of the physician is to assist nature
when it frequently malfunctions. When nature underresponds to disease
the physician is to assist her processes; when nature overresponds, the

physician should tamp down nature's responses, to lessen their "violence."

The sick at home, who were the well to do and could afford a practitioner's fees, summoned and dismissed physicians, surgeons, apothecaries, or female midwives, as the sick person chose. A physician therefore might find himself – no women had yet been admitted to the practice of medicine – summoned before, after, or simultaneously with a competitor, with his concepts and diagnosis and treatment put to the acid test. As the Porters put it, in this setting, because patients were in control of physicians' behavior, there existed only a patient-physician relationship, not a physician-patient relationship (Porter, 1989). The former was exclusively contractual, while the latter would reflect a not yet extant professional relationship.

Like their Hippocratic forbears, physicians left off the care of dying patients. Indeed, Frederick Hoffmann in his *Medicus Politicus* (1749), written in an overtly Christian natural law tradition, makes doing so a matter of obligation for the physician. One would suffer punishing economic consequences if one had a high mortality rate. Better, then, to label the patient incurable, withdraw, and turn matters over to clergy. Gregory rails against this intellectual fraud and the pursuit of self-interest that it signifies and so he calls for the physician to continue to attend the dying. Indeed, he says, "It is as much the business of a physician to alleviate pain, and to smooth the avenues of death, when unavoidable, as to cure diseases" (Gregory, 1772, p. 35). Obviously, some risk to self-interest is entailed in this ethics of caring for the dying, because one's mortality rates will go up and perhaps one's market share down. Gregory does not explicitly address what we now call physician-assisted suicide; neither does he condemn it and he is quick to condemn some things, e.g., sexual abuse of female patients or "sporting" with patients in the Royal Infirmary by using experiments as the first line of treatment.

With the invention and introduction of forceps, physicians added midwifery to their practices. Midwifery joined the medical curriculum at Edinburgh, with Thomas Young teaching while Gregory was there. William Smellie, one of Gregory's friends, was a famous man-midwife. These man-midwives were accused of sexual abuse of female patients, usually by the patient's husband (Porter, 1987). It hardly needs to be noted now that sexual abuse of patients is antiethical to the fiduciary obligation of the professional (McCullough, Chervenak, and Coverdale, 1996), but it needed to be said then.

The Royal Infirmary, as noted above, was established to care for the deserving, working poor. They received free care and there was a teaching ward at the Royal Infirmary of Edinburgh, to which Gregory took his students on rounds, following the example of his own teacher, Rutherford – long, long before Osler took credit for the practice. The sick had first to obtain a ticket of admission from one of the benefactors of the Infirmary and then pass screening by the lay managers of the institution, who selected against patients with "fever" or any other sign of life-threatening illness. Thus market segmentation was invented out of institutional self-interest – more than two centuries ago, by a not-for-profit institution! The lay managers exerted strict control of resources and complained regularly of the overuse of resources and high mortality rates on the teaching ward. The pursuit of institutional self-interest can loom large in such an environment, inhibiting the formation of a profession.

There were regular accusations against the physicians of the Infirmary, who were appointed by the benefactors and served without recompense, i.e., direct recompense. The honor and prestige of such an appointment by Edinburgh's great men had undoubted beneficial effect on one's competition for market share in the private practice of medicine. Physicians were accused of arrogance and mistreatment of patients. The label "incurable" was abused in yet another way, by applying it too quickly, so as to rationalize the use of experiments as the first line of treatment, particularly on the teaching ward. There existed institutional support for an emerging physician-patient relationship and for institutional power over that relationship, but no ethics yet to guide and regulate either.

All of the above problems were occurring in larger contexts. The older, medieval Highland society of clans and their members and of aristocracy and serfs was rapidly giving way to an economically and socially more complex society. As a consequence, it was not so easy, as it were, to know one's place and therefore fall back automatically on accepted behaviors. Scotland, as so much of Europe and elsewhere, experienced the social throes of the end of the age of manners. Moreover, people could put on manners; there was a thriving market in instructing the "lowborn" in how to be gentlemen. Thus emerged a social predator, the physician of false manners who could and did insinuate himself into the good graces of great men and their families and thus win a handsome economic reward.

All of these problems combined to create a crisis of confidence, on the part of the sick, in those who put themselves forward as medical practitioners. "Whom can I trust?," was a question not far from the mind or lips of any sick person in a medical market place driven largely, often exclusively, by physicians' and other practitioners' pursuit of self-interest. Medicine, Gregory feared, would become commercial, a trade or means to the end of the physicians' self-interest, not an art or profession in service of patients. Gregory, deeply under the influence of the fading Highland ideals, railed against commerce in general and commerce in medicine in particular (Gregory, 1765).

Gregory utilizes the great invention of Scottish Enlightenment philosophy, moral sense philosophy, and its central concept, sympathy, to address these and other problems. His topic list includes: truth-telling to the seriously ill; confidentiality, especially with regard to female patients; sexual abuse of female patients; consultations, including negotiating the then very unstable borders between medicine and surgery and between physicians and apothecaries, mistakenly thought by some commentators to involve mere etiquette; abandoning dying patients; the abuse of patients for experimentation; animal experimentation; and the definition and clinical determination of death. Gregory's topic list – which anticipates so much of what we now take to be "new" problems – is not our main concern here. Rather, it is with his method for moral philosophy and its consequence, the invention of the ethical concept of the physician as the moral fiduciary of the patient.

Gregory was steeped in Scottish moral sense philosophy, having studied and accepted Hume's moral philosophy while he was teaching at King's College in the 1750s, during which time he helped to found and played a very active role in the proceedings of the Aberdeen Philosophical Society (Ulman, 1990). Numerous sessions of the Society were devoted to the critical study and discussion of Hume's *A Treatise of Human Nature* (Hume, 1978). Gregory absorbed and accepted the central concept in Hume's moral philosophy, sympathy, which had a technical meaning for Hume that was well understood and accepted in his time.

Gregory was also steeped in Baconian scientific method, which insisted that all hypotheses be tested against observation and experience and revised or discarded as necessary in light of the results. Hume's concept of sympathy has its roots in the new science of the nervous system, neurology, that was then developing, especially with the work of Robert Whytt (1765), whom Gregory succeeded at Edinburgh. The physiologists of the nervous system posited the concept of sympathy to explain how the agitation or disturbance of one organ could result in agitation or disturbance of another organ that was not directly connected to the first. The nerves, they had demonstrated on observation – just as good Baconian scientific method requires – convey the disturbance from one organ to another. The second organ, as it were, reacts in sympathy or fellow feeling with the first.

The physiologists reported on but could not explain a macro-phenomenon of sympathy, inter-personal sympathy. The common examples they noted but could not explain included the "contagions" of yawns from one person to another, or of moods from one person to another, including the projection of moods onto audiences in dramatic productions of the theater. Hume offers an explanation that he, too, thought was demonstrable on the basis of Baconian method. After all, Hume had subtitled the Treatise: *Being an Attempt to Introduce the Experimental Method of Reasoning into Moral Subjects*, which his contemporaries would immediately and unmistakably recognize as a reference to the Baconian scientific method of careful observation and the formulation of causal explanations based on repeated, stable patterns of observation over time (Hume, 1978).

To understand Hume's explanation of sympathy, we need two terms from his technical apparatus, impressions and ideas. Impressions are sensations of which we are immediately aware and originate "in the soul, from the constitution of the body, from the animal spirits, or from the application of objects to the external organs" (Hume, 1978, p. 275). Ideas we form as abstractions from impressions, a conceptualization of them: "By *ideas* I mean the faint images of these [i.e., impressions] in thinking and reasoning ... " (Hume, 1978, p. 1).

Hume explains interpersonal sympathy in the following way. Suppose that one sees another writhing in pain, calling out in what Hume and Gregory confidently term the natural language of pain. From these impressions one forms the idea of another being in pain. This idea, in turn, generates the impression of pain in oneself, the very same pain that the observed individual experiences. This experience of pain is judged bad by natural, instinctual judgment, which then motivates one to relieve the person who is in pain. This physiological process of sympathy is called the double relation of impressions and ideas.

We are all born with the natural capacity of sympathy and we need to develop it properly. Sympathy can be corrupted by insufficient exercise, leading to dissipation, the moral indifference of being unmoved by the pain and suffering of others, which Gregory condemns. He also condemns the over-reaction of sympathy, which leads to loss of emotional control. Properly functioning sympathy permits us to be directly engaged in the affective experience of others in a disciplined, steady fashion.

The virtues of properly functioning sympathy are tenderness and steadiness. Women of learning and virtue, Gregory believes, provide the moral exemplars of these virtues. His particular exemplars were the learned women of the London Bluestocking Circle, especially Mrs. Elizabeth Montagu (Myers, 1990). To use current terminology Gregory genders sympathy feminine, in the sense described by Rosemarie Tong (1993), thus writing the first feminine medical ethics – long before theories of care based on affiliative or relational psychology (More, 1994).

In the following passage from his *Lectures* Gregory presents his core views on sympathy:

> I come now to mention the moral qualities peculiarly required in the character of a physician. The chief of these is humanity; that sensibility of heart which makes us feel for the distresses of our fellow creatures, and which, of consequence, incites us in the most powerful manner to relieve them. Sympathy produces an anxious attention to a thousand little circumstances that may tend to relieve the patient; an attention which money can never purchase: hence the inexpressible comfort of having a friend for a physician. Sympathy naturally engages the affection and confidence of a patient, which, in many cases, is of the utmost consequence to his recovery. If the physician possesses gentleness of manners, and a compassionate heart, and what Shakespeare so emphatically calls "the milk of human kindness," the patient feels his approach like that of a guardian angel ministering to his relief: while every visit of a physician who is unfeeling, and rough in his manners, makes his heart sink within him, as at the presence of one, who comes to pronounce his doom. Men of the most compassionate tempers, by being daily conversant with scenes of distress, acquire in process of time that composure and firmness of mind so necessary in the practice of physick. They can feel whatever is amiable in pity, without suffering it to enervate or unman them. Such physicians as are callous to sentiments of humanity, treat this sympathy with ridicule, and represent it either as hypocrisy, or as the

indication of a feeble mind. That sympathy is often affected, I am afraid is true. But this affectation may be easily seen through. Real sympathy is never ostentatious; on the contrary, it rather strives to conceal itself (Gregory, 1772, pp. 19-20).

Sympathy makes the protection and promotion of the patient's interest – in being rid of pain, suffering, and disease – the physician's primary and motivating concern. Sympathy also puts the pursuit of self-interest into a systematically secondary status. Indeed, self-interest must sometimes be sacrificed, e.g., in accepting the psychologically distressful and unpleasant duty of having to inform a patient who is gravely ill and who has not prepared his estate that he is gravely ill. After all, sympathy will cause one to experience the same sense of foreboding about impending death and lack of preparation for the well-being of one's heirs that the patient will experience and no one wants that. Sympathy thus provides Gregory with the second and third conditions for being a fiduciary listed above.

The first condition, being in a position to know reliably what is in the patient's interest, comes from Gregory's theory of medicine. The title of the 1770 version of his lectures, *Observations on the Duties and Offices of a Physician, and on the Method of Prosecuting Enquiries in Philosophy* (Gregory, 1770), reveals the twin concern for the duties of the physician, medical ethics, and the proper scientific method for medicine, theory of medicine. Gregory, no surprise, argues that the Baconian method should govern the physician's concepts of health and disease, diagnosis – he provides methods for working up patients in his clinical lectures – and treatment. Scientific discipline involves always being open to new information and being willing to change one's thinking in the face of new observations, staples of the Baconian method. Gregory calls this intellectual virtue diffidence or being "open to conviction." With this Gregory puts in place the first condition of being a fiduciary.

When synergized in Gregory's *Lectures* his theory of medicine and his medical ethics produce the concept of the physician as a professional in the ethical meaning of the term. In effect, Gregory forges the conceptual foundations for the social role of the physician as the patient's moral fiduciary. The physician who practices according to the demands of the intellectual virtue of being open to conviction can rightly claim intellectual authority for his clinical judgment. Absent this virtue, clinical judgment – whether of the practitioner, the patient, or the patient's relatives or friends – lacks intellectual authority. Intellectually authoritative clinical judgment provides a clinical perspective on the interests of the patient. In an era in which there were frightful mortality rates from disease and injury, in which the goal of medicine was to "relieve man's estate" by lowering these mortality rates, and in which, as we saw above, there was no stability to clinical judgment, clinical judgment that can claim intellectual authority solves very serious problems for medicine.

This clinical judgment is routinely put in service of patients – rather than in the service of the physician's interest in market share, income, prestige from being an experimentalist, or power, as documented by the Porters (1989) – by properly functioning sympathy. Thus, Gregory argues, medicine should not be practiced as a trade, as a form of commerce in which the pursuit of self-interest rules the roost. Indeed, Gregory found commerce to be anathema to the moral life generally (Gregory, 1765); all the more so in the case of medicine (Gregory, 1772).

By forging the conceptualization of the social role of the physician as a moral fiduciary of the patient and therefore a professional in the ethical sense of the term, Gregory also forged the social role of a patient, i.e., someone who is sick and who can indeed trust the physician and who can place himself or herself with moral confidence under the physician's authority and power. In works on medical ethics before Gregory, notably Hoffmann's, the word 'patient' does not appear. Instead, 'the sick' is used, reflecting the contractual, patient-physician relationship described by the Porters (1989). Gregory uses both terms, indicating a transition from the reality of medicine of his time – in which there was at best a patient-physician relationship, really a "the sick-practitioner" relationship, based only on contractual expectations – to a professional ethics for medicine – in which there would exist a physician-patient relationship as normative because the physician could be counted on by the patient to be the patient's moral fiduciary. None of this existed in English-language medical ethics before Gregory, nor as a matter of routine in the practice of British medicine of the time (McCullough, 1998).

It is important to note that Gregory's medical ethics is secular, in the sense that it appeals to a secular method, the experimental method of Bacon applied in theory of medicine and in medical ethics. In this Gregory departs sharply from the overtly theological natural law *medicus politicus* of Hoffmann. In addition, Gregory meant his medical ethics to be adequate to the social pluralization of his time, with the collapse of

Highland society and the aristocracy and the emergence of a socially and economically more complex society.

Thomas Percival picks up Gregory's concept of the physician as fiduciary in his *Medical Ethics* (Percival, 1803). Percival's book is well known but it is less well known that it represents the first text on the secular, philosophical ethics of medical institutions. As noted above, the royal infirmaries created considerable institutional power, wielded by lay managers, but there was no ethics to guide and judge the use of this power. Essentially, Percival argues that the physician at the bedside in the Royal Infirmary of Manchester, England, should never allow economic considerations to affect clinical judgment in the care of gravely ill patients. He then takes this pure form of the fiduciary role and writes it large for the institution: its ethics come unchanged from the bedside and thus unalloyed to economic considerations.

Since, as we saw above, the trustees and lay managers of the royal infirmaries strictly controlled physicians' access to medications, the management of scarce resources became an ethical issue for Percival. His *Medical Ethics* thus also becomes the first secular, philosophical account of the ethical obligations of the managers of medical institutions. In the following passage, Percival makes this argument:

> The physicians and surgeons should not suffer themselves to be restrained, by parsimonious considerations, from prescribing wine, and drugs even of high price, when required in diseases of extraordinary malignity and danger. The efficacy of every medicine is proportionate to its purity and goodness, and on the degree of these properties, *ceteris paribus* [other things being equal], both cure of the sick, and the speediness of its accomplishment must depend. But when drugs of an inferior quality are employed, it is requisite to administer them in larger doses, and to continue the use of them a longer period of time, circumstances which probably more than counterbalance any savings in their original price. If the case, however were far otherwise, no economy of fatal tendency ought to be admitted into institutions founded on principles of purest beneficence, and which, in this age and country, when well conducted, can never want contributions adequate to their liberal support (Percival, 1803, pp. 74-75).

Note first in this passage that Percival confines himself to the ethics of rationing resources in the care of gravely ill patients, implying the permissibility of doing so with all other patients. Percival argues in two stages. First, he advances a cost-benefit argument (in the second and third sentences), thus showing that one should not hesitate to make economic arguments when they advance the interests of patients and the fulfillment of the physician's fiduciary obligation to protect those interests. Percival thus undermines the assumption that institutional economic interests must always be at odds with the interests of patients, as some now seem to think (Angell and Kassirer, 1996). Second, Percival argues for the unalloyed fiduciary responsibilities of the physician, and of the institution, when cost-benefit arguments fail and the patient is gravely ill. Economics must give way when death is the result, but may properly influence clinical judgment for the less-than-gravely ill patients.

The problem for us is that the realities of the American economy make being a pure fiduciary – however attractive it may seem – no longer a practical, realistic option. Ethics has to start with the world as we find it and not as we imagine it to be, unless we wish to write a medical ethics for where the rubber meets the sky. But, then we should not be surprised when we are ignored.

The present realities mean that our task is to identify the ethically defensible forms of the *economically disciplined* fiduciary physician and medical institution. There will, in my judgment, be more than one such form, hence the plural. However, we can tell with careful analysis and argument which forms of economic discipline undermine the very possibility of fiduciary medicine and these forms of managed practice should be resisted. I turn now to showing how a preventive ethics approach to managed practice can accomplish this goal.

III. A PREVENTIVE ETHICS APPROACH TO THE BUSINESS STRATEGIES OF MANAGED PRACTICE

In this section I consider the following business strategies of managed practice: gag orders; paying physicians under conflict-of-interest schedules; rationing of demand and supply; and population-based concepts of quality. I will show how the ethical concept of the physician as fiduciary and the informed consent process can be effectively deployed in a preventive ethics approach to these powerful business strategies.

A. *Gag orders*

Gag orders have earned this sobriquet because in one way or another they are designed to discourage physicians from telling patients about how their managed care plan works. Consider the following example:

> Physician shall take no action nor make any communication which undermines or could undermine the confidence of enrollees, potential enrollees, their employers, plan sponsors or the public in ＿＿＿ Care, or in the quality of care which ＿＿＿ Care enrollees receive. (New York Times, 1996).

A physician reading such a clause in his or her contract with ＿＿＿ Care would be prudent to conclude that, for example, the patient should not be told about how the plan pays the physician (e.g., on a withhold basis or by capitation) because that could undermine the confidence of the enrollee in the plan. Hence, we now have added 'gag orders' to the medical lexicon.

The ethics of the informed consent process has direct bearing here. It is now well understood that the physician has an affirmative obligation (i.e., an obligation that does not depend on the patient asking for information) to disclose to the patient all reasonable alternatives, the basic benefits and risks of each alternative, and the alternative of non-intervention with its benefits and risks (Faden and Beauchamp, 1986; Wear, 1993). This settled ethics of informed consent generates the following preventive ethics response to gag orders.

The general rule that should guide the physician's disclosure of information in fulfillment of this affirmative obligation is that salient features of clinical judgment and decision making should be shared with the patient so that the patient can replicate the physician's clinical judgment and decisions in the patient's own terms (Wear, 1993). Managed practice strategies are surely salient in clinical judgment and decision making, because such strategies are precisely designed to influence clinical judgment and decision making in the direction of economic discipline. Therefore, the presence and effect of managed practice strategies on clinical judgment and decision making, as well as therefore on practice, ought to be disclosed and explained to the patient. It follows, simply as a logical corollary, that gag orders are ethically impermissible because they contradict the settled ethics of the informed consent and therefore undermine one of the cornerstones of the contemporary ethics of the physician-patient relationship. The physician's negotiating position should be to strike out gag order clauses and managed care organizations should discontinue their use. Institutions employing or contracting with physicians should remove these clauses from their contracts. Seeking statutory relief would lead to legislation properly understood to be in the public interest, rather than to serve the narrow, special-interest legislation for medicine as a merchant guild trying to protect its own interests.

It is not, however, enough to oppose gag orders or simply for MCOs to withdraw them, as US Health Care has recently done, to the misplaced applause of the American Medical Association. Physicians need to insist that institutions fulfill their obligations, as fiduciary providers, to obtain valid consent from subscribers to be cared for under the managed practice strategies of the plan. I dissent vigorously from the view of Mechanic and Schlesinger that this is solely the institution's obligation (Mechanic and Schlesinger, 1996), because the physician is co-fiduciary with the institution and logically, therefore, cannot escape such responsibility.

Each medical institution should discharge its affirmative obligation to explain to potential subscribers the following basic sorts of information, none of which is beyond the comprehension of the lay person of average sophistication. First, the fact that physicians are paid in ways that impose conflicts of interest on them should be explained. The basic concept to be conveyed is that physician judgment and behavior are subject to both positive and negative economic incentives, just like virtually everyone else in the American economy. This concept will be foreign to very few potential subscribers, because all physician payment schemes involve a variation of base salary plus incentives based on performance.

Second, the institution should explain that it also uses non-economic means to regulate clinical judgment and decision making. These are now managed processes and not the unmanaged prerogative of the physician. The physician is no longer free to decide, simply on his or her own, what is "best" for the patient.

Third, the potential subscriber should be told that these two business strategies should be expected to have both direct and indirect effects on his or her care, and that, at present, some of these are unknown. Fourth, the plan should explain to the patient how the plan stacks up against other plans in terms of both process and outcomes measures of quality.

The physician should monitor what the institution is doing in these respects. In other words, as co-fiduciary the physician has an obligation to monitor the activities of the marketing and advertising divisions of the institution. Mechanic and Schlesinger (1996) wrongly want to wall the physician off from such responsibility, to preserve trust. In my judgment, their proposal will create just the opposite because patients will continue to count on their physician's integrity to protect them from lapses of integrity on the part of institutions that can forget that they, too, are fiduciaries of patients. Physicians are in a powerful position to hold institutions' feet to the fire in this matter.

In patient care the physician can then explain to the patient how the managed practice strategies of the institution affect clinical decision making and behavior in the actual context of the patient's present problem. The present situation puts physicians in the unenviable and unwelcome position of being downstream from defective disclosure and so it is no surprise that physicians gag themselves. Explaining what should have been explained by someone else already, especially to a stressed or angry patient – or, in pediatrics all too often these days, an angry parent – is one of the worst ethical traps into which to fall, as every physician knows too well. This is yet another reason why the physician has strong obligations to monitor the institution's fulfillment of its disclosure obligations. Physicians should also accept the institution's monitoring of their disclosure of the particular effects of managed practice strategies as a routine dimension of total quality management.

B. Conflicts of interest in how physicians are paid

Paying people on economic schedules that reward behavior valued by the payer and punishing behavior disvalued by the payer is a very familiar practice in the business world. Indeed, not to be paid on such economic schedules would be uncommon in many businesses. Of course, fee-for-service rewarded behavior that produced revenues for the physician and hospital and discouraged behavior that did not produce revenues. Hence, there was frequent complaint for years from primary care physicians that time spent with patients, especially the time required for preventive medicine, was not compensated and so discouraged. In short, the economics of fee-for-service involved conflicts of interest (Rodwin, 1993). The phrase comes to us from eighteenth-century English where the use of 'interest' was understood to mean self-interest. Thus, a conflict of interest involves a conflict between the physician's self-interest in remuneration, as well as job security and advancement, on the one hand, and the physician's fiduciary obligations to the patient, on the other.

Conflicts of interest were often papered over in fee for service by the language of doing the "best" for patients, i.e., doing everything available and for which payment could be received. Thus, too, the standard of care that we inherit from fee-for-service medicine – doing what is in the "best interests" of the patient – includes a considerable dimension of self-interest. Hence, we should abandon the use of 'best' in reference to the interests of patients, because this locution misleads patients, physicians, and society about what the standard of care should be.

Managed practice thus does not introduce economic conflicts of interest into the private practice of medicine; they have been there for centuries. Gregory, for example, was well aware of this problem and addresses it at some length (Gregory, 1772). What managed practice does introduce is the use of negative economic incentives that are not controlled by physicians. Rather than self-imposed and self-managed (and, therefore, often poorly managed) conflicts of interest in the unmanaged cottage industry model, managed practice imposes conflicts of interest on physicians in the attempt to manage physician judgment, decision making, and behavior. This tool is used because it works in business.

It also works in medical care, or is at least perceived to have worked in the case of the DRG or Prospective Payment System introduced by HCFA in the early 1980s for Medicare payment. This payment system involves the direct imposition of negative economic incentives on hospitals, experienced in the form of a conflict between their self-interest in a balanced budget and survival, on the one hand, and their fiduciary obligations to patients, on the other. Managed practice simply adopts in the private sector a management tool first deployed on a massive scale in the public sector.

Notice that large-scale conflicts of interests as a deliberate management tool were introduced into American medicine by a governmental agency that has fiduciary responsibilities to its beneficiaries, not by supposedly rapacious for-profit MCOs. This is important, because any MCO – not-for-profit and for-profit alike – that hopes to survive for the long run will have to use some version of conflicts of interest in how physicians are paid, because this tool works. People in American culture usually adapt their thinking and behavior in response to economic incentives. Indeed, poorly capitalized, not-for-profit MCOs have a stronger financial interest in using punishing conflict of interest schedules than well capitalized for-profit MCOs.

56   LAURENCE B. MCCULLOUGH

The ethical problem with conflicts of interest in payment schemes is that they can incline or actually lead the physician to put self-interest first, thus sacrificing fiduciary obligations to patients for economic gain or job security and advancement. In a market soon to be very oversupplied with physicians, especially specialists and sub-specialists, the interest in job security may well replace the interest in maximizing remuneration.

The deeper problem here was recognized by the Ancient Greek philosophers, who thought that *akrasia*, or moral weakness, was a constituent element of human nature. All of us are inclined not to fulfill our obligations when doing so imposes significant burdens on or harms to our self-interest. Conflict of interest payment schemes simply and explicitly exploit moral weakness. The preventive ethics response to conflicts of interest in how physicians are paid begins with distinguishing between conflicts of interest that threaten to undermine fiduciary obligations in ways that are, with diligent effort, manageable and those that should be expected to undermine fiduciary obligations because we should not expect people to resist the invitations of *akrasia*. The next step is to undertake an "ethics impact statement," i.e., an ethical analysis of the conflicts of interest in a payment scheme. This should be a group process in which the central question becomes, "Is this a manageable conflict of interest?" Uncertainty should be resolved in favor of the conclusion that the conflict is unmanageable, as a matter of prudence. If the conflict is judged unmanageable, no physician committed to being the moral fiduciary of patients should submit to it and no medical institution committed to being a moral fiduciary should include such conflicts of interest in their payment schemes.

For conflicts of interests judged to be manageable after a rigorous group thinking-through of them, quality control mechanisms must be put into place to prevent drift into unmanageable conflicts of interest that result when people stop paying attention. Conflicts of interest in the payment of fiduciary professionals, it should now be clear, are in all cases ethically unstable. Unmonitored ethical instability leads straight to trouble, just as unmonitored nuclear instability does, as Chernobyl teaches to this day.

Why should physicians and medical institutions make the commitment to being moral fiduciaries? This question must now be addressed, because the proposed preventive ethics strategy depends on how we answer this question. I propose the following. First and foremost, the medical profession belongs to both physicians and society. Society has invested vast sums in the creation of physicians and medical institutions and is therefore a co-owner of the profession. Society has not given it permission to destroy the fiduciary nature of the medical profession. Indeed, society is still unaware of the fact that we have embarked on a social experiment that involves just this risk without society's consent.

Second, fiduciary medicine is better for physicians and patients than its opposite, medicine simply as a non-fiduciary business structured by contracts and product safety laws. I say this because what it takes to sustain oneself as a physician in response to the suffering of human beings cannot be supplied by non-fiduciary business traditions. Fiduciary traditions well sustained also mean that patients can trust their doctors. Absent that trust, the relationship becomes a power struggle that will rapidly exhaust both parties, resulting in raw – i.e., ethically unconstrained – power of physicians. Raw power is extremely toxic to the user. Americans know exactly how to respond to it, with the instrumentalities of systematic government regulation. This is almost certainly a formula for mediocrity, which will drive out those committed to excellence in medicine, an important ingredient in sustaining fiduciary professions.

C. Rationing demand and supply

In American medicine, from the payer's point of view, to which medical institutions must respond in an economically rational fashion or else become extinct, there is no scarcity of supply. Thus, scarcity of demand, i.e., rationing of demand, is created as a way to force down the excessive supply and thus eliminate its costs from medical care. In addition, rationing of demand addresses the wide, documented variability in the use of medical services with no effect on outcome. This wide variability is, Gregory would say, precisely what one would predict the unmanaged cottage industry of medicine to produce and solving this problem becomes one of the central economic tasks of managed practice. In other words, one of the goals of managed practice is to introduce greater rationality into physician demand for services. After all, physicians, not patients, control that demand.

The main tool for achieving a more rational level of demand and thus eliminating the costs of oversupply is the regulation of clinical judgment, decision making, and behavior. This form of rationing is not intrinsically unethical. Rather, rationing below a level of demand that ought to define the standard of care – either as a result of careful scientifically based studies or rigorous consensus judgment – is ethically unjustified. This is equally true of rationing of demand and rationing of supply. However, rationing of demand is the goal of the regulation of clinical judgment, decision making and behavior by such techniques as practice guidelines, critical pathways, case management, and utilization review.

The first step in a preventive ethics approach to rationing of both demand and supply involves a little-appreciated dimension of informed consent, namely, consent by physicians to the rationing of demand. Deming (1986) teaches that one should not hope for success by imposing change; rather one must persuade those who need to change to do so and the way to do this is to actively involve them. This makes good business sense, especially in American culture and even more so with ego-strong people who become physicians. My point about consent is different: fiduciaries need to carefully evaluate rationing strategies in terms of their possible impact on fiduciary obligations and the informed consent process, we already know, does this work. Here, physicians must be especially on guard for the insidious effects of conflicts of interests in payment on their definition of the standard of care. Boundary issues are especially important here, i.e., what should be the proper province of primary, secondary, and tertiary providers, severity of illness for admission to the hospital or critical care unit, and discharge planning.

The standard of care for these and other matters should be defined in beneficence-based terms. At the very least this means that a clinical intervention should have a significant probability of being life-saving or of preventing disease, injury, handicapping, or pain and suffering. Such clinical intervention should also have a low or manageable probability of causing serious, far-reaching, and irreversible disease, injury, handicap, or pain and suffering. These judgments should not be a matter of clinical impression but either of evidence-based clinical judgment or a most rigorous consensus-based process for making well-informed clinical judgments.

The second step is, once a rationing scheme passes initial muster via the consent process, it should be monitored for its actual effect on the fulfillment of fiduciary obligations of physicians to their patients. Like conflicts of interests, untended rationing schemes that pass initial muster should be expected to drift into rationing schemes that undermine fiduciary obligations.

In short, for rationing and for conflicts of interest, ethics adds an important dimension to total quality management and continuous quality improvement: protection of the fiduciary nature of the medical profession from unmonitored – and therefore negligent – collapse. In my judgment, this dimension of quality management has received insufficient emphasis in the quality literature. Ethics challenges quality managers to add a distinctive new feedback loop to the quality management process.

Third, patients must consent to care under the rationing scheme. The institution, as explained above, should obtain consent on enrollment for the general use of such strategies. The physician then has the correlative obligation to explain to the patient how particular rationing schemes will affect the patient's care and how those effects will be monitored.

Fourth, it should be explained to patients, when there is high confidence that the rationing scheme passes ethical muster, that refusal to accept the rationing scheme is one of the patient's options. However, the economic consequences for the patient should be spelled out. An economically disciplined fiduciary is not obliged to fund a patient's undisciplined preferences for something above a reasonable standard of care; this obligation falls properly on the patient. Thus, payment for out-of-network services when in-network services have been put through this preventive ethics strategy can justifiably be made the contractual financial obligation of the patient alone.

D. Population-based measures of quality

Increasingly, quality management uses population-based measures. This is as it should be in the scientific practice of medicine because there is no science of individuals and their responses to diseases and their medical management. That clinical judgment could ever in a scientifically disciplined way be about individual patients remains one of the enduring, but nonetheless false, myths of cottage industry, fee for service private practice of medicine – reinforced by the infectious disease model of medicine.

Population-based measures of outcomes focus on biomedical outcomes: mortality, morbidity, and objectively measured function (e.g., minutes on a treadmill with defined changes in cardiac function). There is nothing inherently wrong about this and it is just how beneficence-based judgment should work. The problem is that the psychosocial dimensions of beneficence are omitted. Moreover, the burdens on others than the patient of disease and its management are not measured. Finally, patients' preferences are systematically discounted.

Frank Chervenak and I have argued elsewhere that these factors combine to cause a resurgent paternalism (Chervenak and McCullough, 1995). Paternalism involves a restriction of patients' autonomy for the purpose of securing benefit, in this case, for a population of patients. The resurgent paternalism of population-based definitions of quality involve discounting patients' preferences, including those captured under a psychosocial concept of beneficence and those directed with reducing caregiving burdens on family members, in favor of biomedically defined outcomes.

The preventive ethics approach here begins by recognizing that the economic discipline being imposed on American medicine by payers is such that any attempt to resist this resurgent paternalism by introducing patient preferences into the definition of quality is doomed to fail. Allowing patient preferences in will simply lead to uncontrolled variability in patient care and Deming (1986) is right to point out that uncontrolled variability in either a production of service process is the main engine of uncontrolled costs.

The next step, therefore, is to obtain consent from patients for this resurgent paternalism. After all, consent will produce ethically authorized constraints on patient autonomy, which are not in any way ethically troublesome. Again, this consent should be obtained in a general way on enrollment and then renewed with specificity in patient care.

Now, some might object that this preventive ethics strategy violates considerations of justice. I think not, unless justice requires funding with the resources of others the scientifically or economically undisciplined preferences of patients. Patients' preferences in this context are all positive rights and positive rights come with limits. The ethical arguments concern what those limits should be. Unconstrained patients' preferences may cause private employers to reduce other benefits or fire people. I know of no theory of justice that gives anyone the positive right to introduce injury of this sort into the lives of others without their consent, especially as a consequence of exercising intellectually undisciplined preferences. Indeed, justice may require patients to exercise only intellectually disciplined preferences, a form of economic discipline patients have yet to learn. In the public sector, increasing taxes to cover additional Medicare and Medicaid expenditures would have the same deleterious effects, raising the same ethical issues. In addition to justice-based obligations to constrain one's preferences as a patient, there may also be concerns in justice, e.g., employment security or even full employment (what ever happened to justice and full employment?), that outweigh the funding of patients' preferences to services that they or their families could provide for themselves, albeit with no little sacrifice in many cases.

IV.CONCLUSION

We are busy at work now and for the future replicating the conditions of eighteenth-century British medicine that Gregory found, rightly in my judgment, to be ethically very problematic, indeed, for the well being of patients and the intellectual and moral integrity of medicine. That is, physicians, medical institutions, private employers, government entities, patients, and society are all making free choices in response to the shift of medicine from a cottage industry to a large-scale institutional undertaking. In the process of doing so we are putting medicine as a profession in its ethical sense, as the moral fiduciary of patients, at risk of extinction. The concept of medicine as an ethical profession is a young idea in the context of the history of ideas and so we cannot expect it to sustain itself. Rather, this is our main task, now and for the foreseeable future. I have proposed here some preventive ethics strategies with the aim of identifying the good, hard work that awaits us in the twenty-first-century practice of medicine, using powerful conceptual tools from the history of medical ethics.

Baylor College of Medicine
Houston, Texas

REFERENCES

Angell, M, Kassirer, J.P.: 1996, 'Quality and the medical marketplace – following elephants,' *New England Journal of Medicine* 335, 883-885.
Baker, R. (ed.): 1995, *The Codification of Morality: Historical and Philosophical Studies of the Formalization of Western Medical Morality in the Eighteenth and Nineteenth Centuries. Volume Two: Anglo-American Medical Ethics and Medical Jurisprudence in the Nineteenth Century*, Kluwer Academic Publishers, Dordrecht, The Netherlands.
Black, H.C.: 1979, *Black's Law Dictionary*, West Publishing Company, Minneapolis, Minnesota.

Buchan, W.: 1769, *Domestic Medicine*, Balfour, Auld & Smellie, Edinburgh.

Chervenak, F.A., McCullough, L.B.: 1995, 'The threat to autonomy of the new managed practice of medicine,' *Journal of Clinical Ethics* 6, 320-323.

Deming, W.E.: 1986, *Out of Crisis*, Massachusetts Institute of Technology, Center for Advanced Engineering Study, Cambridge, Massachusetts.

Faden, R.R., Beauchamp, T.L.: 1986, *A History and Theory of Informed Consent*, Oxford University Press, New York.

Gregory, J.: 1743, 'Medical notes,' Aberdeen University Library, MS 2206/45.

Gregory, J.: 1765, *A Comparative View of the State and Faculties of Man Compared with those of the Animal World*, J. Dodsley, London.

Gregory, J.: 1770, *Observations on the Duties and Offices of a Physician, and on the Method of Prosecuting Enquiries in Philosophy*, W. Strahan and T. Cadell, London. Reprinted in L.B. McCullough (ed.), *John Gregory's Observations and Lectures on the Duties, Offices, and Qualifications of a Physician*, Kluwer Academic Publishers, 1998, pp. 93-159.

Gregory, J.: 1998, *Lectures on the Duties and Qualifications of a Physician*, W. Strahan and T. Cadell, London. Reprinted in L.B. McCullough (ed.), *John Gregory's Observations and Lectures on the Duties, Offices, and Qualifications of a Physician*, Kluwer Academic Publishers, pp. 161-245.

Hoffmann, F.: 1749, *Medicus Politicus, sive Regulae Prudentiae secundum quas Medicus Juvenis Studia sua et Vitae Rationem Dirigere Debet*, in Frederici Hoffmanni, *Operum Omnium Physico-Medicorum Supplementum in Duas Partes Distributum*, apud Fratres de Tournes, Genevae.

Hume, D.: 1978, *A Treatise of Human Nature*, 2nd. ed., P.H. Nidditch (ed.), The Clarendon Press, Oxford. Based on 1739-1740 edition.

Lawrence, C.J.: 1975, 'William Buchan: Medicine laid open,' *Medical History* 19, 20-35.

Lochhead, M.: 1948, *The Scottish Household in the Eighteenth Century: A Century of Scottish Domestic Life*, The Moray Press, Edinburgh.

McCullough, L.B.: 1996, 'Reification and synergy in clinical ethics and its adequacy to the managed practice of medicine,' *Journal of Medicine and Philosophy* 21, 1-6.

McCullough, L.B.: 1998, *John Gregory (1724-1773) and the Invention of Professional Medical Ethics and the Profession of Medicine*, Kluwer Academic Publishers, Dordrecht, The Netherlands.

McCullough, L.B.: 1998, (ed.), *John Gregory's Observations and Lectures on the Duties, Offices, and Qualifications of a Physician*, Kluwer Academic Publishers, Dordrecht, The Netherlands.

McCullough, L.B., Chervenak, F.A.: 1994, *Ethics in Obstetrics and Gynecology*, Oxford University Press, New York.

McCullough, L.B., Chervenak, F.A., Coverdale, J.F., 1996, 'Ethically justified guidelines for defining sexual boundaries between obstetrician-gynecologists and their patients,' *American Journal of Obstetrics and Gynecology* 175, 496-500.

Mechanic, D., Schlesinger, M.: 1996, 'The impact of managed care on patients' trust in medical care and their physicians,' *Journal of the American Medical Association* 275, 1693-1697.

More, E.S.: 1994, '"Empathy" enters the profession of medicine,' in E.S. More and M.A. Milligan (eds.), *The Empathic Practitioner: Empathy, Gender, and Medicine*, Rutgers University Press, New Brunswick, New Jersey, pp. 19-39.

Myers, S.H.: 1990, *The Bluestocking Circle: Women, Friendship, and the Life of the Mind in Eighteenth-Century England*, Clarendon Press, Oxford.

New York Times: 1996, 'Word for Word/H.M.O Contracts,' *New York Times* September 22, 1996, Week in Review, 7 (National Edition).

Percival, T.: 1803, *Medical Ethics, or a Code of Institutes and Precepts, Adapted to the Professional Conduct of Physicians and Surgeons*, Printed by J. Russell, for J. Johnson, St. Paul's Church Yard & R. Bickerstaff, Strand, London.

Porter, D., Porter, R.: 1989, *Patient's Progress: Doctors and Doctoring in Eighteenth- Century England*, Stanford University Press, Stanford, California.

Porter, R.: 1987, 'A touch of danger: The man-midwife as sexual predator,' in G.S. Rousseau and R. Porter (eds.), *Sexual Underworlds of the Enlightenment*, Manchester University Press, Manchester, pp. 206-232.

Risse, G.: 1986, *Hospital Life in Enlightenment Scotland: Care and Teaching at the Royal Infirmary of Edinburgh*, Cambridge University Press, Cambridge.

Rodwin, M.: 1993, *Medicine, Money, and Morals: Physicians' Conflicts of Interest*, Oxford University Press, New York.

Starr, P.: 1982, *The Social Transformation of American Medicine*, Basic Books, New York.

Tong, R.: 1993, *Feminine and Feminist Ethics*, Wadsworth Publishing Company, Belmont, California.

Ulman, H.L.: 1990, *The Minutes of the Aberdeen Philosophical Society*, Aberdeen University Press, Aberdeen, Scotland.

Wear, S.: 1993, *Informed Consent: Patient Autonomy and Physician Beneficence within Clinical Medicine*, Kluwer Academic Publishers, Dordrecht, The Netherlands.

Whytt, R.: 1765, *Observations on the Nature, Causes and Cure of those Disorders Which Have Been Commonly Called Nervous, Hypochondriac or Hysteric,* 2nd. ed., corrected, Balfour, Edinburgh.

E. HAAVI MORREIM

SAVING LIVES, SAVING MONEY:
SHEPHERDING THE ROLE OF TECHNOLOGY

I. OVERVIEW

Whether or not it is the single greatest driver in health care cost escalation
(Aaron and Schwartz, 1984 and 1985), technology poses formidable
challenges to achieving rational expenditure limits. As those who pay for
health care try to rein in spending, they must make the best use of
available resources in the face of ever-increasing costs that range from
new genetic tests and treatments, to laser surgeries, to multi-organ
transplants, to a wide array of drugs and devices (Abraham, *et al.*, 1995;
Goldsmith, 1992, 1993 and 1994; Gulick, *et al.*, 1997; Hammer, *et al.*,
1997; Hanania, *et al.*, 1995).

Historically, new technologies have tended to diffuse rapidly – often
before their best uses have been identified.[1] Even when new technologies
reduce per-patient costs, as with laparoscopic surgery, overall costs can
rise rather than fall as indications for its use expand (Garber, 1994, p.
117; ECRI, 1995b; Escarce, Chen and Schwartz, 1995). There are several
reasons for such rapid technologic development and diffusion. Although
third-party payers typically purport to cover only "medically necessary"
services, coverage for new technologies has often been surprisingly swift.
For new drugs and devices, Food and Drug Administration (FDA)
approval as "safe and effective" has commonly been equated with
"necessary" and thereby covered.[2] Surgical and medical procedures,
which do not require FDA evaluation (Steinberg, Tunis, and Shapiro,
1995, p. 145), have become reimbursable upon achieving even marginal
physician acceptance.[3]

Conversely, insurers have usually declined to pay for research, or for
technologies deemed experimental, for reasons of both cost and quality.[4]
And under this same umbrella of not subsidizing research, most payers
have also shunned paying for technology assessment (TA). After all,
since FDA-approved treatments can be prescribed however a physician
wishes, the profession's accepted practices were not deemed
experimental. Although payers are now veering away from these policies,
they were the norm for many years.[5] So long as they prevailed, these

*Stephen Wear, James J. Bono, Gerald Logue and Adrianne McEvoy (eds.), Ethical Issues in
Health Care on the Frontiers of the Twenty-First Century, 65–112.*
© 2000 *Kluwer Academic Publishers. Printed in Great Britain.*

policies of covering virtually all uses of approved or accepted drugs, devices, and procedures, while generally refusing to subsidize even the research that might establish their best uses, has spawned a number of somewhat perverse results.

First, by virtually guaranteeing a market for new products once approved, such policies helped to generate enormous sums of money for research and development, thereby spawning a wide range of new technologies that in turn raised the overall cost of health care.[6] Some new technologies have actually reshaped definitions of disease: treatments for problems like childlessness, for instance, tend to make it a medical, not just a social ailment (ten Have, 1995).

Second, TA to discover new products' and procedures' best uses, or even their most efficient production modes (Garber, 1992), was actually contrary to producers' interests. So long as FDA approval plus professional acceptance ensured good sales, it would be foolish for manufacturers to do research that could ultimately reduce sales. Similarly, fee-for-service rewarded physicians and hospitals for maximizing services, not for studying which ones to delete. Thus, for providers and producers TA was not just unnecessary, but expensive and self-defeating.

Third, TA did of course occur – unsystematically and unscientifically, by trial-and-error and cumulative anecdote. Physicians would get acquainted with a new technology, such as magnetic resonance imaging, by using it widely. Absent any significant threat of harm to patients, a test or treatment could become accepted practice with no clear scientific evidence to support many of its uses.[7] Interestingly, such rapid proliferation of technology can actually preclude adequate assessment. For example, courts' willingness to mandate bone marrow transplant for women with breast cancer has made the procedure so easily available that researchers conducting a controlled study had difficulty recruiting enough patients willing to have their treatment randomized (Kolata, 1995, 1999; Steinberg, Tunis and Schapiro, 1995; U.S. General Accounting Office, 1996).

Hence an irony: payers' refusal to fund formal scientific research in fact subsidized vast amounts of informal, unscientific TA while also fueling an empire of corporate research and development. In the process, physicians saw an obligation to emphasize the latest technology. After all, medicine prides itself on being scientific. Since most of the scientific research focused on cutting-edge tests and treatments, then a profession promoting scientific care must provide these, the most well-documented interventions. Older treatments and home remedies could be dismissed with "there is no evidence ...." Medicine became the science of the expensive.

The problem is not restricted to new technologies. Many medical routines are the product, not of careful science, but of reimbursement patterns, local habits, malpractice fears, or product marketing (Avorn, Chen and Hartly, 1982; Burnam, 1987; Holoweiko, 1995; Wong and Lincoln, 1983; Woosley, 1994). Too often, tests and treatments once thought to be necessary later turn out to be useless or even harmful.[8] The result has been wide, often inexplicable, variations in medical practices.[9]

Against this background, this article explores three major areas of challenge: scientific, ethical, and legal. Section II addresses the scientific issues. After discussing current problems in outcomes studies, technology assessment, and health services research, this section proposes a way to address these challenges. Section III turns to ethical challenges as they arise for three sorts of technologies: high-cost, unproven new technologies; high-cost technologies whose value has been scientifically substantiated; and ordinary interventions for commonplace maladies. Understanding the ethical issues that arise as health plans attempt to distribute these kinds of technology fairly will require distinguishing four sorts of justice: distributive, formal, contractual, and contributive. Section III then proposes some preliminary resolutions to these ethical concerns.

However, any such remedies are moot if they can not be effectively enforced. That is, it is useless for health plans to draw ethically reasonable limits on technology if those limits will not be honored in practice. Thus, Section IV explores the ways in which the legal system has fueled the development and rapid diffusion of technology and identifies the legal issues that must be resolved before health plans can enforce limits on technology coverage. In particular, health plans must be established on a sounder contractual basis than many currently are. Instead of presuming that each health plan must cover all and only "medically necessary" interventions, and then quietly inserting widely varying definitions of necessity, health plans should instead offer particular packages of services that describe benefits considerably more clearly than current plans typically do. When patients can make informed choices among clearer options, the legal system is better able to enforce contractual limits, and to tame the sometimes excessive potential for tort liability placed on health care providers.

II. SCIENCE AND TECHNOLOGY ASSESSMENT

A. *Challenges*

Health care's current economic climate is starkly different from just a few years ago. Employers often contract with managed plans that provide all services for a fixed sum; providers' revenues are stagnant if not declining; manufacturers of drugs and devices can no longer count on sales simply because a new product has been approved (Browning, 1995; Fins, 1994); and new surgical procedures may not be covered unless well-supported by scientific evidence.

At long last, the needed research is underway. Pharmacoeconomic studies, for instance, inquire which drugs work best for what indications, at what cost (Browning, 1995). Guidelines and critical pathways suggest which treatments are appropriate, and how long patients should be hospitalized (Pearson, Goulart-Fisher, and Lee, 1995). Health services research helps to determine how health care systems should more effectively be organized (Gray, 1992; Forman and McClennan, 1994; Gabel, 1998). This research is important, but there are significant problems.

Many of these studies are commissioned or undertaken by manufacturers, employers, insurers, MCOs, and entrepreneurial firms whose commercial product is clinical guidelines sold to insurers, MCOs, and employers (Nelson, Quiter and Solberg, 1997). While many of these firms still refuse to help pay for research conducted independently, such as by academic medical centers, they hold a clear financial interest in the results of the TA and outcomes studies they sponsor on their own, thus opening the door to conflicts of interest and significant opportunities for bias.[10]

For instance, post-marketing research on FDA-approved products has no official requirements for scientific rigor, so long as manufacturers do not use the results in official marketing (Hillman, *et al.*, 1991). In the worst examples, some of these studies are simply marketing devices, with no scientific merit whatever.[11] Other studies look more like science, yet lack acceptable methodology,[12] while still others may be corrupted by researchers' personal conflicts of interest (Brody, B., 1995; Gosfield, 1994; Stelfox, *et al.*, 1998). More broadly, outcomes studies as a whole suffer from a lack of standardized methodologies – what counts as an outcome, which costs are tallied, and the like.[13] And each methodology has distinct (dis)advantages, for instance as administrative data bases (hospital billing records) permit great breadth and quantity of data, but are often littered with gaps and inaccuracies (Hornberger and Wrone, 1997; Iezzoni, 1997; Ray, 1997). And because so many organizations undertake their studies independently, it is difficult for any single project to be large enough to achieve statistical significance.[14]

Even an excellent methodologic design can be compromised during implementation. Some drug manufacturers use for-profit research organizations that cost less than academic medical centers, but may not be directly supervised by the manufacturer, thereby potentially diminishing the company's ability to verify the quality of the research (Browning, 1995; Grandinetti, 1997; Kent, 1996). Bias can also enter when a firm refrains from undertaking a study that might show its product poorly, or refrains from comparing its product to the best competitor.[15] Or sometimes, excellent research whose conclusions do not serve their sponsors' interests may never see the light of day.[16]

On the bright side, there are many high-quality TA and outcomes studies, and they are helping health plans and providers to put their limited resources to the best uses. Yet even the best research has limits (Eddy, 1993; Pearson, Goulart-Fisher and Lee, 1995). Tightly controlled studies with narrow eligibility criteria may not apply well to the more complex patients so typical in the ordinary clinical setting.[17] Studies with clear statistical significance may have little clinical significance (Gifford, 1996). And most studies only collect data on a limited array of outcomes, such as mortality or tumor shrinkage, and do not include other important factors, such as a treatment's effects on quality of life,[18] or its relationship to patient preferences (Hayward, *et al.*, 1995; King, *et al.*, 1994; Feinstein and Horowitz, 1997).

Guidelines creation is a second, related area of concern. Insurers, MCOs, and corporate employers may base their clinical guidelines on outcomes studies or, absent sufficient scientific evidence on expert consensus (Eddy, 1993). Clearly, the choice of which studies or experts to use can be biased by the goals the guidelines are expected to achieve. In other cases, pressures to create a guideline may be strong even if existing studies do not point to clear scientific conclusions (Woolf and Lawrence, 1997). Moreover, guidelines created by entrepreneurial firms are usually proprietary, meaning that data, analyses, and even the conclusions are usually unavailable to examination by the public or the scientific community (American College of Physicians, 1994b, p. 425). Payers may simply rely on the Merck manual, Medicare guidelines, "an administrator who 'asked friends who are doctors,' or an insurance company's employee-physician (usually not a specialist in the field in question) who reads textbooks and discusses the issue with other insurance company physicians" (Holder, 1994, p. 19). Needless to say, such dubious origins can produce equally dubious guidelines (Anders, 1994).

In sum, quality outcomes studies are crucial if health care plans are to draw medically, economically, and ethically intelligent limits on the use of technologies. Yet in addition to the methodological and other problems noted above, a "free-rider" phenomenon poses a major obstacle to adequate research: each player, whether an insurer, MCO, manufacturer, employer, or provider, has an interest that proper studies be done, yet each would prefer someone else to pay the tab. A further drawback is the possibility that a given study might show one's own product to be less than desirable, or that an expensive treatment is preferable – even though in the long run such studies can be expected to work to the advantage of everyone.[19]

B. Responding to the Scientific Challenges

One attractive response would be to create one or a few independent, non-profit agencies to collect contributions from those who need high-quality outcomes science and TA, including MCOs and insurers, manufacturers, corporate employers, physician organizations, and perhaps even patient groups. Such an entity could disburse funds for studies that can be scientifically credible as well as socially, medically, and economically more useful. Woosley, for instance, proposes a dozen or so Centers for Education and Research in Therapeutics (CERTs) to study clinical outcomes of various drugs, to be funded by pharmaceutical companies, governments, MCOs, and the like (Woosley, 1994). Other commentators propose variations on this theme (Kong and Wertheimer, 1998; McGivney, 1992; Neumann, *et al.*, 1996). Although such an entity(s) could be government-based, private agencies might be considerably less subject to the political pressures that recently closed the government's Office of Technology Assessment and precipitated major cutbacks in the scope of the Agency for Health Care Policy and Research (Deyo, *et al.*, 1997; Kahn, 1998).

A central funding approach could offer several advantages. First, the quality of outcomes science and guideline construction might improve considerably if studies are funded independently of any particular manufacturer, MCO, or other interested party. Grants could be made instead to researchers, such as those in academic centers, who specialize in high-quality science.[20] Further, such a funding entity would be better able to coordinate various studies, so that methodologies might be more consistent among studies on the same basic topic. Only then can knowledge be cumulative over time. At least as important, collective funding of larger studies would be better able to include sufficient numbers of research subjects to achieve statistical significance. The broader the recruitment base and the broader the funding, the sooner it is possible to reach credible conclusions.

Second, consolidated research efforts should ultimately save substantial sums of money, both in research and in the delivery of care. Top-quality outcomes science is more likely than poor-quality research to discover what works and what doesn't, and less likely to arrive at ill-founded conclusions that ultimately backfire with poorer health outcomes and greater costs. At the same time, the manufacturers, MCOs, and the other firms that are already concurrently spending large sums of money to reinvent the same wheel may be able to spend less by pooling resources. Consolidation of research efforts can also ameliorate the "free-rider" problem in which each party seeks benefit, but with others bearing the cost (Bennett, et al., 1995; Gabel, 1998).

Third, resource consolidation among multiple parties would also reduce the conflicts of interest that arise when a lone entity pursues its own TA or pharmacoeconomics. For instance, an employer with a self-funded plan may initially want to avoid a new technology that could raise costs. But it will benefit from high-quality, cost-effective care that can help patients return to active lives. The drug or device manufacturer can suffer an initial setback if one of its products is shown not to be as effective or cost-effective as a competing product; but such information can more clearly delineate that company's directions for future research and lead to products that serve everyone's needs better. When these goals are added to physicians' desire to help patients and to patients' need for quality, affordable care, the common denominator is the pursuit of high-quality science to find out which interventions work best for what conditions, at what economic and medical costs. Fact is ultimately better than fiction for everyone.

Encouragingly, some coordination of research is already beginning to occur. A large number of major corporations have pooled funds to create the Foundation for Accountability (FAcct), an organization that is conducting outcomes studies on a variety of conditions. The research looks not only at standard morbidity and mortality as many such studies do, but also investigates quality of life, return to normal functions of living, and other matters important to a broader view of medical outcomes.[21] In an analogous effort, the Managed Care Outcomes Study, funded by six MCOs and the National Pharmaceutical Council, recently led to publication of a study indicating that excessively stringent formulary limits tend perversely to increase patient visits to outpatient offices, emergency rooms, and hospitals, and to raise overall costs of care.[22] In the same vein, the HMO Research Network coordinates twelve research organizations located within integrated health care organizations (Durham, 1998, p. 114).

Such joint efforts are encouraging, but an even broader pooling of funds would be better. If a large number of MCOs, insurers, employers, manufacturers, and providers all contribute to a neutral agency, such as a private, non-profit foundation, whose sole mission is to conduct or allot grants for high-quality research and guidelines construction, it is reasonable to expect even better results, with fewer distorting biases, sooner than they might otherwise be reaped.

III. ETHICAL ISSUES

Ethical challenges primarily concern issues of justice, as we try to determine which patients should have access to which technologies, at whose expense. These issues, in turn, emerge more clearly if technologies are subdivided into three rather different realms.

The first, most obvious realm regards high-cost but still unproven technologies, such as new surgical and medical procedures, and off-label uses of already-approved drugs and devices.[23] Unapproved new drugs and devices will not generally be found in this area, since they require FDA certification as being safe and effective before they can enter general use; hence they may be new and high-cost, but not unproven. New medical and surgical procedures, in contrast, do not require FDA approval (Steinberg, Tunis, and Shapiro, 1995; U.S. General Accounting Office, 1996). And once a drug or device is approved, physicians are generally permitted to prescribe it as they wish. Further scientific studies for additional FDA approval are required only if the manufacturer wishes to advertise additional uses for its product. High-dose chemotherapy with bone marrow transplant, for instance, uses FDA-approved drugs, but can be innovative when applied to conditions for which its value has not been established. Thus, although the procedure's effectiveness is well documented for diseases such as acute leukemia in remission, recurrent neuroblastoma, resistant non-Hodgkin's lymphoma, and advanced Hodgkin's disease (ECRI, 1995a, p. 10), the evidence regarding other conditions, such as breast cancer, is much less well-documented (Antman et al., 1999; Gradishar, 1999; Jeffrey and Waldhold, 1999a, 1999b; Rowlings et al., 2000; Waldholz, 1999, 2000). Other examples in this first area of technology include new surgeries, such as liver-bowel transplants or other multi-organ transplants.

The second realm of technologies encompasses tests and treatments whose value is scientifically well-substantiated, but which are very costly. In this area one inquires whether the bare fact that something works means that it must be covered regardless of cost. Alglucerase, for example, is of great value for people with Gaucher's disease, but its cost can range as high as $400,000 a year.[24] Analogously, drug combinations including protease inhibitors for HIV infection can approach $20,000 per year. These regimens work wonderfully for many patients, and can actually reduce costs by averting hospitalizations for opportunistic illness. But in other cases, side-effects may prompt nonadherence among some patients, a development that could, in turn, spawn new, protease-resistant forms of HIV. Given that many patients with HIV are unable to work and must rely on Medicaid, these new costs pose sobering potential burdens for publicly funded health plans.[25]

The third realm covers the ordinary care of commonplace maladies. With traditional research emphasizing exotic new developments at the frontiers of medicine, much of what physicians routinely do has not been well-studied – a phenomenon that partly explains the wide variations in physicians' practices and the need for guidelines development.[26] On the one hand, health plans are often right to prefer evidence over physicians' habits as justification for paying out substantial sums of money. These habits can be based not so much on time-tested science, as on such non-medical factors as malpractice fears, the beliefs of local leaders, reimbursement patterns, and the like (Burnam, 1987; Wong and Lincoln, 1983). Too often, tests and treatments once thought to be necessary later turn out to be useless or downright harmful.[27] Still, physicians can not cease caring for patients simply because they must use judgment instead

of journal articles. When the needed science is simply lacking, or when
the science does not fit the individual patient, the question arises how
much flexibility physicians and patients should be granted to do whatever
they wish, potentially at great cumulative cost.

A. *Identifying the Justice Issues*

Perhaps the most difficult ethical challenges surrounding the management
of technologies pit the needs and interests of individuals against those of
the broader population needing health care services – they are classic
issues of justice.

From one perspective, medicine is devoted to the welfare of individual
patients. Much of contemporary medical ethics emphasizes physicians'
duties to promote each patient's welfare, not just above the physician's
own interests, but also above societal interests and even those of other
patients (Hall, 1994; Levinsky, 1984; Pellegrino and Thomasma, 1988). If
the patient does not have his own physician as an advocate, he may have
no one. Further, an overt denial of a treatment that offers at least some
chance of saving a life or averting a major harm is often seen as moral
callousness, an abandonment that may harm those who issue the denial
almost as much as those who are denied. For some commentators, this
hazard is so serious that resource limits, however necessary, must be
drawn covertly (Calabresi and Bobbitt, 1978; Miles, 1992).

Beyond this, it may be reasonable to argue that the definitions of
"safe" and "effective" can legitimately soften as a patient's condition
worsens. After all, if death is virtually certain without treatment, then it is
difficult to call a treatment "dangerous" that might actually prolong life.
And when no other treatment offers even a remote chance of survival, a
new treatment might be "effective" even if its chance of success is
miniscule.[28]

Accordingly, if we believe that each individual is intrinsically
precious, and not valued only because of his usefulness to society, then it
is important to honor that specialness by making at least a reasonable
effort to save endangered lives (Jonsen, 1986). Even when it is impossible
to help everyone in need it is still possible to honor symbolically the
value of each individual by making a serious effort to help the limited
number of persons whose needs become openly known. Such moral
symbolism may seem inadequate if one is tallying the totality of needs

against the total who are helped. But it is not trivial, and can honor an important value whose complete fulfillment is impossible.

From a very different perspective, however, acknowledging the importance of individuals does not require society or any particular health plan to spend endless amounts of money pursuing infinitesimal chances for dying people. Nor does it mandate that health plans bankrupt themselves buying costly, last-hope treatments for a very few subscribers. Justice requires due consideration for the broader population, comprised of individuals whose needs and entitlements are no less real than those of the identified person in special need. In this context, four notions of justice are useful to consider.

Distributive justice is probably bioethics' most familiar conception of justice. It calls for a fair distribution of benefits and burdens across society. In health care, selecting the criteria to guide such fairness can spawn vigorous discussion, such as whether allocation of limited resources should be based on equal shares for everyone, or allocated according to the individual's need, his prior contributions to society, or his personal merit – or whether instead some sort of market distribution should prevail.[29] And one can debate other priorities: acute care versus preventive care, rehabilitation, long-term custodial care, and so forth. Though distributive justice is clearly important, other notions of justice must also figure in any attempt to reconcile the needs and demands of individuals with those of society at large or, of special interest for this article, other members of the same health plan.

Formal justice points out that similar cases must be treated similarly (Beauchamp and Childress, 1994, p. 328). If a bone marrow transplant is granted to a woman with breast cancer, despite the lack of substantial scientific support for its efficacy (ECRI, 1995a, pp. 1-12; Antman *et al.*, 1999; Rowlings *et al.*, 2000); or if liver-bowel transplant is mandated as a benefit in a state's Medicaid program even when nearly 60% of those receiving this new treatment have died;[30] or if intensive care must be available to support the life of an anencephalic infant (*Matter of Baby K*, 1993; 1994); the implications do not just concern people with breast cancer or short-bowel syndrome or anencephaly. Any health plan that provides these treatments must make similarly costly, comparably unproven or marginally effective treatments available to equally desperate patients with other illnesses. Determining precisely what counts as 'similar' will, of course, be controversial. But whatever the definition, the similarities are bound to have a broad and costly scope. Beyond this, if each individual is important, then that importance does not disappear just because the individual's name has not appeared in headlines. Hence, formal justice requires consistency in the application of resource criteria and in the exercise of discretion. Once the full implications of formal justice are taken into account, the total costs of generosity and compassion for one suffering individual can potentially be crushing (Morreim, 1995d).

Contractual justice holds that each individual should receive whatever level of care he actually bought, neither more nor less.[31] It is a principle of fairness and promise-keeping in economic transactions. When a health plan has identified the benefits it does and does not provide, contractual justice requires that the plan not hand out extra benefits, simply because some particular individual urgently wants, needs, or demands more than he paid for.[32] Such "generosity" would be unfair to fellow subscribers by depleting collective resources. On another level it is even unfair to the individual. If health plans can not be counted on to honor the limits as well as the benefits of their contracts, it becomes impossible for individuals to translate their personal resource values into a concrete choice of a health plan whose explicit benefits and limits fit those values. 'Generosity' that changes the character of the plan's benefit scheme, and ultimately its premium prices, does not give the buyer what he paid for.

As discussed below in Section IV, a serious problem arises from the vagueness and ambiguity typical of health plan contracts (Bergthold, 1995; Havighurst, 1995, p. 15, 110 ff). The less specifically a plan defines beneficiaries' entitlements, the more difficult it is to determine precisely what is required by contractual justice. While ambiguities are commonly interpreted in favor of the insured, taken to the extreme such a presumption could effectively destroy contractual integrity. Hence, contractual justice in itself carries its own version of the individuals-vs-population tension.

Contributive justice begins with the fact that, because at least some linguistic ambiguity is inevitable, and because no contract can cover every contingency, judgement and discretion are unavoidable in the administration of a health plan.[33] The exercise of such discretion has implications for all the other people served by that health plan. Formal justice, as noted, insists that every act of discretion favoring a particular individual be duplicated for all others similarly situated. Contributive justice asks how the remaining people in the plan, whose contributions create the plan's resources and who then rely on that plan for their own care, will be affected. Generosity, after all, has very real opportunity costs. A hospital might provide an exotic, news-making treatment such as surgical separation of conjoined twins (Thomasma, *et al.*, 1996). Yet that same institution may be laying off nurses, dieticians, and physical therapists – with clear costs to other patients via reduced services for their comfort, recovery, function, perhaps even survival. Similarly, a Medicaid MCO that must use its fixed funds for liver-bowel transplants or the latest drugs for HIV has no choice but to find other places to cut back on the services it provides.[34] The 'easy' decision to provide extraordinary help to the identified individual is easy only as long as one ignores the implicit but real effects on everyone else in the health plan.

B. Resolving the Ethical Challenges

The tensions between individuals and populations raise important questions of justice in the three realms of technology listed at the outset: when to provide still-unproven new technologies; how quickly to disseminate costly technologies whose value is proven; and how best to utilize the ordinary technologies whose best uses have not yet been identified by credible outcomes science. Even if it is imperative to honor the moral significance of suffering individuals, it is likewise mandatory to observe justice for the larger population.[35]

1. Ordinary care
As boundaries are drawn on the uses of health care technologies, needy individuals can enjoy a moral focus in several ways. As a prerequisite, it seems imperative that all citizens have access to at least some basic level of care – a point that will be more assumed than argued here. Morally it is considerably less troubling to deny someone an exotic, unproven treatment when this individual will still receive good basic care, than when the alternative is little or no care. If adequate health is a precondition for most of life's important goals and activities, and if ordinary care actually does preserve and restore health, as advertised, then surely all citizens should have assured access. The best uses of routine treatments need to be guided by careful outcomes research, of course, to ensure wise resource use and to avoid iatrogenic injuries. And that research should, as noted above, be the product of high-quality science. This realm is the most fundamental level of health care. Further, the same universal access that honors the moral importance of the

individual also serves the broader group by promoting a vigorous, healthy, productive population.

2. Costly, unproven treatments

Miracles are not morally mandatory, and health plans should not be compelled to honor the importance of individuals by showering them with exotic, unverified heroics that may more likely waste money or cause harm than achieve their intended purpose. Even if such efforts symbolically honor these people as valued individuals, there are other, more defensible ways to show that deference.

One of those alternatives is to grant wider access to promising new technologies via controlled trials. If the needy individual is in a scientific trial, then although some patients may be randomized to receive placebo, no one can complain that access is completely denied. Health plans should recognize that it is in their interest to help fund such research. It is better for them to pay for serious research at an early stage than to be forced, as many currently are, to pay for unproven treatments that proliferate rapidly, unsystematically, and unscientifically via the pressures of litigation and intense public demand. Arguably, one reason why many patients have been so desperate, and judges have been so liberal in granting access to new technologies like bone marrow transplant for breast cancer, is that so few of those who might have been candidates for the treatment have had access to trials.[36]

For patients who are clearly beyond curative treatments and trials, whether from fatal illness or from irreversible disability, greater emphasis on palliative care and on making the best of the patient's remaining life also seems imperative. The intrinsic moral value of each individual can be significantly honored in these less dramatic but highly important forms of support. It is an area too neglected by medical science and too often ignored by health plans.

Granting access to unproven technologies, though only through research trials, serves the broader population as well as needy individuals. Barring access to promising new technologies has tended to trigger public (and sometimes judicial) outrage, thus perversely spawning uncontrolled access. The result has been not just wasted resources, but harms occur to both current and future patients via a failure to learn as quickly as possible whether and for what indications a new treatment works.[37]

3. Costly-but-proven care

Within this realm it is important to distinguish between situations for which no comparable alternative is available, and those for which there is a viable alternative. When a treatment has been shown to offer a reasonable likelihood of substantial benefit for seriously ill patients – markedly to prolong life or significantly to improve its quality – and when no alternative offers comparable benefit, the treatment should ordinarily be covered. As noted above, high-cost treatments for which the magnitude or likelihood of benefit are marginal or unproven are not morally mandatory. But where benefits are clear and large, and there is no alternative, such treatments arguably ought to be covered even if they are costly. The operational definitions of "reasonable likelihood" and "substantial benefit" will of course be debatable. But at least some cases will be clear enough, and the point here is that, in those cases, costs should not ordinarily be a decisive obstacle to care.

The reasons are both moral and practical. Morally, society's "rule of rescue" says that it is at least *prima facie* desirable to rescue endangered life, at least if the rescue is likely to be successful and if the hazards and costs are not undue (Jonsen, 1986; Morreim, 1994). Refusing to provide interventions of known and substantial help, where the cost is not prohibitive, would seem to be a clearly unacceptable abandonment. And where a technology is costly but proven, the first prerequisite for the rescue imperative is clearly met: *ex hypothesi*, these rescues are likely to be successful because of these interventions' documented effectiveness. They benefit those who are currently ill and reassure the broader population, any of whom might need such help in the future.

The practical side concerns the level of burden. So long as health plans are able to spread risk over a fairly broad population, the actual financial impact of clearly beneficial treatments will ordinarily not, at least for now, be undue. Indeed, a major reason that some of the treatments in this group are so costly is that the illnesses are "orphan diseases." Because they are by definition rather rare, it is unlikely that enough treatment will be provided to permit the price reductions that usually come with increasing sales volume. But reciprocally, that same rarity means that even if treatment costs are high, the relatively few who receive them will not ordinarily make an intolerable financial dent.

Admittedly, this reasoning has caveats. Although the basic principle may be relatively clear, interpretation and implementation can vary. Even if health plans should generally cover alglucerase for Gaucher's disease, for instance, plans might differ about the precise conditions for which the drug is indicated. At this time, although the benefit is clear for certain kinds of patients, further research is needed to determine just when to initiate treatment, and what dosages are best for which situations (NIH, on Gaucher Disease, 1996). On these details plans may legitimately differ. Further, as costly new technologies proliferate, such as genetically-based diagnostics and therapeutics, their impact on the overall cost of care could become so substantial that their coverage or noncoverage could cumulatively make a large difference. At that point, plans might legitimately differ in their coverage. Such variations will be discussed in more detail below.

In one more practical consideration, perhaps a bit more generosity in the area of well-proven technologies that lack any viable alternative may actually conserve resources in the long run. Society's strong rule of rescue can prompt rescues in the heat of passion, sometimes costing exorbitant amounts for improbable outcomes. Stronger assurances to cover worthwhile rescues, in this realm of costly-but-proven technologies, might help to defuse the tendency to leap into more dubious efforts.[38]

In contrast, where alternative treatments offer roughly comparable benefits at markedly lower cost, a basic health plan should generally be obligated only to cover the least costly among them. Partly this is because a society's obligation to its less fortunate members is not limitless. Even if it is agreed that an affluent society should ensure health care alongside other necessities such as food and shelter, it would be difficult to argue that such assistance must encompass optimal health care regardless of cost, particularly since the high cost of a maximal commitment in health care would surely impair society's ability to help in other important areas.

C. Implementing the Resolutions

A discussion about implementing these technology limits begins with the point that, since it has already been proposed that costly unproven technologies should be available only through controlled trials, the remaining task concerns the other two realms: routine, relatively low-cost care, and costly-but-proven interventions. For both practical and theoretical reasons, these two should be handled differently. Arguably, patients should have considerable freedom to control the details of ordinary care, while explicit resource management should focus mainly on the relatively few situations that engender the highest costs.

1. Routine Care
Perhaps in theory, health plans might create detailed guidelines to indicate precisely which intervention – which antibiotic, how many x-rays – should be provided under what conditions. However, although guidelines may inform choices, a complete account of every detail of care is neither plausible nor desirable.

It is implausible for several reasons, beginning with the fact that the vast majority of health care encounters are relatively low-cost: "85% of Americans spend less than $3000 a year on medical care, and 73% have less than $500 a year in claims."[39] High-quality scientific studies to cover every permutation of every medical contingency would be prohibitively costly and logistically impossible. Given the myriad co-morbidities and biological idiosyncrasies that ordinary patients have, scientific studies would have to be enormous in order to provide the right instructions for every patient. And given the steady emergence of new technologies and new uses of existing technologies, such studies would need constant updating.

Applying such guidelines to literally every situation would be insufferably cumbersome, with even the simplest lab test or prescription potentially delayed by a process of guideline searches, permissions and appeals – a cascade of administrative costs second only to the research costs forsworn a moment ago. It would surely be the worst kind of utilization review nightmare, of which so many physicians already complain bitterly. And under such constraints, widespread "gaming the system" for interventions that go beyond the guidelines would quite likely be a major problem. To complete the nightmare, that battery of rules would provide only "nickel-and-dime" savings.

Such detailed rules would also be undesirable, not just implausible. Personal values permeate medical decisions. Even though a person might want generally to subscribe to a given level of care, for instance, the discomforts of a particular drug taken for a chronic illness might warrant a switch to a costlier alternative. Some of the common hypertension medications, for instance, have side-effects such as fatigue or impotence that affect relatively few individuals. Those people may want alternatives. Equally important, it is difficult if not impossible for physicians and patients to develop mutually trusting relationships if their every decision must be monitored by a third party. A significant measure of clinical discretion – the freedom to do whatever makes the best sense in a given situation – is essential if physicians are to be able to negotiate with their patients to reach mutually acceptable treatment regimens (Morreim, 1998).

In sum, good medical care cannot be provided without resources, and the use of those resources requires judgment and flexibility – not intrusive, costly micromanagement of routine details. Still, these lesser interventions are cumulatively even if not individually costly. Utterly unrestricted access, at the plans' expense, is not a viable option any more than pervasive controls. Perhaps the most plausible alternative would permit physicians and patients together to make their own decisions, but under some financial consequences. Such consequences are essential, given a basic economic dynamic. Because every medical decision is also a spending decision, and because health plans are created to help individuals absorb the high costs of such medical care, health plans have only three responses when physicians and patients make spending decisions. They can either pay the bill without question (no longer an option); or they can dictate medical decisions by dictating spending decisions (often intrusive, and not a viable option for routine care, as noted above); or they can permit physicians and patients to make their own decisions by shifting to them some of the financial consequences.[40]

Directing such consequences primarily at physicians, unfortunately, can create formidable conflicts of interest. A superior approach would place at least some financial incentives on patients themselves. The arguments for this are discussed elsewhere and need not be belabored here.[41] Patients will have neither economic nor medical control unless they experience in some way some economic consequences for their medical (spending) choices. Medical savings accounts, and variants of them, are one way to ensure immediate access to care without barriers or penalties for being ill, while still rewarding patients for considering the cost-worthiness of the care they request (Morreim, 1995a). Self-control replaces external control. And with the great majority of decisions left up to physicians and patients, health plans need not be so preoccupied with the daily details of care. Administrative savings could be substantial if cost constraints focus instead on the limited realm of higher-cost care.

2. *Costly Care*

While most people consume rather little in health care resources during a given year, conversely most of the health care budget is devoted to a relatively few people with high-level needs.[42] Practicality and justice both suggest that systematic resource policies be applied to this arena of serious chronic illness, and major acute illness and injury.

Practically, it is here that careful science can and should identify the best uses of diagnostic and therapeutic technologies. By focusing on a relatively limited array of conditions, such research can more reasonably be expected to produce credible conclusions. And where better science is describing the most effective uses of various technologies, the resulting guidelines can be enforced more consistently, permitting greater predictability of health plan expenditures. Furthermore, the savings from adhering to carefully drawn guidelines in this realm can be substantial, by reducing both the waste and the suffering that can occur when high-cost technologies are applied imprudently.

Justice also requires rather faithful adherence to the guidelines drawn at this level. Although virtually all health care decisions can involve cost-value tradeoffs, the economic and thereby medical consequences for everyone in the health plan are considerably greater at this high-cost level than for routine ailments. Hence, it is particularly at this level, featuring major illnesses, major costs, and major trade-offs, where a resource philosophy should be articulated and embodied in a set of guidelines.[43] At that point, fairness to everyone requires rather strict enforcement.

Formal justice requires that whatever benefits are made available to one person must be offered to everyone else who is similarly situated. Therefore, deviating from protocols to provide one desperate person a costly treatment not otherwise covered by his health plan may require similar deviations for others equally desperate.

Contractual justice requires enforcing commitments. If a person has freely and knowledgeably chosen a health plan with a specific resource philosophy and explicit guidelines and limits, then it is imperative to respect the contractual expectations of all the others who have signed onto that same plan, by implementing its provisions and limits in good faith. People should receive fully the benefits of their bargain, but they should not receive more than they paid for, at others' expense.

Concomitantly, *contributive* justice requires that ambiguities and requests for flexibility be resolved by looking to the overall spirit of the health plan's resource philosophy, whether it be a "lean-and-mean" plan, or "nothing but the best," or something in between.

Enforcing health plans' limits also honors individuals' autonomy. When one chooses one level of health care rather than another, he has explicitly made cost-value trade offs that permit him to receive certain health benefits, and to forego others, in exchange for a specified price. If health plans dishonor those limits and hand out extra benefits – ultimately raising costs for their subscribers – they thwart the individual's decision to leave money available for other priorities, whether his children's education, his retirement, or his next new car.

It is one thing to assess the ethics of health plans' obligations toward their members. It is another thing to ensure that plans can effectively enforce their limits. As discussed below, legal structures have tended to foster a too-rapid diffusion of technology, and to pose significant obstacles to plans' efforts to draw limits.

IV. LEGAL ISSUES

A. Sources of Difficulty

Two areas of law have spurred the rapid proliferation of unexamined technologies: contract and tort.

1. Contract

Subscribers receive health care benefits by contracting with health plans. Ordinarily, judges must interpret a contract by its plain language.[44] However, three factors invite judges to favor subscribers over health plans: adhesion, ambiguity, and arbitrariness.

First, a number of courts have viewed health care contracts as contracts of adhesion. In a contract of adhesion, a stronger party dictates the terms of an agreement, presenting a take-it-or-leave-it offer with little opportunity to negotiate specific terms. The weaker party is vulnerable, purchasing the service out of personal need rather than from a desire to profit, so that "leaving it" is not a viable option.[45] Under such conditions, courts can be quick to side with the vulnerable party when contract benefits are in dispute. In a variety of cases, courts have cited this feature of health plan contracts to award patients disputed (and sometimes scientifically very dubious) technological interventions (Ferguson, Dubinsky and Kirsch, 1993; Kalb, 1990).

Health care often carries another adhesory element: beyond an inability to negotiate internal terms of the health plan, the subscriber often can not even choose the plan itself. Most Americans receive health coverage through the workplace. Of businesses that provide it at all, 84% offer only one plan.[46] Although recent years have seen some change, many judges have found it difficult to bind subscribers to the limits of contracts they did not choose (Morreim, 1995d).

Second, in any contract, ambiguities are construed against the drafter. This doctrine, also called *contra proferentum* (Havighurst, 1995, p. 182) is based on the fairness principle that, since the party writing the agreement had the opportunity to make the wording clear, its failure to do so should not harm the party who lacked this opportunity.[47] Unfortunately, vagueness is unavoidable in health care contracts. To specify which interventions are covered, an insurer would have to describe all of medicine: what diagnostic and therapeutic interventions are indicated, under exactly what conditions. Clearly this is impossible since, in medical services, flexibility and individuality of care are indispensable. Accordingly, health plans have usually described benefits rather generally, defining medical necessity mainly by deferring to the medical profession's accepted practices (Bergthold, 1995; Havighurst, 1995, p. 15, 110 ff.), and defining experimental treatments, if at all, rather vaguely.[48] Since accepted practices (and payers' perceptions of them) vary widely, so can benefits determinations.

Third, judges frown on arbitrariness. Where ambiguity concerns the wording of the agreement, arbitrariness focuses on the human processes of implementing and interpreting whatever is written. Unfortunately, the inevitable vagueness in health plan contracts virtually guarantees some measure of arbitrariness as different administrators decide which medical interventions are necessary, unnecessary, or experimental. In a further entrée for arbitrariness, many health plans' definition of medical necessity is now shifting away from the traditional reliance on professional customs, toward written guidelines. However, those guidelines vary from one plan to the next.[49] And because they are often proprietary and undisclosed, there is little opportunity to determine, in advance, just what kinds of care each plan is likely to consider necessary for which medical indications. So long as plans purport to cover all medically necessary care, and then base necessity judgments on varying and undisclosed guidelines (Holder, 1994; Holoweiko, 1995), an appearance of arbitrariness is inevitable.

In sum, adhesion, ambiguity, and arbitrariness provide ample opportunities for courts to grant subscribers a generous contract-based access to new technologies. Effective, enforceable limits have been difficult to draw.

2. Tort

Tort law has also fueled diffusion of technology, mainly through the standard of care to which physicians are held in the malpractice context. First, the standard of care is mainly based on the profession's prevailing practices, which historically have quickly expanded to embrace new technologies. Partly this has been a self-fulfilling prophecy: physicians feel that they must use the latest technologies, even if not of proven value, lest they be sued. The greater the apprehension, the faster the proliferation, and the more the technology actually becomes prevailing practice and thereby the malpractice standard of care.[50]

Second, physicians must provide the same standard of care to all patients whom they accept, regardless of the patient's (in)ability to pay. Where prevailing practices include costly technology, tort standards essentially expect physicians to commandeer other people's money and property to provide that care even for an indigent patient.[51] To some extent courts have reinforced this view, as seen, for instance, in *Wickline v. State of California* and *Muse v. Charter Hospital Winston-Salem, Inc.*

B. Responding to the Legal Challenges: Contract

Myriad decisions are required for any health plan to determine which interventions will be covered for what indications. These are made more difficult by a serious tension. On the one hand, health plans typically purport to cover all and only "medically necessary" care, requiring a determination whether each specific intervention, in a particular situation, is "necessary" or "unnecessary." It purports to be a black-and-white determination with no hint of gray. On the other hand, treatments often differ from one another, not by any clear necessity or disutility, but on a range of probabilities and merit. The costlier, wider-spectrum antibiotic is a bit more likely to eradicate the infectious organism; a finer-grained diagnostic test such as magnetic resonance imaging may be somewhat likelier to detect an unusual or unexpected condition; the genetically synthesized clot lysis drug may be slightly more likely than its cheaper alternative to save the life of a heart attack patient.

In other words, medical interventions are not "effective" or "ineffective" *per se*. Each is likely, or not, with some probability, to reach a specified goal, which in turn is more or less desirable, according to some particular values or viewpoint. Antibiotics can be more or less effective in eradicating various kinds of bacteria, depending on each organism's sensitivity or built-up resistance. (They are completely ineffective against viruses, yet physicians have often prescribed them for patients with viral illnesses – very effectively – toward a different sort of goal: to placate a demanding patient.)

Analogously, a given intervention is rarely "(un)safe," simpliciter. Each presents a given risk for certain outcomes or side-effects that are more or less (un)desirable, depending on whose preferences are applied. A surgeon may refuse to operate on a patient with a very high mortality risk, for example, but to the patient for whom the procedure represents a last hope, the risk may be entirely acceptable. Indeed, the value of most medical interventions is highly contingent on which goal is being sought, and those goals can differ widely among various patients and providers.

"Experimental" care is similar. Though some tests and treatments are utterly new and untested, virtually any medical technology can be in a state of flux. Major and minor modifications of drugs and devices, and new uses of existing products, mean that, in some sense, much of medical care is experimental to some degree. Thus, a given technology is rarely either "experimental" or "accepted" *per se*, but rather has been supported by science and/or experience, to a greater or lesser degree, for this or that particular use.[52] Hence, it is virtually impossible for those writing health plan contracts to define "experimental" in a clear and credible way.[53]

Accordingly, instead of pretending to cover all and only "necessary" care, and presuming that necessity is a straightforward matter of science, health plans should finally acknowledge that many benefits designs and coverage decisions are normative, far more than medical (Glassman, Jacobson, and Asch, 1997). The real question is whether the prospective improvement to be provided by treatment is sufficiently likely, desirable, and substantial to merit the medical risks and financial costs.

Those costs include not just direct costs, but opportunity costs, the things that must be foregone because finite funds have been spent on one use rather than on alternatives. Though these value questions permeate the life and death issues so familiar in bioethics, they figure even more prominently in the quality-of-life services that comprise the bulk of ordinary medical care – services ranging from pain relief for arthritis and childbirth, to symptom reduction for self-limited illnesses, to cataract removal or acne treatment (Morreim, forthcoming). People can legitimately differ on the importance of such care, depending on how they value health care relative to other goods in life, from housing to education.

In sum, "medically useful" does not entail "mandatory purchase." In a more coherent, defensible health care system, the concept of medical necessity should be discarded[54] in favor of benefit packages at varying levels. Although this article takes it as given that all people should have access to a basic, reasonable level of care, choices should be available beyond that point. Some people will prefer a lower-cost plan that covers only those diagnostic and treatment options whose value is well-documented and the least costly among reasonable alternatives. Others might want to upgrade to costlier, less cost-efficient, less well-proven kinds of care. Overall, the less clear and substantial the value of the intervention is, the more its inclusion in the health plan should be a matter of choice.

Such tiered approaches, alluded to in Section III, have been defended by a number of commentators (Hall and Anderson, 1992; Havighurst, 1995; Kalb, 1990) and need only be briefly described here. "Basic" care, the kind to which all citizens arguably should have assured access, encompasses interventions that are clearly effective in promoting the central goals of medicine: preserving life; preventing illness, injury, and premature death; relieving unnecessary pain and suffering; preserving or improving function for major life activities, ameliorating disabling conditions.[55] As noted above, when only one intervention exists that can reliably achieve such a goal, it should be considered basic. If more than one treatment is available with roughly comparable effectiveness, then basic care might include only the less costly.

Tiers of care might differ in other ways. First, plans might vary in the extent to which they offer access to research trials. Although health plans would help themselves as well as their members by contributing more systematically to carefully selected research protocols, basic health plans might limit their participation to the most promising trials, and permit only a limited number of patients to be in a given trial at any one time. Costlier plans might offer broader access to approved trials, and still richer plans might permit access to trials that are based more on theoretical promise than on research-based evidence. Thus, novel surgeries might be found only in the richer plans until they have built a reliable track record.

Second, plans may differ on the threshold of evidence required before a test or treatment is deemed sufficiently "effective" toward a desired outcome. The strength and quality of scientific evidence on behalf of a conclusion can vary considerably. David Hadorn identifies several possible standards of proof: (1) "Reasonably expected to provide significant net health benefit;" (2) "Reasonably well demonstrated to provide significant net health benefit;" and (3) "Clearly demonstrated to provide significant net health benefit" (Hadorn, 1992, p. 20). On such a scheme, leaner plans might demand a stronger burden of proof before adding a new treatment, while richer plans might accept any treatment that is sufficiently promising, until evidence shows it to be ineffective (Eddy, 1993). A multi-organ transplant, for instance, might be excluded by a basic plan if it showed less than a fifty percent chance of one-year survival.[56] Leaner plans might also exclude treatments that are not clearly for a medical problem, such as growth hormone for non-growth-hormone-deficient children (Bercu, 1996; Cuttler, et al., 1966).

Third, plans could vary regarding interventions that are only marginally different in their effectiveness and safety. If a costly new drug offers only a small added benefit, such as an improved success rate of two percent or less, a leaner plan might presume to use the less costly version and restrict the new one, perhaps for limited indications or with special justification. Thus, for instance, as science permits diagnosis of genetic predispositions toward breast cancer or ovarian cancer, leaner plans might demand a high rather than low likelihood that the trait would produce disease, before covering prophylactic treatment.[57]

Fourth, health plans might differ on matters of transient comfort and convenience (as distinct from long-term quality of life). Thus richer plans could include the convenience of costlier, once-a-day drugs (Eddy, 1992; Jacobson and Rosenquist, 1988) while leaner ones might permit low-osmolar radiographic contrast dyes only for people at risk for allergic reaction to the older, less comfortable ones (Ellis, et al., 1996; Jacobson and Rosenquist, 1988).

The clarity of choices among such tiers could be considerably enhanced by limiting the number of guideline-tiers from which to choose, perhaps to just four or five levels. Such a move has precedent. Recognizing that excessive variety can become confusing if not paralyzing to intelligent choice,[58] the federal government made basic changes in the so-called "Medi-gap" insurance that many older citizens buy to cover expenses not covered by Medicare. These plans had become so numerous, and varied so widely, that comparison-shopping was virtually impossible. Accordingly, federal legislation limited insurers to just ten basic levels of medi-gap insurance. A given insurer might offer all ten, or only selected options. But each of the ten coverage-levels must be the same, no matter which insurer offers it. Plans then compete on other factors, such as quality of service and cost.[59]

By analogy, rather than the current situation, in which virtually every health plan bases benefits decisions its own idiosyncratic set of guidelines, subscribers could make more informed, intelligent decisions by choosing among a few basic tiers of care.[60] Those tiers might be created via the independent research agency described above, in section II-B, with the respective guideline-sets available to any health plan that wanted to offer that tier of care. Competition could then be based on quality of providers, efficiency of service, price, and the like.

Such guidelines would be available to public examination and critique – unlike the proprietary guidelines currently hidden from patients and physicians. That secrecy not only inhibits thoughtful medical scrutiny and improvement, but also poses a significant obstacle to effective contracting: if the buyer has no real opportunity to learn what he has bought, it is more difficult for courts to enforce the contract's resource limits.

A limited number of guideline-sets produced by an independent agency might also help resolve another nagging legal problem, stemming from the requirement to construe contract ambiguities against the drafter. So long as health plans create their own contracts, and write or adapt their own guidelines, courts must continue to construe ambiguities against the plans. Because ambiguities are inevitable, plans often lose. And so, ultimately, do the plan members. When courts award costly, untested treatments that, by implication, must likewise be awarded to everyone else with a similarly desperate desire for a comparably unproven remedy, plans are hard-pressed to limit their costs or thereby their premiums. In contrast, if guidelines were designed by an independent, nonprofit entity, their ambiguities could no longer be construed against the health plan, because the plan did not draft them. Litigation against the plan would instead focus on whether its agents interpreted the guidelines fairly and in a manner consistent with the overall spirit of that tier's resource philosophy.[61]

Note that this recommendation, in which people would choose their health plan on the basis of how much they want to invest for what level of health care services, does not entail that an individual must know and understand every element of each tier's guidelines. Such detailed comprehension is not required for valid contracting, any more than contracts to buy an automobile require the buyer to understand how the car is put together in every mechanical respect. Rather, the consumer must have a basic description of the resource philosophy at each tier, its values concerning effectiveness, risks, convenience, and such (Havighurst, 1995, p. 201). And the guidelines themselves should be available for whoever wants more detailed information.

These improvements in the contracting process seem essential if health plans are to be able to draw effective limits. And only if those limits are enforceable can they provide a real opportunity for subscribers to make choices that permit them to determine what priority they will accord to health care compared to the other things they value. Admittedly, it can be difficult to live with the real consequences of plans saying "no, you may not have this intervention, because you did not buy it." A society accustomed to an "artesian well of money" that poses no barrier to the latest and greatest technologies will surely take offense at some of those denials. But there is no alternative. Endless expenditures that proliferate untested treatments and costs must now give way to a sober recognition of limits. Further, courts have shown a real willingness to enforce contractual limits when the plans' terms are reasonably clear, and when subscribers had a *bona fide* opportunity to choose among alternatives (Havighurst, 1995, pp. 161-164).

Fortunately, opportunities for real choice are emerging. Many employers are moving away from traditional "defined benefits" toward "defined contributions" in which the firm establishes a specific sum for employees' health *coverage* and then makes several choices available. The defined contribution covers basic plans, while those wanting costlier plans pay extra. The range of choices grows markedly when groups of employers come together in "purchasing pools" that lower costs via group leverage.[62] Such pools include Pacific Business Group on Health, a collaboration of 27 large West-coast firms that together spend some $3 billion to cover 25 million employees, dependents, and retirees (Schauffler and Rodriguez, 1996; Robinson, 1995; Luft, 1995); the

Health Insurance Plan of California, a California consortium of small private employers covering some 100,000 lives (Robinson, 1995; Luft, 1995; Shewry, *et al.*, 1996; the Buyers' Health Care Action Group, a collection of 24 large firms in the Twin Cities area (Murata, 1996; O'Brien, 1996; Slomski, 1996; Wetzell, 1996; Report to the Board of Trustees, AMA, 1996); and the Federal Employees Health Benefits Program (American College of Physicians, 1996; Butler and Moffit, 1995).

C. Responding to the challenge: Tort law

When resource issues are managed through explicit contract rather than through vague expectations, important corresponding adjustments can be made in the malpractice standard of care. This issue has been discussed elsewhere, (Morreim, 1997; Morreim, 1992; Strosberg, *et al.*, 1992) and so needs only to be summarized here.

As noted above, courts traditionally expect physicians to deliver a roughly uniform standard of care. As medical care has grown to encompass costly technologies, physicians have essentially been expected to commandeer other people's money and property on behalf of their patients. Fears that a failure to use the latest technology could precipitate tort liability have prompted many physicians, in a self-fulfilling prophecy, to integrate new technologies quickly into the standard of care to which, in turn, they have been held accountable in tort. In days of generous funding, this expectation posed little problem. But as payers and health plans have tightened the resource reins, a jurisprudential challenge has arisen in which physicians stand to be held personally liable for others' resource decisions.

The resolution must be to distinguish between expertise issues and resource disputes. The former concerns the knowledge, care, and skill physicians bring to their performance in diagnosing and treating illness. These are matters under physicians' control, to which they can rightly be held accountable. In contrast, resource obligations are largely a function of contracts between health plans and their members. While the former should properly be addressed under tort as a standard of care issue, the latter is suitably managed under contract. Once this separation is achieved, physicians may feel considerably less pressure to drive up technology proliferation out of malpractice fears.

V. CONCLUSION

The creation, diffusion, and utilization of new technologies pose enormous challenges in an era of constrained resources. Dramatic rescues can symbolically honor the preciousness of each individual, but ultimately can also imperil other citizens' access to basic care. A more intelligent approach has become imperative.

As argued in this essay, we need first to develop a better scientific basis on which to decide which technologies really work, and for what indications. This in turn needs greater research cooperation among the many players who need such information: employers, health plans, drug and device manufacturers, and even patients.

Second, we must recognize the deeply normative character of the choices surrounding emerging technologies. We must acknowledge that medicine is not so scientific as we suppose, that not all interventions are of equal worth, that "effective" need not mean "medically necessary" or "mandatory purchase," and that resources provided for one person carry costs for others that must be recognized. Accordingly, distributive, formal, contractual and contributive justice all urge us to acknowledge that enforcing limits is not only acceptable, it is morally essential. Those limits need not be one-size-fits-all, but rather should permit individuals to participate in choosing the limits that will affect this important area of their lives.

Finally, it will do little good to recognize a moral imperative to limit technology if the legal system is unwilling to enforce such limits. Through a wider array of choices, clearer contracting, and more realistic apportioning of tort liability, it should be possible to achieve more rational, fair, and defensible ways of sharing the costly technologies that can do so many people so much good – at such a very high price.

College of Medicine, University of Tennessee
Memphis, Tennessee

NOTES

[1] Gray *et al.*, (1993, p. 1521) note that "new procedures are infrequently evaluated in randomized trials, and the trials that are conducted rarely include cost analyses." See also Grimes 1993; Garber, 1992; Kessler, Kessler and Myerburg, 1995; NIH Technology

Assessment Panel on Gaucher Disease, 1996; Buto, 1994 p. 138; Reiser 1994; United States General Accounting Office, 1996.

[2] State Medicaid programs, for instance, have been required to reimburse all new drugs for a six-month period after introduction, after which point it may be difficult to withdraw them from general availability to people who have grown accustomed to using them. See Soumerai, Ross-Degnan, Fortess, and Abelson, 1993, p. 234.

[3] Garber, 1994, pp. 121-122. Indeed, once a treatment received approval, it could be very difficult to remove it from the list, even if its value was later clearly disproved. See also Buto, 1994, p. 138.

[4] See Anderson, Hall, Steinberg, 1993, p. 1636; Hall, Anderson, 1992, p. 1638. As one court said, "subscriber premiums should not have to pay for procedures which are purely experimental or investigative or subsidize every scientist stirring a magic potion in some laboratory at the top of a mountain with lightning flashing about" (Holder, 1994, p. 20).

Another commentator has argued that "private health insurance pays for medical technologies and interventions that are proven to be safe, effective, and medically necessary. Asking private insurers to defray the costs of clinical trials that seek to answer safety and efficacy questions about emerging therapies is, therefore, inherently contradictory... . extending coverage policies to include expensive unproven technologies (ones that cannot be expected to cure but may offer some marginal clinical benefits) increases the cost of covering the treatments that are known to be 100% effective (i.e., appendectomies)" (Cova, 1992, p. 744). See also Eddy, 1997; Buto, 1994; Peters and Rogers, 1994, p. 477.

[5] One of the earlier changes came when insurers began to challenge oncologists using new combinations of approved drugs for chemotherapy. See Chalmers, 1988 and Antman, Schnipper, and Frei, 1988.

[6] Reciprocally, where payment for new technologies is not relatively sure, research and development is likely decrease. Steinberg, Tunis, Shapiro, 1995, p. 150.

It can be difficult for insurers to determine whether and when physicians actually are using new approaches. The replacement of conventional surgery with new laparoscopic techniques for cholecystectomy (gall-bladder removal), for instance, was not initially visible to insurers. So long as there was no separate billing code for the laparoscopic technique, the procedure would be billed under the conventional code for that surgery. Partly as a result of this, laparoscopic cholecystectomy diffused rapidly prior to published studies documenting any superiority. The laparoscopy device itself had been approved, and so its emerging uses were left up to the medical community. See Buto, 1994, p. 138-39.

[7] McGivney (1992, p. 743) suggests that estimates about misuse of technologies may significantly underestimate the problem. "When rates of inappropriate utilization of 15%-30% are discussed, two other important factors are missing from the computation. First, there are technologies (e.g., external carotid-internal carotid artery bypass, Garren gastric bubble, gastric freezing) that should never have been introduced into general practice; thus, the rate of inappropriate utilization is 100%. Second, there are specific devices (e.g., magnetic resonance imaging) that may be applied for an approved indication; yet they are applied uselessly due to insufficient power of the individual device (e.g., magnet) or inadequate technical considerations. Thus, the cost of the inappropriate utilization of technology may well exceed present estimates."

See also Reiser, 1994; Garber, 1994; Jensen, et al., 1994; Kent, et al., 1994; American College of Physicians, 1994a; Durenberger and Foote, 1994.

[8]   Jensen *et al.*, 1994; Deyo, 1994; Ewigman *et al.*, 1993; Berkowitz, 1993; Winslow, 1996; Goodwin and Goodwin, 1984.

[9]   Wennberg, 1986; Chassin, *et al.*, 1986; Wennberg, *et al.*, 1987; Chassin, *et al.*, 1987; Wennberg, 1990; Leape, *et al.*, 1990; Leape, *et al.*, 1989; Wennberg, 1991; Cleary, *et al.*, 1991; Fisher, *et al.*, 1992; Greenfield, *et al.*, 1992; Welch, *et al.*, 1993; Miller, *et al.*, 1994; Detsky, 1995; Guadagnoli, *et al.*, 1995; Pilote, *et al.*, 1995.

[10] "A major issue is that the primary source of funding for this research is often the primary financial beneficiary of positive study results. Unfortunately, even valid studies done under the best of circumstances may be suspect" (Task Force on Principles for Economic Analysis of Health Care Technology, 1995, p. 62). See also Kassirer and Angell, 1994; Evans, 1995; Garber, 1994; Cho and Bero, 1996; Neumann, *et al.*, 1996.

[11] One such "study," for instance, was ostensibly intended to assess the efficacy and tolerability of [the company's drug] in controlling mild-to-moderate hypertension. The sponsor used its sales force to recruit 2500 office-based 'investigators' who were frequent prescribers of drugs in the therapeutic class in question. Each investigator was to enroll 12 patients (for a total enrollment of 30,000) and was offered reimbursement of $85 per patient enrolled, or $1050 per physician. The 'study' was not capable of achieving even the modest objectives stated. There was no control group, and the study was not blinded. There was thus no possibility that it would generate useful data on efficacy and little likelihood that it would produce data on safety other than the potential for detecting a rare adverse event. Kessler, *et al.*, 1994, p. 1351.

[12] One pharmaceutical company, for instance, compiled an extensive registry listing patients who have had a particular illness (e.g., heart attack) and what treatments they received with what outcome. The registry paid physicians to provide considerable information about the use of drugs manufactured by the company, but not about patients who received alternate treatments. As a result, the data is not only not the product of any controlled trials, it does not even include comparative information. Though it has the look of science and is often cited by physicians, it is thus not seriously scientific (King, 1996). Similarly, a study of clot lysis in patients who have suffered heart attack might, if funded by the manufacturer of one particularly expensive drug, incorporate methodological or analytic techniques that might tend to favor that drug (Brody, 1995).

As another example, many of the studies of high-dose chemotherapy with autologous bone marrow transplant (HDC/ABMT) for breast cancer also show sobering scientific deficits. For instance, treatment-related deaths (death within one month after transplant) were frequently disregarded; these patients were reported as 'unevaluable' because they 'did not survive long enough to exhibit a clinical response.' In many cases, omitting these patients led to higher response rates. Eliminating early deaths is inappropriate (particularly when they may have been caused by the treatment) and not standard for trial design or analysis (ECRI, 1995).

Other methodological problems were rampant. Some studies lacked controls entirely, others included only those patients who had already shown they were responsive to chemotherapy, and still other studies neglected to keep track of key patient characteristics, such as the number of metastatic sites, or estrogen receptor status.

[13] Soumerai, *et al.*, 1993; Balas, *et al.*, 1998; Kassirer and Angell, 1994; Evans, 1995; Epstein, 1995.

Economic analysis in particular may require "unique methodologic choices, such as which types of costs to include (direct, indirect, intangible, induced), which perspective to apply (that of society, payer, provider, patient), which design to adopt (cost-identification, cost-benefit, cost-effectiveness, cost-utility), from where to obtain costs (indemnity database, managed care or capitated database, hospital cost systems, Medicare, Medicaid), and whether to collect resource consumption data prospectively or retrospectively through various modeling techniques" (Task Force on Principles for Economic Analysis of Health Care Technology, 1995, pp. 61-62).

In designating endpoints of study, one study might take survival alone as the mark of success, while another might focus on survival without major neurological deficits. See Whyte, et al., 1993 and Allen, et al., 1993.

"These guidelines activities are uncoordinated, however, and different agencies sometimes issue different guidelines on the same topic. Occasionally, their recommendations conflict. At least some of these differences are probably attributable to the vastly different methods used to create the guidelines. Methodological differences include the types of people selected to be in the expert group, methods used to collect and synthesize evidence, methods used to structure the group discussion and arrive at consensus, and degree to which recommendations are linked directly with the evidence behind them" (Power, 1995, 205). See also Holoweiko, 1995 and Jones, 1995.

[14] "Even when measures are aggregated, statistical power may be insufficient to detect a significant improvement in outcomes. For example, in a modestly sized health plan with 25,000 members, one might estimate that 150 persons will have diabetes and use insulin. Even if all 150 participated in a program that reduced the number of diabetes-related complications by 50% through improved diet, exercise, and appropriate use of insulin, the statistical power would be too low to document the program value compared with complication rates with a previous program" (Epstein and Sherwood, 1996, p. 833). See also Durham, 1998.

[15] As Furberg notes regarding the question whether drugs called calcium agonists (CAs) should be used for hypertension:

> The documentation on efficacy and safety is limited to small, short-term clinical trials, typically a few hundred patients treated for 2 to 3 months. Pharmaceutical companies have used skillful marketing based on concepts and mechanisms of action to promote these agents and have avoided for more than a decade calls for large-scale, randomized clinical trials to determine the effect of these drugs on major cardiovascular disease end points. Clinicians have chosen to prescribe CAs without proper evidence of health efficacy and long-term safety (1995, p. 2157).

[16] In one recent case, a study showing that inexpensive thyroid medications were just as effective as the leading, costly brand was abruptly pulled from publication. The study's sponsor – the manufacturer of the costlier leading drug – had reserved the right to approve or refuse any publications prior to press (Rennie, 1997; King, 1996).

[17] "The distinction between effectiveness and efficacy is critical to disease management; treatments with proven efficacy do not always perform as well under conditions of typical clinical practice." Treatments must be shown to have efficacy before they will be accepted, but the "evaluation generally requires a priori hypotheses, randomization (to eliminate selection bias and confounding), homogeneous patients at high risk for the outcome, experienced investigators who follow a protocol, a comparative measure such as a placebo (if ethical), and intensive follow-up to ensure compliance. Under these circumstances, if a treatment proves to be better than a placebo (or a comparative measure), one can be reassured

that the treatment can work. However, questions may remain about the ability of the treatment to work adequately in a broader range of patients and in usual practice settings in which both pts and providers face natural barriers to care" (Epstein and Sherwood, 1996, p. 883). See also Feinstein and Horwitz, 1997; Wells and Sturm, 1995; Editorial, 1995.

[18] A study of the results of newborn intensive care might presume that "intact survival" will only be counted if the patient has an I.Q. over 85, and does not suffer from cerebral palsy or visual or auditory impairment (Gifford, 1996, p. 42). See also Tannock, 1987.

[19] Suppose, for instance, that two organizations are deciding whether to fund a cost-effectiveness study: "If both fund the study, then they will split the cost evenly. If both decide not to fund the study, then no economic study is performed. However, if one decides to fund the study and the other says no, then the agency that says yes is obligated to fund the entire study." The result is a prisoner's dilemma. "Letting the other organizations fund the economic study is optimal from the individual point of view, but it is inefficient from the viewpoint of society as a whole" (Bennett, et al., 1995, p. 2460).

[20] As a further advantage, such funding might help academic medical centers that currently are seeing major diminutions in their revenues for conducting research and training.

[21] FAcct is compiling research and guidelines on a number of conditions, including asthma, breast cancer, coronary artery disease, depression, diabetes and low-back pain. See Keister, 1995 and Terry, 1996.

[22] "Both in general and for patients with similar levels of disease severity, the more limited the formulary, the *higher* the patient prescription count and doses ... , and the *higher* the number of ambulatory office visits, ED visits, and hospitalizations per patient per year " (Horn, et al., 1996 , p. 259).

[23] Buto, 1994, p. 137; See Reiser's (1994) concept of crossover technologies, p. 132.

[24] Note, some issues remain unclear, such as when to begin treatment. NIH Technology Assessment Panel on Gaucher Disease, 1996; Garber, 1992; Holoweiko, 1995.

[25] Gulick, et al., 1997; Hammer, et al., 1997; Bartlett, 1996; Deeks, et al., 1997; *Wall Street Journal*, 1996; Shelton, 1996.

[26] Wennberg, 1986; Chassin, et al., 1986; Wennberg, et al., 1987; Chassin, et al., 1987; Wennberg, 1990; Leape, et al., 1990; Leape, et al., 1989; Wennberg, 1991; Cleary, et al., 1991; Fisher, et al., 1992; Greenfield, et al., 1992; Welch, et al., 1993; Miller, et al., 1994; Detsky, 1995; Guadagnoli, et al., 1995; Pilote, et al., 1995.

[27] Magnetic resonance imaging for lower back problems has been commonplace, but studies show it has little value except in a limited range of cases. See Jensen, et al., 1994; Deyo, 1994. Similarly, ultrasonography is commonly used as a screening tool during pregnancy, but recent research findings suggest it is only of use in certain higher-risk situations. See Ewigman, et al., 1993; Berkowitz, 1993. See also Winslow, 1996. Reciprocally, efficacious therapies are sometimes rejected because they do not 'make sense' in light of accepted theories. See Goodwin and Goodwin, 1984.

[28] The arguments parallel those in the so-called futility argument. An intervention is neither effective nor ineffective *per se*; it is (in)effective toward some particular goal. Though many people would call "futile" a treatment that can only preserve a permanently vegetative existence, for instance, vitalists would call that same treatment "medically necessary" so long as it can keep that individual alive. See Morreim, 1995c. The two sides have a fundamental difference in their goals: life of a certain quality, versus any life at all. Similarly, persons in otherwise good health, for instance someone contemplating a knee or hip replacement, want a good chance of success before accepting the risks of such a procedure. However, a person

who is otherwise doomed may take any prospect of survival as his goal, in which case a treatment offering even a thin thread of hope might be seen as "safe" and "effective" enough to warrant a try.

[29] Beauchamp and Childress, 1994. Similarly, deciding how broadly to grant access to health care, and for what reason, has raised theories ranging from the utilitarian, to libertarian, communitarian, egalitarian, and fair opportunity. See Beauchamp and Childress, 1994, pp. 334-341.

[30] Brandie Hinds, a three-year-old child with short bowel syndrome due to mid-gut volvulus with gangrene, had been nourished via total parenteral nutrition (TPN), a liquid nutrition usually administered by catheter into a large vein. Because this form of feeding ultimately destroys the liver, Hinds needed a liver transplant. However, because the continuing use of TPN would eventually destroy any transplanted liver, Hinds's physicians sought also to transplant bowel in hopes of returning her to normal nutrition. Hinds was a Medicaid patient, and the state's Medicaid office regarded the combined transplant procedure to be experimental and thus not covered by her health plan, which was one of the dozen MCOs in the state's TennCare managed competition plan. This type of surgery had been performed for only a few years, at only a small number of institutions, and 58% of all patients had died. *Hinds v Blue Cross and Blue Shield of Tenn*, No. 3:95-0508, M.D.TN, 12/28/95.

[31] Eddy notes that coverage criteria are important for the plan "to act in the best interest of its entire membership. If selected individuals can get coverage for particular treatments that do not meet the criteria agreed to in the contract, they are siphoning resources from the pool available to provide care to other members" (Eddy, 1996, p. 651). Note that although government programs such as Medicare and Medicaid are not technically contracts between governments and beneficiaries, the moral principles of contract nevertheless apply. People should receive the specific benefits to which they are legitimately entitled, but they should not expect to receive more.

[32] An important caveat here arises from the fact that, in many cases, individuals did not choose their own health plans; their employers or government did. As noted above, it is much more difficult to hold individuals to the terms of contracts that were drawn on their behalf, but without their consultation. This challenge will be addressed further just below.

[33] Formal, contractual, and contributive justice are further described in Morreim, 1995d.

[34] One consequence, for instance, might be that a Medicaid MCO pays such minimal fees for specialist physicians, that few specialists are willing to sign on with the plan. For instance, in Tennessee's managed competition Medicaid plan – TennCare – one of the TennCare MCOs allegedly paid so little to specialists, that for a period of time the MCO had no contracts with some types of specialists, and very few contracts with other types of specialists. For example, it could not attract any dermatologists in Memphis and Shelby County to sign up; similarly, it could not arrange for allergists in Nashville and Davidson County. Although the MCOs have an obligation to provide such care, in some cases it has been met by requiring patients to drive significant distances (Derks, 1995). The more that such plans with limited budgets must provide exotic care to a few individuals, the more that deficits of various kinds are likely to arise for all the other members of the plan.

Issues of contributive justice figured prominently in the recent case of *Hinds v. Blue Cross and Blue Shield of Tenn*, No. 3:95-0508, M.D.TN, 12/28/95 In this case, the Tennessee state Medicaid office determined that the liver-bowel transplant requested for a three-year-old suffering from short-bowel syndrome was experimental and thus not covered. As the matter proceeded to litigation in federal district court, the state argued against mandating the surgery, arguing that "the public interest is not served by granting of an injunction. 'Services upon demand' rather than a proper package of Medicaid benefits is inconsistent with managed care for the maximum number of citizens. TennCare participants have a right to medically necessary services, not largely untested and risky procedures not approved or accepted by the medical community" (Defense brief, submitted to federal district court, M.D.TN., 8/17/95, p. 16.).

The federal district court disagreed. The chief judge argued that the procedure is not experimental, since 42% of its patients are still alive and because no other treatment is available for this needy individual: "safety and effectiveness are relative concepts: as the potential benefit of a technology increases, the level of acceptable risk also increases" (12/28/95, at p. 18). The court went on:

> The injunction will not harm others. Defendants submit that a small bowel transplant is extremely expensive and will place the TennCare program under an enormous economic burden. While the Court is cognizant of the need to contain health care costs in order to distribute benefits among the widest number of recipients, the law is clear that necessary medical treatment must not be sacrificed in an effort to contain costs. ... The public interest is served by ensuring that the laws governing Medicaid and TennCare benefits are followed. The public interest is also served by allowing Brandie to obtain the benefits to which she is entitled under these laws (12/28/95, at p. 18).

[35] Practicality suggests that any revisions regarding the evaluation and best uses of technology should probably retain intact the two major structures by which new technologies are created and introduced in this country. Hence, this article presumes that manufacturers will continue to undertake basic research and development of new drugs and devices, and that the Federal Drug Administration will continue to rule on their safety and efficacy. Though there may be room for improvement, these structures seem to work reasonably well, and a major overhaul of that system seems unlikely.

[36] Admittedly, broader access to trials will not entirely resolve the problems that arise when desperate individuals ask for unconventional treatments. Some situations, such as techniques for surgically separating conjoined twins who share major organs, are simply too rare for scientific trials ever to collect enough cases to achieve statistical significance. However, such cases would be a tiny and considerably less troublesome minority once the more common cases involving demands for new technology are resolved. Arguably, they should be resolved within the overall spirit of the health plan. Plans that advertise low-cost premiums in exchange for conservative care would arguably violate contributive justice by exercising such off-plan "generosity" with subscribers' money. On the other hand, plans that offer a much broader access to less well-proven, less cost-effective treatments, might be able to justify a few such cases of unusual expense. See text for further discussion of tiers among health plans.

[37] Quite likely, for instance, if bone marrow transplant for breast cancer had been carefully studied at a much earlier date, women today would be the beneficiaries via a more scientific understanding of which types of breast cancer, if any, it actually helps. Instead, the early studies attesting to its value were fraught with sobering defects. For instance, treatment-related deaths (death within one month after transplant) were commonly disregarded on the ground that these patients were 'unevaluable,' failing to survive long enough to show clinical response. The frequent result was a probably-factitious finding of favorable response rates. Other methodological problems were rampant. Some studies lacked controls entirely, others

included only those patients who had already shown they were responsive to chemotherapy, and still other studies neglected to keep track of key patient characteristics, such as the number of metastatic sites, or estrogen receptor status. ECRI. High-dose chemotherapy with autologous bone marrow transplantation and/or blood cell transplantation for the treatment of metastatic breast cancer. Healthy Technology Assessment Information Service, 1995, p. 7. Of five studies released in 1999, four indicated that high-dose chemotherapy with bone marrow transplant was no better for breast cancer than conventional chemotherapy. A fifth study, done in South Africa, suggested some benefit. However, several months later, as scientists looked at this study more closely in an effort to replicate its results, the principal investigator admitted to having falsified some of the data "in a 'foolish desire' to make his research more 'acceptable' to the scientific meeting sponsored by the American Society of Clinical Oncology" (Waldholz, 2000, B2). See Antman, *et al.*, 1999; Gradishar, 1999; Jeffrey and Waldhold, 1999a, 1999b; Kolata, 1999; Rowlings, *et al.*, 2000; Waldholz, 1999, 2000.

[38] Daring and unprecedented separations of twins conjoined at major organs, for instance, can cost astronomical amounts of money while promising little if any prospect of success. See Thomasma, *et al.*, 1996.

[39] Editorial, 1994.

[40] For further discussion of this economic dynamic, see Morreim, 1998.

[41] See Morreim, 1995a and Morreim, 1995b.

[42] As noted above, 85% of us spend less than $3000 a year on medical care, while 73% are under $500 a year (Editorial, 1994). Alternately described, "[t]he most expensive 1 percent of the U.S. population accounts for 30 percent of all health spending, and the least expensive 50 percent of the population accounts for only 3 percent of health spending" Blumberg and Nichols, 1996, p. 36.

[43] Moreover, if lower-price decisions are freely made by patients and physicians under an MSA form of accountability, then higher-cost care is the only area left for health tiers to differ.

[44] *Loyola University of Chicago v. Humana Ins. Co.*, 1993; *Fuja v. Benefit Trust Life Ins.*, 1994; *McGee v. Equicor-Equitable HCA Corp.*, 1992; *Harris v. Mutual of Omaha Companies*, 1993; *Arrington v. Group Hospitalization & Medical Service*, 1992; *Barnett v. Kaiser Foundation Health Plan, Inc.*, 1994; *Goepel v Mail Handlers Benefit Plan*, 1993; *Nesseim v. Mail Handlers Benefits Plan*, 1993; *Farley v. Benefit Trust Life Ins. Co.*, 1992; *Harris v. Blue Cross Blue Shield of Missouri*, 1993.

[45] "A contract of adhesion is a contract which has some or all of the following characteristics: the parties to the contract were of unequal bargaining strength; the contract is expressed in standardized language prepared by the stronger party to meet his needs; and the contract is offered by the stronger party to the weaker party on a 'take it or leave it' basis." *Morris v Metriyakool*, 1984 (opinion of Justice Ryan). See also *Madden v. Kaiser Foundation Hospitals*, 1976; Anderson, *et al.*, 1993; Gottsegen, 1981.

[46] Blendon, *et al.*, 1995; American College of Physicians, 1996.

Another study concluded that 78% of firms offer just one plan, although often with multiple products, such as a POS version of an HMO. See Berenson, 1997.

From another vantage point, a study looking at workers rather than employers found that 48% of employees have only one health plan available, while 23% have only two plans to choose from, and 12% have three plans to choose from. Etheredge, *et al.*, 1996.

[47] "Ambiguities in the plan should be resolved against the insurer," *Bucci v Blue Cross-Blue Shield of Connecticut*, 1991. "If a court finds that an insurance policy is ambiguous, ... an

ambiguous policy will be construed in favor of the insured," *Katskee v. Blue Cross/Blue Shield*, 1994.

[48] Holder, 1994; Holoweiko, 1995a; Steinberg, *et al.*, 1995.

Even when contracts are worded quite precisely, judges sometimes interpret liberally. In a phenomenon called "judge-made" insurance law, judges may stretch clear contractual language to favor desperate subscribers seeking their only, even if very small, chance to survive. Abraham, 1981; Ferguson, *et al.*, 1993. Indeed, courts need not become involved at all where insurers agree to provide benefits under the mere threat of litigation. Peters and Rogers, 1994; Anders, 1994a; Holoweiko, 1995. In other cases, federal or state governments mandate the use of a new technology, obviating even the need to threaten litigation. For example, nearly half a dozen states plus the federal government have mandated HDC/ABMT for breast cancer (Holoweiko, 1995; ECRI, 1995a). "On September 10, 1994, the United States Office of Personnel Management (OPM) 'mandat[ed] immediate coverage of HDC/ASCR for all diagnoses for which it is considered standard treatment and, in addition, specifically for breast cancer, multiple myeloma, and epithelial ovarian cancer' for the approximately 200 insurers participating in the Federal Health Benefits program, which covers nine million federal employees," *id.*

[49] For further discussion see Morreim, 1997.

[50] In some cases a technology becomes standard even though its value is doubtful – electronic fetal monitoring and fetal ultrasound in low-risk pregnancy have been cited as examples. D.A. Grimes suggests, regarding electronic fetal monitoring: "Based on uncontrolled case-series reports and cohort studies with historical controls, this expensive and intrusive technology was widely adopted and disseminated during the 1970s. When the requisite randomized controlled trials were finally done, the consensus was striking: routine electronic fetal monitoring confers no demonstrable benefit to the fetus, yet poses a significantly increased risk of operative delivery (*e.g.*, cesarean delivery or forceps) for the woman. Even for high-risk fetuses, evidence of the benefit of electric fetal monitoring over traditional auscultation is lacking. After two decades of use, electric fetal monitoring has not been shown to be superior to intermittent auscultation with a stethoscope." Grimes, 1993. Recent studies have suggested that fetal ultrasound in low-risk pregnancy does not improve fetal outcomes. Ewigman, *et al.*, 1993; Berkowitz, 1993.

[51] For a considerably more detailed discussion of these points, see Morreim, 1997; Morreim, 1987.

[52] Reiser discusses this phenomenon of evolving uses of technologies under the term "crossover technologies" (Reiser, 1994).

[53] Havighurst, 1995; Holoweiko, 1995; Holder, 1994.

In one case, a court discarded the 5-factor criteria used by Blue Cross/Blue Shield, its Technical Evaluation Criteria (TEC) "Summarized, the criteria are (1) government regulatory approval; (2) evidence which permits conclusions as to the effect on patient health; (3) demonstrated improvement of the patient's health; (4) demonstration of medical benefit at least equal to that offered by established alternative treatment; and (5) improvement other than in investigational settings," *Bucci v. Blue Cross-Blue Shield of Connecticut*, 1991. This particular court rejected these criteria on the ground that they were too subjective, instead looking as whether there is a "reasonably substantial, qualified, responsible, relevant segment of the medical community which accepts the procedure as properly within the range of appropriate medical treatment, under the circumstances of the case, as judged by the standards of the community" (764 F. Supp. 732). Critics might wonder whether the latter standard is significantly less subjective.

Medicare standards differ somewhat. "The clearest articulation of the considerations that go into determining whether a particular service is experimental is found in a letter Medicare uses to explain to his clients and providers why a service is ineligible for reimbursement: 'In making such a decision [whether to provide payment for a particular service], a basic consideration is whether the service has come to be generally accepted by the professional medical community as an effective and proven treatment for the condition for which it is being used. If it is, Medicare may make payment. On the other hand, if the service is rarely used, novel or relatively unknown, then authoritative evidence must be obtained that it is safe and effective before Medicaid may make payment'" (Miller by *Miller v. Whitburn*, 1993; citing *Rush v Parham*, 1980). This policy mainly looks for some level of consensus in the medical community, requiring evidence is only if the test or treatment is rare, novel or unknown.

[54] Challenges to "medical necessity" in benefit language are also endorsed by other commentators. See Eddy, 1996; Bergthold, 1995; Glassman, *et al.*, 1997.

[55] Beauchamp and McCullough, 1984; Callahan, *et al.*, 1996. See also Jonsen, *et al.*, 1992. These authors list the goals of medicine as: (a) health promotion/disease prevention; (b) relieving symptoms, pain, suffering; (c) curing disease; (d) preventing untimely death; (e) improving functional status, maintaining compromised status; (e) educating, counseling patients; (f) avoiding harm to patients during care.

[56] See, e.g., the case of Brandie Hinds, *supra* note 30.

[57] For further discussion of this issue see Glazier, 1997; see also *Katskee v. Blue Cross Blue Shield of Nebraska*, 1994.

[58] "People can efficiently process and use only five or six variables or pieces of data in each decision. With more information, a person's ability to use that information declines" (Hibbard, *et al.*, 1997).

[59] Pub. L. No. 101-508, 4358, 104 Stat. 1388 (codified at 42 U.S.C. 1395ss (1994)); see also Farrell, 1997.

[60] Note also that a limited number of tiers and guidelines would make intelligent resource decisions far easier for physicians, who currently must negotiate their way around a stunning diversity of rules and regulations for patients' widely varying health plans.

[61] Havighurst recommends focusing litigation on the procedural aspects of implementing the contract in just this way (1995).

[62] Schauffler and Rodriguez, 1996; Robinson, 1995; Butler and Moffit, 1995; Luft, 1995. Purchasing pools have been strongly endorsed by the American College of Physicians, which argues that "the mutual desire of physicians and patients for the maintenance of high-quality, accessible, affordable health care is clearly best served by the empowerment of individual persons in the marketplace. The College believes that this empowerment can be achieved through purchasing pools" (American College of Physicians, 1996).

REFERENCES

Cases and statutes

Arrington v. Group Hospitalization & Medical Services, 806 F.Supp. 287, 290 (D.D.C. 1992).

Barnett v. Kaiser Foundation Health Plan, Inc., 32 F.3d 413 (9th Cir., 1994).

Bucci v. Blue Cross-Blue Shield of Connecticut, 764 F. Supp. 728, 731 (D.Conn. 1991).

Doe v. Group Hospitalization & Medical Services, 3 F.3d 80 (4th Cir. 1993).

Farley v. Benefit Trust Life Insurance. Co., 979 F.2d 653 (8th Cir. 1992).

Fuja v. Benefit Trust Life Insurance. Co., 18 F.3d 1405, 1412 (7th Cir. 1994).

Gee v. Utah State Retirement Bd, 842 P.2d 919, 920-21 (Utah App. 1992).

Goepel v. Mail Handlers Benefit Plan (No. 93-3711, 1993 WL 384498 (D. N.J. 9/24/93)).

Harris v. Blue Cross Blue Shield of Missouri, 995 F.2d 877 (8th Cir. 1993).

Harris v. Mutual of Omaha Companies, 992 F.2d 706, 713 (7th Cir. 1993).

Hinds v. Blue Cross and Blue Shield of Tennessee, No. 3:95-0508, M.D.TN, 1995.

Katskee v. Blue Cross/Blue Shield, 515 NW2d 645, 647 (Neb. 1994).

Loyola University of Chicago v. Humana Insurance Co., 996 F.2d 895 (7th Cir 1993).

Madden v. Kaiser Foundation Hospitals, 552 P.2d 1178 (Cal. 1976).

Matter of Baby K, 832 F. Supp. 1022 (E.D. Va. 1993).

Matter of Baby K, 16 F. 3d. 590 (4th Cir. 1994).

McGee v. Equicor-Equitable HCA Corp., 953 F.2d 1192 (10th Cir. 1992).

McLeroy v. Blue Cross/Blue Shield of Oregon, Inc., 825 F.Supp. 1064 (N.D. Ga. 1993).

Miller by Miller v. Whitburn, 10 F3d 1315, 1320 (7th Cir. 1993) (citing Rush v Parham, 625 F.2d 1150, 1156 (5th Cir. 1980)).

Morris v. Metriyakool, 344 NW 2d 736, 756 (Mich. 1984) (opinion of Justice Ryan).

Muse v. Charter Hospital Winston-Salem Inc., 452 S.E.2d 589 (N.C.App. 1995).

Nesseim v. Mail Handlers Benefits Plan, 995 F.2d 804 (8th Cir., 1993).

Sarchett v. Blue Shield of California, 729 P. 2d 267 (Cal. 1987).

Thomas v. Gulf Health Plan, Inc., 688 F. Supp. 590 (S.D. Ala. 1988).

Wickline v. State of California, 192 Cal. App. 3d 1630 (1987).

Aaron H.J., Schwartz W.B.: 1984, *The Painful Prescription: Rationing Hospital Care*, Brookings Institution, Washington D.C.

Aaron, H.J., Schwartz, W.B.: 1985, 'Hospital cost control: A bitter pill to swallow,' *Harvard Business Review* 64, 160-167.

Abraham E., Wunderink R., Silverman H., Perl T.M., Naraway S., Levy H., Bone R., Wenzel R.P., Balk R., Allred R., Pennington J.E., Wherry J.C.: 1995, 'Efficacy and safety of monoclonal antibody to human tumor necrosis factor in patients with sepsis syndrome,' *Journal of the American Medical Association* 273, 934-941.

Abraham, K.S.: 1981, 'Judge-made law and judge-made insurance: Honoring the reasonable expectations of the insured,' *Virginia Law Review* 67, 1151-1191.

Allen, M.C., Donohue, P.K., Dusman, A.E.: 1993, 'The limit of viability – neonatal outcome of infants born at 22 to 25 weeks' gestation,' *New England Journal of Medicine* 329, 1597-1601.

American College of Physicians: 1994a, 'Magnetic resonance imaging of the brain and spine: A revised statement,' *Annals of Internal Medicine* 120, 872-875.

American College of Physicians: 1994b, 'The oversight of medical care: A proposal for reform,' *Annals of Internal Medicine* 120, 423-431.

American College of Physicians: 1996, 'Voluntary purchasing pools: A market model for improving access, quality, and cost in health care,' *Annals of Internal Medicine* 124, 845-853.

Anders, G.: 1994a, 'More insurers pay for care that's in trials,' *Wall Street Journal*, February 15, B-1.

Anders G: 1994b, 'Limits on second-eye cataract surgery are lifted by major actuarial firm,' *Wall Street Journal* 12/15/94, B-6.

Anderson G.F., Hall M.A., Steinberg E.P.: 1993, 'Medical technology assessment and practice guidelines: Their day in court,' *American Journal of Public Health* 83, 1635-1639.

Antman, K.H., Heitjan, D.F., Hortobagyi, G.N.: 1999, 'High-dose chemotherapy for breast cancer,' *Journal of the American Medical Association* 282, 1701-1703.

Antman, K., Schnipper, L.E., Frei, E. III.: 1988, 'The crisis in clinical cancer research: Third-party insurance and investigational therapy,' *New England Journal of Medicine* 319, 46-48.

Avorn, J., Chen, M., Hartley, R.: 1982, 'Scientific versus commercial sources of influence on the prescribing behavior of physicians,' *American Journal of Medicine* 73, 4-8.

Balas, E.A., Kretschmer, R.A.C., Gnann, W., *et al.*: 1998, 'Interpreting cost analyses of clinical interventions,' *Journal of the American Medical Association* 279, 54-57.

Bartlett, J.G.: 1996, 'Protease inhibitors for HIV infection,' *Annals of Internal Medicine* 124, 1086-1088.

Beauchamp, T.L., Childress, J.F.: 1994, *Principles of Biomedical Ethics*, 4th edition, Oxford University Press, New York.

Beauchamp, T.L., McCullough, L.B.: 1984, *Medical Ethics: The Moral Responsibilities of Physicians*, Prentice-Hall, Englewood Cliffs, New Jersey.

Bennett, C.L., Smith, T.J., George, S.L., Hillner, B.E., Fleishman, S., Niell, H.B.: 1995, 'Free-riding and the prisoner's dilemma: Problems in funding economic analyses of phase III cancer clinic trials,' *Journal of Clinical Oncology* 13, 2457-2463.

Bercu, B.B: 1996, 'The growing conundrum: Growth hormone treatment of non-growth hormone deficient child,' *Journal of the American Medical Association* 276, 567-568.

Berenson, R.A.: 1997, 'Beyond competition,' *Health Affairs* 16(2), 171-180.

Bergthold, L.A.: 1995, 'Medical necessity: Do we need it?' *Health Affairs* 14(4), 180-190.

Berkowitz, R.L.: 1993, 'Should every pregnant woman undergo ultrasonography?' *The New England Journal of Medicine* 329, 874-875.

Blendon, R.J., Brodie, M., Benson, J.: 1995, 'What should be done now that national health system reform is dead?' *Journal of the American Medical Association* 273, 243-244.

Blumberg L.J., Nichols L.M.: 1996, 'First, do no harm: Developing health insurance market reform packages,' *Health Affairs* 15(3), 35-53.

Brody, B.: 1995, *Ethical Issues in Drug Testing, Approval, and Pricing*, Oxford University Press, New York.

Browning, E.S.: 1995, 'Change in health care shakes up the business of drug development,' *Wall Street Journal*, March 28, A-1, A-6.

Burnum, J.F.: 1987, 'Medical practice a la mode,' *The New England Journal of Medicine* 317, 1220-1222.

Butler, S.M., Moffit, R.E.: 1995, 'The FEHBP as a model for a new medicare program,' *Health Affairs* 14(4), 47-61

Buto, K.A.: 1994, 'How can Medicare keep pace with cutting-edge technology?' *Health Affairs* 13(3), 137-140.

Calabresi, G., Bobbitt, P.: 1978, *Tragic Choices*, Norton & Co, New York.

Callahan, D., *et al.*: 1996, *The Goals of Medicine: Setting New Priorities*, unpublished manuscript, Hastings Center.

Chalmers, T.C.: 1988, 'Third-party payers and investigational therapy (letter),' *New England Journal of Medicine* 319, 1228.

Chassin, M.R., Brook, R.H., Park, R.E., *et al.*: 1986, 'Variations in the use of medical and surgical services by the Medicare population,' *The New England Journal of Medicine* 314, 285-290.

Chassin, M.R., *et al.*: 1987 'Does inappropriate use explain geographic variations in the use of health care services?' *Journal of the American Medical Association* 258, 2533-2537.

Cho, M.K., Bero, L.A.: 1996, 'The quality of drug studies published in symposium proceedings,' *Annals of Internal Medicine* 124, 485-489.

Cleary, P.D., Greenfield, S., Mulley, A.G., *et al.*: 1991, 'Variations in length of stay and outcomes for six medical and surgical conditions in Massachusetts and California,' *Journal of the American Medical Association* 266, 73-79.

Cova, J.L.: 1992, 'A swift response to a "modest" proposal,' *Journal of the National Cancer Institute* 84, 744-745.

Cuttler, L., Silvers, B.J., Singh, J., Marrero, U., Finkelstein, B., Tannin, G., Neuhauser, D.: 1966, 'Short stature and growth hormone therapy: A national study of physician recommendation patterns,' *Journal of the American Medical Association* 276, 531-537

Deeks, S.G., Smith, M., Holodniy, M., Kahn, J.O.: 1997, 'AHIV-1 protease inhibitors: A review for clinicians,' *Journal of the American Medical Association* 277, 145-53.

Derks, S. A.: 1995, 'Specialists may be sparse for enrollees in Access plan,' *Comm. App.*, May 9, B-1, B-2.

Detsky, A.S.: 1995, 'Regional variation in medical care,' *New England Journal of Medicine* 333, 589-590.

Deyo, R.A., Psaty, B.M., Simon, G., Wagner, E.H., Omenn, G.S.: 1997, 'The messenger under attack – intimidation of researchers by special interest groups,' *New England Journal of Medicine* 336, 1176-1180.

Deyo, R.A.: 1994, 'Magnetic resonance imaging of the lumbar spine: Terrific test or tar baby?' *New England Journal of Medicine* 331, 115-116.

Durenberger, D.F., Foote, S.B.: 1994, 'Technology and health reform: A legislative perspective,' *Health Affairs* 13(3), 197-205.

Durham, M.L.: 1998, 'Partnerships for research among managed care organizations,' *Health Affairs* 17(1), 111-122.

ECRI: 1995a, 'High-dose chemotherapy with autologous bone marrow transplantation and/or blood cell transplantation for the treatment of metastatic breast cancer,' *Healthy Technology Assessment Information Service: Executive Briefings*, February.

ECRI: 1995b 'Xenograft transplantation: Science, ethics, and public policy,' *Health Technology Assessment News*, September-October, 1-12.

ECRI: 1995c, 'Pallidotomy and thalamotomy for Parkinson's disease,' *Health Technology Assessment News*, June, 1-10.

Eddy, D.M.: 1992, 'Applying cost-effectiveness analysis: The inside story,' *Journal of the American Medical Association* 268, 2575-2582

Eddy, D.M.: 1993, 'Three battles to watch in the 1990's,' *Journal of the American Medical Association* 270, 520-526.

Eddy, D.M.: 1996, 'Benefit language: Criteria that will improve quality while reducing costs,' *Journal of the American Medical Association* 275, 650-657.

Eddy, D.M.: 1997, 'Investigational treatments: How strict should we be?' *Journal of the American Medical Association* 278, 179-185.

Editorial: 1994, 'Consumer-first health care,' *Wall Street Journal*, July 21, A-12.

Editorial: 1995, 'Tomorrow's doctoring: Patient, heal thyself,' *The Economist* February 4, 19-21.

Ellis, J.H., Cohan, R.H., Sonnad, S.S., Cohan, N.S.: 1996, 'Selective use of radiographic low-osmolality contrast media in the 1990's,' *Radiology* 200, 297-311.

Epstein, A.: 1995, 'Performance reports on quality – prototypes, problems, and prospects,' *New England Journal of Medicine* 333, 57-61.

Epstein, R.S., Sherwood, L.M.: 1996, 'From outcomes research to disease management: A guide for the perplexed,' *Annals of Internal Medicine* 124, 832-837.

Escarce, J.J., Chen, W., Schwartz, S.: 1995, 'Falling cholecystectomy thresholds since the introduction of laparoscopic cholecystectomy,' *Journal of the American Medical Association* 273, 1581-1585.

Etheredge, L., Jones, S.B., Lewin, L.: 1996, 'What is driving health system change?' *Health Affairs* 15(4), 93-104.

Evans, R.G.: 1995, 'Manufacturing consensus, marketing truth: Guidelines for economic evaluation,' *Annals of Internal Medicine* 123, 59-60.

Ewigman, B.G., Crane, J.P., Frigoletto, F.D., LeFevre, M.L., Bain, R.P., McNellis, D., Radius Study Group: 1993, 'Effect of prenatal ultrasound screening on perinatal outcome,' *The New England Journal of Medicine* 329, 821-827.

Farrell, M.G.: 1997, 'ERISA preemption and regulation of managed health care: The case for managed federalism,' *American Journal of Law & Medicine* 23, 251-289.

Feinstein, A.R., Horwitz, R.I.: 1997, 'Problems in the "evidence" of "evidence-based" medicine,' *American Journal of Medicine* 103, 529-535.

Ferguson, J.H., Dubinsky, M., Kirsch, P.J.: 1993, 'Court-ordered reimbursement for unproven medical technology: Circumventing technology assessment,' *Journal of the American Medical Association* 269, 2116-2121.

Fins, J.J.: 1994, 'Prescription for health care reform: A page from the formulary,' *P & T*, August, 750, 753-759.

Fisher, E.S., Welch, H.G., Wennberg, J.E.: 1992, 'Prioritizing Oregon's hospital resources: An example based on variations in discretionary medical utilization,' *Journal of the American Medical Association* 267, 1925-1931.

Forman, H.P., McClennan, B.L.: 1994, 'Health services research in radiology: Opportunities and imperatives,' *AJR* 163, 257-261.

Furberg, C.D.: 1995, 'Should dihydropyridines be used as first-line drugs in the treatment of hypertension? The con side,' *Archives of Internal Medicine* 155, 2157-2161.

Gabel, J.R.: 1998, 'On drinking with your competitors after five: Research collaboration in the real world,' *Health Affairs* 17(1), 123-127.

Garber, A.M.: 1992, 'No price too high?' *New England Journal of Medicine* 327, 1676-1678.

Garber, A.M.: 1994, 'Can technology assessment control health spending?' *Health Affairs* 13(3), 115-126.

Gifford, F.: 1996, 'Outcomes research and practice guidelines: Upstream issues for downstream users,' *Hastings Center Report* 26(2), 38-44.

Glassman, P.A., Model, K.E., Kahan, J.P., Jacobson, P.D., Peabody, J.W.: 1997, 'The role of medical necessity and cost-effectiveness in making medical decisions,' *Annals of Internal Medicine* 126, 152-156.

Glassman, P.A., Jacobson, P.D., Asch, S.: 1997, 'Medical necessity and defined coverage benefits in the Oregon health plan,' *American Journal of Public Health* 87, 1053-1058.

Glazier, A.K.: 1997, 'Genetic predispositions, prophylactic treatments and private health insurance: Nothing is better than a good pair of genes,' *American Journal of Law & Medicine* 23, 45-68.

Goldsmith, J.C.: 1992, 'The reshaping of health care,' *Healthcare Forum Journal* 34(4), 34-41.

Goldsmith, J.C.: 1993, 'Technology and the end to entitlement,' *Healthcare Forum Journal* 36(5), 16-23.

Goldsmith, J.; 1994, 'The impact of new technology on health costs,' *Health Affairs* 13(3), 80-81.

Goodwin, J.S., Goodwin, J.M.: 1984, 'The tomato effect: Rejection of highly efficacious therapies,' *Journal of the American Medical Association* 251, 2387-2390.

Gosfield, A.G.: 1994, 'Clinical practice guidelines and the law: Applications and implications,' in C.B. Callaghan (ed.), *Health Law Handbook*, Thomson Legal Publishing, Inc., Deerfield I.L. (reprinted in, and with pagination of, NHLA's Legal Issues Related to Clinical Practice Guidelines), pp. 59-95.

Gottsegen, S.W.: 1981, 'A new approach for the interpretation of insurance contracts – *Great American Insurance Co. v. Tate Construction Co.*,' *Wake Forest Law Review* 17, 140-152.

Gradishar, W.J.: 1999, 'High-dose chemotherapy,' *Journal of the American Medical Association* 282, 1378-1380.

Grandinetti, D.A.: 1997, 'Add fun and profits to your practice: Do research,' *Medical Economics* 74(25), 67-79.

Gray, B.H.: 1992, 'The legislative battle over health services research,' *Health Affairs* 11(4), 38-66.

Gray, D.T., Fyler, D.C., Walker, A.M., Weinstein, M.C., Chalmers, T.C., et al.: 1993, 'Clinical outcomes and costs of transcatheter as compared with surgical closure of patent ductus arteriosus,' *New England Journal of Medicine* 329, 1517-1523.

Greenfield, S., Nelson, E.C., Subkoff, M., et al.: 1992, 'Variations in resource utilization among medical specialties and systems of care: Results from the medical outcomes study,' *Journal of the American Medical Association* 267, 1624-1630.

Grimes, D.A.: 1993, 'Technology follies: The uncritical acceptance of medical innovation,' *Journal of the American Medical Association* 269, 3030-3033.

Guadagnoli, E., Hauptman, P.J., Avanian, J.Z., Pashos, C.L., McNeil, B.J., Cleary, P.D.: 1995, 'Variation in the use of cardiac procedures after acute myocardial infarction,' *New England Journal of Medicine* 333, 573-578.

Gulick, R.M., Mellors, J.W., Havlir, D., et al.: 1997, 'Treatment with indinavir, zidovudine, and lamivudine in adults with human immunodeficiency virus infection and prior antiretroviral therapy,' *New England Journal of Medicine* 337, 734-739.

Hadorn, D.C.: 1992, 'Chapter 2: Necessary-care guidelines,' in D.C. Hadorn (ed.) *Basic Benefits and Clinical Guideline*, Westview Press, Boulder.

Hall, M.A.: 1994, 'Rationing health care at the bedside,' *New York University Law Review* 69 (4-5), 693-780.

Hall, M.A., Anderson, G.F.: 1992, 'Health insurers' assessment of medical necessity,' *University of Pennsylvania Law Review* 140, 1637-1712.

Hammer, S.M., Squires, K.E., Hughes, M.D., et al.: 1997, 'A controlled trial of two nucleoside analogues plus indinavir in persons with human immunodeficiency virus infection and CD4 cell counts of 200 per cubic millimeter or less,' *New England Journal of Medicine* 337, 725-733.

Hanania, E.G., Kavanagh J., Hortobagyi G., Giles R.E., Champlin R., Deisseroth A.B.: 1995, 'Recent advances in the application of gene therapy to human disease,' *The American Journal of Medicine* 99, 537-552l.

Havighurst, C.C.: 1995, *Health Care Choices: Private Contracts as Instruments of Health Reform*, The AEI Press, Washington, D.C.

Hayward, R.S.A., Wilson, M.C., Tunis, S.R., Bass, E.B., Guyatt, G., *et al.*: 1995, 'Users' guides to the medical literature: VIII. How to use clinical practice guidelines; A. Are the recommendations valid?' *Journal of the American Medical Association* 274, 570-574.

Health Technology Assessment Information Service: 1995, Executive Briefings, February.

Hibbard, J.H., Jewett, J.J., Legnini, M.W., Tusler, M.: 1997, 'Choosing a health plan: Do large empoyers use the data?' *Health Affairs* 16(6), 172-180.

Hillman, A.L., Eisenberg, J.M., Puly, M.V., Bloom, B.S., Glick, H., Kinosian, B., Schwartz, J.S.: 1991, 'Avoiding bias in the conduct and reporting of cost-effectiveness research sponsored by pharmaceutical companies,' *New England Journal of Medicine* 324, 1362-1365.

Holder, A.R.: 1994, 'Medical insurance payments and patients involved in research,' *IRB* 16 (1-2), 19-22.

Holoweiko, M.: 1995, 'When an insurer calls your treatment experimental,' *Medical Economics* 72(17), 171-182.

Horn, S.D., Sharkey, P.D., Tracy, D.M., Horn, C.E., James, B., Goodwin, F.: 1996, 'Intended and unintended consequences of HMO cost-containment strategies: Results from the managed care outcomes project,' *The American Journal of Managed Care* 2, 253-264.

Hornberger, J., Wrone, E.: 1997, 'When to base clinical policies on observational versus randomized trial data,' *Annals of Internal Medicine* 127, 697-703.

Iezzoni, L.I.: 1997, 'Assessing quality using administrative data,' *Annals of Internal Medicine* 127, 666-674.

Jacobson, P.D., Rosenquist, C.J.: 1988, 'The introduction of low-osmolar contrast agents in radiology,' *Journal of the American Medical Association* 260, 1586-1592.

Jeffrey, N.A., Waldholz, M.: 1999a, 'Studies are likely to question breast-cancer therapy,' *Wall Street Journal*, March 24, B-1, B-4.

Jeffrey, N.A., Waldholz, M.: 1999b, 'Oncologists to speed release of results of breast-cancer studies,' *Wall Street Journal*, March 11, B-2.

Jensen, M.C., Brant-Zawadzki, M.N., Obuchowski, N., Modic, M.T., Malkasian, D., Ross, J.S.: 1994, 'Magnetic resonance imaging of the lumbar spine in people without back pain,' *New England Journal of Medicine* 331, 69-73.

Jones, L.: 1995, 'Does prevention save money?' *American Medical News* 17, 20-22.

Jonsen, A.: 1986, 'Bentham in a box: Technology assessment and health care allocation,' *Law, Medicine, and Health Care* 14, 172-174

Jonsen, A.R., Siegler, M., Winslade, W.J.: 1992, *Clinical Ethics*, 3rd ed., McGraw-Hill, New York.

Kahn, C.N.: 1998, 'The AHCPR after the battles,' *Health Affairs* 17(1), 109-110.

Kalb, P.E.: 1990, 'Controlling health care costs by controlling technology: A private contractual approach,' *Yale Law Journal* 99, 1109-1126.

Kassirer, J.P., Angell, M.: 1994, 'The Journal's policy on cost-effectiveness analyses,' *New England Journal of Medicine* 331, 669-670.

Keister, L.W.: 1995, 'With health costs finally moderating, employers' focus turns to quality,' *Managed Care* 10, 20-24.

Kent, C.: 1996, 'Can clinical research thrive (or survive)?' *American Medical News* 3, 44.

Kent, D.L., Haynor, D.R., Longstreth, W.T., Jr., Larson, E.B.: 1994, 'The clinical efficacy of magnetic resonance imaging in neuroimaging,' *Annals of Internal Medicine* 120, 856-871.

Kessler D.K., Kessler K.M., Myerburg R.J..: 1995, 'Ambulatory electrocardiography: A cost per management decision analysis,' *Archives Internal Medicine* 155, 165-69.

Kessler, D.A., Rose, J.L., Temple, R.J., Schapiro, R., Griffin, J.P.: 1994, 'Therapeutic-class wars – drug promotion in a competitive marketplace,' *New England Journal of Medicine* 331, 1350-1353.

King, R.T., Jr.: 1996, 'How a drug firm paid for university study, then undermined it,' *Wall Street Journal*, April 25, A-1, A-6.

King, S.B., Lembo, N.J., Weintraub, W.S. *et al.*: 1994, 'A randomized trial comparing coronary angioplasty with coronary bypass surgery,' *New England Journal of Medicine* 331, 1044-1050

Kolata, G.: 1995, 'Women rejecting trials for testing a cancer therapy,' *New York Times*, Feb 15, C1.

Kolata, G.: 1999, 'Hope for sale: Business thrives on unproven care, leaving science behind,' *New York Times*, October 3, A-1.

Kong, S.X., Wertheimer, A.I.: 1998, 'Outcomes research: Collaboration among academic researchers, managed care organizations, and pharmaceutical manufacturers,' *American Journal of Managed Care* 4, 28-34.

Leape, L.L., Park, R.E., Solomon, D.H., Chassin, M.R., Kosecoff, J., Brook, R.H.: 1989, 'Relation between surgeons' practice volumes and geographic variation in the rate of carotid endarterectomy,' *The New England Journal of Medicine* 321, 653-657.

Leape, L.L., Park, R.E., Solomon, D.H., Chassin, M.R., Kosecoff, J., Brook, R.H.: 1990, 'Does inappropriate use explain small-area variations in the use of health care services?' *Journal of the American Medical Association* 263, 669-672.

Levinsky, N.G.: 1984, 'The doctor's master,' *New England Journal of Medicine* 311, 1573-1575.

Lipson, D.J., De Sa, J.M.: 1996, 'Impact of purchasing strategies on local health care systems,' *Health Affairs* 15(2), 62-76.

Luft, H.S.: 1995, 'Modifying managed competition to address cost and quality,' *Health Affairs* 15(1), 23-38.

McGivney, W.T.: 1992, 'Proposal for assuring technology competency and leadership in medicine,' *Journal of the National Cancer Institute* 84, 742-744.

Miles, S.: 1992, 'Medical futility,' *Law, Medicine and Health Care* 20, 310-315.

Miller, M.G., Miller, L.S., Fireman, B., Black, S.B.: 1994, 'Variation in practice for discretionary admissions,' *Journal of the American Medical Association* 271, 1493-1498.

Morreim, E.H.: 1987, 'Cost containment and the standard of medical care,' *California Law Review* 75(5), 1719-1763.

Morreim, E.H.: 1989, 'Stratified scarcity: Redefining the standard of care,' *Law, Medicine and Health Care* 17, 356-367.

Morreim, E.H.: 1992, 'Rationing and the law,' in M.A. Strosberg, J.M. Wiener, R. Baker, and I.A. Fein, (eds.) *Rationing America's Medical Care: The Oregon Plan and Beyond.* Brookings Institution, Washington, D.C., pp. 159-184.

Morreim, E.H.: 1994, 'Of rescue and responsibility: Learning to live with limits,' *Journal of Medicine and Philosophy* 19, 455-470.

Morreim, E.H.: 1995a, 'Diverse and perverse incentives in managed care; bringing the patient into alignment,' *Widener Law Symposium Journal* 1, 89-139.

Morreim, E.H.: 1995b, 'The ethics of incentives in managed care,' *Trends in Health Care, Law and Ethics* 10(1-2), 56-62.

Morreim, E.H.: 1995c 'Futilitarianism, exoticare, and coerced altruism: The ADA meets its limits,' *Seton Hall Law Review* 25, 101-149.

Morreim, E.H.: 1995d, 'Moral justice and legal justice in managed care: The ascent of contributive justice,' *Journal of Law, Medicine, and Ethics* 23, 247-265.

Morreim, E.H.: 1997, 'Medicine meets resource limits: Restructuring the legal standard of care,' *University of Pittsburgh Law Journal* 59(1), 1-95.

Morreim, E.H.: 1998, 'Revenue streams and clinical discretion,' *Journal of the American Geriatrics Society* 46(3), 331-337.

Mortenson, L.E.: 1989, 'Insurers target chemotherapy payments,' *Wall Street Journal*, May 11.

Murata, S.K.: 1996, 'Here come big changes in your patients' insurance,' *Medical Economics* 73(7), 185-190.

Nelson, A.F., Quiter, E.S., Solberg, L.I.: 1998, 'The state of research within managed care plans: 1997 survey,' *Health Affairs* 17(1),128-138.

Neumann, P.J., Zinner, D.E., Paltiel, A.D.: 1996, 'The FDA and regulation of cost-effectiveness claims,' *Health Affairs* 15(3), 54-71.

NIH Technology Assessment Panel on Gaucher Disease: 1996, 'Gaucher disease: current issues in diagnosis and treatment,' *Journal of the American Medical Association* 275, 548-553

O'Brien, C.L.: 1996, 'Direct contracting: Potential legal and regulatory barriers. *Minnesota Medicine* 79, 21-25.

Pearson, S.D., Goulart-Fisher, D., Lee, T.H.: 1995, 'Critical pathways as a strategy for improving care: Problems and potential,' *Annals of Internal Medicine* 123, 941-948.

Pelligrino, E, and Thomasma, D.C.: 1988, *For the Patient's Good*, Oxford University Press, New York.

Peters, W.P., Rogers, M.C.: 1994, 'Variation in approval by insurance companies of coverage for autologous bone marrow transplantation for breast cancer,' *New England Journal Of Medicine* 330, 473-477.

Pilote, L., Califf, R.M., Sapp, S., Miller, D.P., Mark, D.B., Weaver, D., Gore, J.M., Armstrong, P.W., Ohman, M., Topol, E.J., for the GUSTO-1 Investigators: 1995, 'Regional variation across the United States in the management of acute myocardial infarction,' *New England Journal of Medicine* 333, 565-572.

Power, E.J.: 1995, 'Identifying health technologies that work,' *Journal of the American Medical Association* 274, 205.

Ray, W.A.: 1997, 'Policy and program analysis using administrative data banks,' *Annals of Internal Medicine* 127, 712-718.

Reiser, S.J.: 1994, 'Criteria for standard versus experimental therapy,' *Health Affairs* 13(3), 127-136.

Rennie, D.: 1997, 'Thyroid storm,' *Journal of the American Medical Association* 277, 1238-1243.

Report to the Board of Trustees, AMA: 1996, 'Direct contracting with employers: A strategy to increase physician involvement in the current health care market – an update to B of T Report' 27 (A-95). *B of T Report I-95*.

Robinson, J.C.: 1995, 'Health care purchasing and market changes in California,' *Health Affairs* 14(4), 117-130.

Rowlings, P.A., Williams, S.P., Antman, K.H., *et al.*: 2000, 'Factors correlated with progression-free survival after high-dose chemotherapy and hematopoietic stem cell transplantation for metastatic breast cancer,' *Journal of the American Medical Association* 282, 1335-1343.

Schauffler, H.H., Rodriguez, T.: 1996, 'Exercising purchasing power for preventive care,' *Health Affairs* 15(1), 73-85.

Shelton, D.L.: 1996, 'Drugs offer hope – at a price,' *American Medical News*, April 15, 3, 22.

Shewry, S., Hunt, S., Ramey, J., Bertko, J.: 1996, 'Risk adjustment: The missing piece of market competition,' *Health Affairs* 15(1), 171-181.

Slomski, A.J.: 1996, 'Here they come: Price-conscious patients,' *Medical Economics* 73(8), 40-46.

Soumerai S.B., Ross-Degnan D., Fortess E.E., Abelson J.: 1993, 'A critical analysis of studies of state drug reimbursement policies: Research in need of discipline,' *Milbank Quarterly* 71(2), 217-252.

Steinberg, E.P., Tunis, S., Shapiro, D.: 1995, 'Insurance coverage for experimental technologies,' *Health Affairs* 14(4), 143-158.

Stelfox, H.T., Chua, G., O'Rourke, K., Detsky, A.S.: 1998, 'Conflict of interest in the debate over calcium-channel antagonists,' *New England Journal of Medicine* 338, 101-106.

Tannock, I.F.: 1987, 'Treating the patient, not just the cancer,' *New England Journal of Medicine* 317, 1534-35.

Task Force on Principles for Economic Analysis of Health Care Technology: 1995, 'Economic analysis of health care technology: A report on principles,' *Annals of Internal Medicine* 122, 61-70.

ten Have, J.A.M.J.: 1995, 'Medical technology assessment and ethics: ambivalent relations,' *Hastings Center Report* 25(5), 13-19.

Terry, K.: 1996, 'Can functional-status surveys improve your care?' *Medical Economics* 73(14), 126-144.

Thomasma, D.C., Muraskas, J., Marshall, P.A., Myers, T., Tomich, P., O'Neill, J.A.: 1996, 'The ethics of caring for conjoined twins: The Lakeberg twins,' *Hastings Center Report* 26(4), 4-12.

United States General Accounting Office: 1996, 'Health insurance: Coverage of autologous bone marrow transplantation for breast cancer,' April, GAO/HEHS-96-83.

Waldholz, M.: 1999, 'For breast cancer, marrow transplants seem no better than standard drugs,' *Wall Street Journal*, April 16, B-8.

Waldholz, M.: 2000, 'Doctor admits faking data on cancer therapy,' *Wall Street Journal*, February 7, B-2.

Wall Street Journal: 1996, 'States move to ration promising AIDS drugs,' August 22.

Welch, W.P., Miller, M.E., Welch, H.G., Fisher, E.S., Wennberg, J.E.: 1993, 'Geographic variation in expenditures for physicians' services in the United States,' *The New England Journal of Medicine* 328, 621-627.

Wells, K.B., Sturm, R.: 1995, 'Care for depression in a changing environment,' *Health Affairs* 14(3), 78-89.

Wennberg, J.E., Freeman, J.L., Culp, W.J.: 1987, 'Are hospital services rationed in New Haven or overutilized Boston?' *The Lancet* 1, 1185-1188.

Wennberg, J.E.: 1986, 'Which rate is right?' *The New England Journal of Medicine* 314, 310-311.

Wennberg, J.E.: 1990, 'Outcomes research, cost containment, and the fear of rationing,' *The New England Journal of Medicine* 323, 1202-1204.

Wennberg, J.E.: 1991, 'Unwanted variations in the rule of practice,' *Journal of the American Medical Association* 265, 1306-1307.

Wetzell, S.: 1996, 'Consumer clout,' *Minnesota Medicine* 79(2), 15-19.

Whyte, H.E., Fitzhardinge, P.M., Shennen, A.T., *et al.*: 1993, 'Extreme immaturity: Outcome of 568 pregnancies of 23-26 weeks gestation,' *Obstetrics and Gyenecology* 82, 1-7.

Winslow, R.: 1996, 'Study questions safety and cost of heart device,' *Wall Street Journal*, September 18, B-1, B-12.

Wong, E.T., Lincoln, T.L.: 1983, 'Ready! Fire! ... Aim!' *Journal of the American Medical Association* 250, 2510-2513.

Woolf, S.H., Lawrence, R.S.: 1997, 'Preserving scientific debate and patient choice: Lessons from the consensus panel on mammography screening,' *Journal of the American Medical Association* 278, 2105-2108.

Woosley, R.L.: 1994, 'Centers for education and research in therapeutics,' *Clinical Pharmacological Therapy* 55, 249-255.

PART II

THE HUMAN GENOME PROJECT

JAMES J. BONO

THE HUMAN GENOME, DIFFERENCE, AND DISEASE:
NATURE, CULTURE, AND NEW NARRATIVES FOR
MEDICINE'S FUTURE

> "Brothers," I said, "who through a hundred thousand
> perils have made your way to reach the West,
> during this so brief vigil of our senses
>
> that is still reserved for us do not deny
> yourself experience of what there is beyond,
> behind the sun, in the world they call unpeopled.
>
> Consider what you come from: you are Greeks!
> You were not born to live like mindless brutes
> but to follow paths of excellence and knowledge."
>
> "... we had entered through the narrow pass –
> when there appeared a mountain shape, darkened
> by distance, that arose to endless heights."
>
> "Our celebrations soon turned to grief."
>
> Dante, *The Inferno*
> (XXVI.112-120; 132-136)

In what may yet prove a cautionary *exemplum* for postmodern medicine, the poet Dante here retells the story of Homer's Odysseus and his men in a tragic mode that transforms the epic hero's famous cunning (*metis*) and pride – *hubris* – from means of survival to fatal flaw and agent of destruction. Knowledge – so *our* own story goes – is there, just beyond the horizon, for us to experience and make our own. But when is the pursuit of knowledge licit? And when do exhortations to "follow the paths of knowledge" turn from wise counsel to overreaching, from virtue to damning vice?

We are not likely, in the late twentieth century, to draw boundaries between licit knowledge and the *hubris* of fraudulent counsel where Dante did, nor even to frame questions of knowledge, power, and the boundaries of thought and action in the same morally universalistic language as did our poet. Far more seductive to modern biomedicine than Dante's representation of the Pillars of Hercules as foreboding threshold

Stephen Wear, James J. Bono, Gerald Logue and Adrianne McEvoy (eds.), Ethical Issues in Health Care on the Frontiers of the Twenty-First Century, 115–126.
© 2000 *Kluwer Academic Publishers. Printed in Great Britain.*

to a dangerous *terra incognita* where the likes of Dante's pilgrim might lose their way (and their souls), is that of Francis Bacon, in the frontispiece to his *Instauratio magna* (London, 1620), where they become a gateway to a grand adventure of discovery, to New Worlds to be studied, conquered, and colonized in the wake of Cristobal Colón. Just as Columbus sought to map the world that lay beyond the great expanse of water beyond the Straits of Gibraltar,[1] and Bacon's merchants of light sought to map the vast, uncharted territory of nature beyond the confines of the Old World,[2] so, too, has modern science sought to map worlds beyond ordinary human apprehension – from the daunting immensity of an expanding universe to the unimaginable finitude of elementary particles and superstrings.

Alongside such seemingly neverending quests, the desire to map the human genome seems positively within our mortal grasp, an imaginable extension of our culture's Baconian quest for power through knowledge. Yet, as Bacon – the champion of modern "experimental and natural history" – himself would caution us, neither knowledge, or power, are ends in themselves:

> Lastly, I would address one general admonition to all; that they consider what are the true ends of knowledge, and that they seek it not either for pleasure of the mind, or for contention, or for superiority to others, or for profit, or fame, or power, or any of these inferior things; but for the benefit and the use of life; and that they perfect and govern it in charity. For it was from lust of power that the angels fell, from lust of knowledge that man fell; but of charity there can be no excess, neither did angel or man ever come in danger by it.[3]

In its nobler aspirations, much of modern biomedicine does embrace the "end" lauded by Bacon – to produce knowledge and techniques "for the benefit and the use of life" – as part of its legitimating professional and social rhetoric. Biomedicine even attempts to make good on its claims by delivering the goods as well. This point can be maintained even in the face of equally apparent evidence that the lure of agonistic competition, profit, fame, power, and the temptation to use scientific knowledge to assert or enhance "superiority to others" remain powerful factors in the complex motivations that draw individual scientists to cutting-edge biomedical and technoscientific research.[4]

The Human Genome Project has itself been marketed to Congress and the public as promising perhaps the ultimate technoscientific fix: it is the

Holy Grail of biomedicine through which humans may at long last conquer disease and reveal the secrets of aging and life itself.[5] By mapping the genome and deciphering the coded information contained in it, the dream unleashed by the originary vision of the founders of molecular biology with their "molecular view of life"[6] – the dream of grasping, altering, manipulating, even "downloading" and "translating"/"transferring" the "Code of Codes" – now promises to become a lived reality that will shape "man's" future.[7] To the extent that this dream can be realized even partially in practice, the Human Genome Project would seem to offer the prospect of a truly Baconian biomedical regime "for the benefit and the use of life." This very prospect, however, makes it all the more urgent that such claims be subject to careful scrutiny. The true limits of biomedical knowledge, and of the bioinformatic and bioengineering technologies based upon such knowledge, must be critically appraised lest the costs – not to mention the lure of individual fame, profit, and even power – prove too great a price to pay for the benefits actually received. Will a new medicine emerge that respects its fiduciary responsibilities to patients and to society? Will that medicine offer relief from pain, suffering, and the more devastating ravages of disease and aging that correspond to the needs, wants, and hopes of actual patients? Or will that medicine create new personal and social choices and tensions, new possibilities for the operation, and therefore the exercise, of power that themselves bear a social cost that may prove problematic, if not divisive?[8]

Here careful ethical analysis – especially historically informed ethical analysis – has an important, perhaps decisive, role to play in allaying fears of, and encouraging circumspect policies to prevent, unwanted problems, including any brave new world scenario. The essays by Eric Juengst, Dorothy Nelkin, and Diane Paul, together with the overview and comments of Jonathan Moreno, enable us to see beyond the hype to many of the unspoken assumptions and critical issues raised by the prospect of genetic therapies and the vision of a new world of medicine and cyborg bodies made possible by the Human Genome Project and the bioinformatic and bioengineering technologies it may well spawn and will certainly encourage. In addition, these essays point to the important role of narratives in biomedicine that I shall briefly elaborate in an attempt to point toward what I feel is the need for *new narratives* for medicine's new future.

Jonathan Moreno's essay concisely highlights the potential contributions of the other essays in this section to ethical and policy issues inherent in discussions of the current and future state of genetic therapies and genetic engineering. He also summarizes salient aspects of each essay, so I shall refrain from duplicating his efforts here. Instead, I shall simply note certain aspects of each of the three essays that I believe are critical to recognizing important assumptions and problems that inhere in the Human Genome Project and in the hype surrounding claims for its future benefits to humankind. My intention is, then, to point toward the critical implications of these studies and to prepare the way for an interpretation of this new biomedicine as a historically and culturally situated project of the early twenty-first century.

Diane Paul's "PKU and Procreative Liberty: Historical and Ethical Considerations," provides the first crucial step in unpacking the claims and assessing the prospects associated with "genetic" therapies and, ultimately, the Human Genome Project. On the one hand, Paul's essay begins by situating the problem of PKU and its role in the production of "mental retardation" in the context of the eugenics movement of the first half of the twentieth century. This discussion serves as a useful reminder of the close connection between disease, social stigmatization, politics, cultural values, and discourses of amelioration in Western society and, especially, medicine. Hence, it also serves a cautionary purpose warning us to look askance at programs for the amelioration of diseases, especially where they talk of eliminating "defects" in a population. Such programmatic discourses – to judge from earlier eugenic programs or earlier still "contagionist" or "environmentalist" discourses – often mask less overt sociopolitical agendas centering on "groups" or "populations" and the reinterpretation of difference as pathology.[9]

On the other hand, Paul's essay does yeomen's work in providing a first-rate case study that points to the disconnect between identification of a disease as genetic and treatment, what she calls the "therapeutic gap." Moreover, Paul shows how, in the case of PKU, the success of prevention can translate into lives fraught with frustration, difficult choices, often barely tolerable circumstances, and even downright unhappiness. Genetic knowledge, and even successful genetic screening programs, do not necessarily translate into unmitigated success stories, whether those stories are cast as heroic medical narratives or as personal struggles for "normalcy" in daily life. As of today, there is little reason to believe that the identification of genes and genetic factors in disease will, in the future, lead to dramatically improved medical or personal scenarios. The burden of proof seems to belong to the proponents of more optimistic claims for the benefits of genetic knowledge and technologies.

Given the sobering realities of successful genetic screening in the case of PKU, Dorothy Nelkin's "From Promises of Progress to Portents of Peril: Public Responses to Genetic Engineering" provides important perspectives on just what stories are circulating in Western societies about genes, disease, and human prospects for life in the rapidly approaching new century. Popular stories, even news stories, often exaggerate the current state of biotechnology inferring the imminent development of cloning technologies or cures for illnesses that remain the stuff of the imagination and imagined futures. Some stories are inveterately optimistic, painting pictures of a future in which humans control, if not dominate, nature. Others are more restrained, even nightmarish or paranoid, in projecting a future in which individuals are stigmatized by the intractable presence of defective genes and traits. But the point that I should most like to stress from Nelkin's account is the presence and circulation of many of the same stories in and among scientists' own accounts of their work. Molecular biologists in particular seem to have caught themselves in their own metaphoric webs, which then shape the very narratives that drive their own research. At certain points, then, the boundaries between popular and scientific discourses tend to blur and even merge. I shall return to the consequences of this proliferation of narratives below.

Eric Juengst's "Concepts of Disease After the Human Genome Project" gives further evidence to the impact upon biomedicine in general and molecular biology in particular of the circulation of such stories. His analysis is further enriched by a historical perspective, like Paul's, that stresses the indebtedness of genetic views of disease to older "constitutional" and even "humoral" views and to deterministic and eugenic styles of biomedical thinking characteristic of an earlier period. Indeed, "new" genetic concepts of disease represent in Juengst's analysis the retelling of older stories, associated in many cases with older concepts of disease, in a new idiom. Thus, carriers of infectious disease – like Typhoid Mary – become transposed into the modern carriers of defective or pathogenic genes, with similar themes of stigmatization, victimization, innocence vs. culpability, guilt, shame, and inevitability, to name but a few, shared by both sets of illness-narratives. Since the link between "genes" and disease is not nearly as evident (despite both scientific and popular hype) as that between pathogenic microbes carried by individuals and disease, the upshot of the new genetic narratives is the unwarranted but very real multiplication and proliferation of individuals labeled "pathological" and hence potentially "defective." As Juengst suggests, such genetic narratives encourage "more people's 'genetic individuality' to be understood as 'sources of disease.'" As such narratives proliferate, more and more individuals become subject to stigmatization, threatened not only by its social consequences but by the refusal of medical coverage by insurance carriers for such predispositions to specific diseases. Such stigmatization only reinforces the reclassification of difference as aberrant, further encouraging a social ethos in which difference is regarded with suspicion and suspicion may all too easily be transmuted into hatred for the alien and the pathological.

Juengst's rejection of this insidious slippage of meaning, his contestation of prevalent models and concepts of disease in the discourse of genetics and the Human Genome Project, his insistence that we rethink concepts of disease after the HGP, and his call for a "new genetic lexicon," are signally important contributions to the ethics and philosophy of medicine. For my part, I should simply like to comment, however briefly, upon the significance of the historical and cultural place of "difference" in Western understandings of nature and, specifically, in the constitution of biomedical knowledge and objects.

Two tropes dominate the Human Genome Project and, more generally, the discourse of molecular biology: the tropes of questing and of universality. The search for the "Code of Codes," ultimately for the "Holy Grail" of the completely sequenced and mapped human genome, casts modern molecular biology in the heroic mode of Homer's Odysseus, Virgil's Aeneas, and a host of Western epic heroes. Expectations – both popular and scientific – run high with the pervasive use of such language, and with them the growing assumption that the fulfillment of such a noble quest will result not in a mere incremental addition to our ever-changing fund of scientific knowledge, but, far more powerfully, if grandiosely, in the ultimate discovery of the language of nature – the language of life – itself! As Lily Kay has persuasively argued, molecular biology's transformation of the problems of heredity, protein synthesis, intracellular "communication," and a host of related problems into processes of information transfer, translation, replication, and so forth invested fundamental biological phenomena with an aura of transhistorical, if not transcendent, significance. Life itself was now a code, a language. Its mastery represents humankind's ultimate intellectual triumph, the ability to correctly transcribe the Book of Life – by implication the usurpation by humans of what had once been the Divine prerogative to total knowledge of God's Book of Nature. Indeed, in what may stand culturally as the ultimate act of heroic *hubris* – a fulfillment of Adam's illicit overreaching and of the haughty ambitions of Babel's architects – the quest to rewrite the Book of Life – to alter the fundamental codes constituting the language of life – claims to have found its fundamental tools for rewriting the genetic script in the inscriptional technologies of bioengineering and bioinformatics![10]

Underwriting this quest is our second trope: universality. For such grandiose claims – claims that purportedly transform important and fundamental scientific work in molecular biology into quasi-theological discoveries of epic proportions – depend upon an assumption that what one finds in DNA, and what scientists will uncover as they reveal to the world the heretofore hidden script of the human genome, is a *universal language*. Historically, the quest for a universal language was closely aligned with the search for the lost Adamic language and with subsequent speculation about the relationship of the metaphor of the Divine Book of Nature to God's creative Word, to the original (Adamic) language of man, and to existing, historical languages. Fundamentally, the variety and differences found in both nature and in human languages were regarded as either temporary manifestations of God's created order or as corruptions of a singular and originary model. That is to say, the variety of natural forms were actually manifestations of a unitary and unified Divine plan to which, in the end, they would return; similarly, linguistic differences were corruptions of a singular and original universal language.[11]

Molecular biology and its *wunderkind*, the Human Genome Project, represent but the latest, secularized version of this age-old quest for unity, universality, and transcendent knowledge.[12] For it is precisely in so far as the human genome is conceptualized as a common and universal scriptural inheritance shared by all of humankind that it derives its immense cultural power, occupies such a formidable place in the contemporary social imaginary, and can claim such potential for ameliorating the very timeless ills that have plagued mankind.[13] The cost exacted by such power is great: it demands, as Eric Juengst so rightly insists, that "difference" be recast as pathology, or, at the very least, as potential pathology. It is this reading of the trope of universality as normative that has so dominated the narratives of molecular biology, as it did much of pre- and early-modern Western thought and science. Thus, the stories molecular biology tells are rife with plots that associate "variety" with defects, subtly enforcing the identification of uniform, universal, or common genomic structures with the "normal."[14] As Juengst, for example, points out, "with genome research, the 'defects' identified are more often structurally sound variations of the gene, part of a range of differentially benign mutant forms that make up the 'polymorphism' of the gene" (p. 138). In the slippage from "polymorphism" to "defect" we see repeated earlier narratives of universality and unity associated with tropes like the "Book of Nature" and the "Book of Life." We also risk reenacting eugenic narratives in which arbitrarily and narrowly chosen exemplars of the "normal" lead to stigmatization of differences as abnormal and, at their extreme, to social projects aimed at eliminating difference masked as medical campaigns to conquer and eliminate disease and defects. At its most benign, this set of genomic tropes and narratives encourages the homogenization of American culture, where, through the futuristic, *Gattaca*-like, projection of genetically altered designer-babies and similar feats of molecular biotechnologies, differences are trumped by standardized ideals of intelligence, size, behavioral traits, and the like. In such a fantasized future, bodies will go the way of retail merchandizing: away from the difference and variety of the local, toward the relentless presence of the same – genomics meets the suburban mall!

Let me end with a plea. We need to scrutinize carefully and replace such constraining and socially volatile narratives in molecular biology. We need, I suspect, to rethink the very meaning of "difference" in ways that capture the vitality and creativity to which difference gives rise, that resists the temptation to read difference as defect. In this rethinking, we can take our lead both from feminist scholars who have taught us so much about the problematics of difference,[15] and from practitioners of the new science studies, such as Hans-Jörg Rheinberger, for whom science is, wondrously, a "machine for making the future."[16] If we wish, like Odysseus, to "follow the paths of knowledge," we would do well not to ignore the byways of difference. Marching in unison along a straight and narrow path leads, as we have too often and tragically learned in the twentieth century, to the illusions and injustices of the same: even perhaps to that overreaching and damning vice that so troubled Dante. A healthy respect for difference may well prove to be the future for medical ethics and humanities as we cross the Herculean Pillars of a new millenium.

University at Buffalo
Buffalo, New York

NOTES

[1] For a fascinating discussion of Columbus and the New World in the context of wider European cultural projects, including science, see Pagden, 1993. See also de Certeau, 1988, for connections between mapping, travel, knowledge, and power and early modern European culture.

[2] Bacon's quest for "scientific" knowledge – for "experimental and natural histories" – was closely tied to travel and exploration. Thus, the frontispiece to his *Instauratio magna*, which as noted earlier depicts ships venturing beyond the Pillars of Hercules, or Strait of Gibraltar, prominently highlights the prophetic words of Daniel, "Multi pertransibunt et augebitur scientia:" "many shall go to and fro, and knowledge will be increased." See especially, Webster, 1975, and Bono, 1995, 199-246.

[3] See Bacon, 1980, p. 16.

[4] See Paul Rabinow's discussion (1996) of the recent and contemporary culture of biotechnology.

[5] See, in general, works and edited volumes by Cranor, 1994; Keller, 1995, 1994, 1992; Kevles and Hood, 1992; and Lewontin, 1993. For an introduction to the "political" history of the human genome and to the history of gene mapping and sequencing, see, respectively, Kevles's and Judson's essays in Kevles and Hood, 1992.

[6] For the early history of the molecular revolution in biology, see Kay, 1993, and also Judson, 1979.

[7] See a number of important essays, and a forthcoming book, by Lily Kay, 1995, 1997, 1998, forthcoming, for critical historical and conceptual analysis of the metaphors and narratives of molecular biology and their implications for the Human Genome Project and postmodern views of the body and identity. See also, Doyle 1994, 1992, and Lewontin, 1993. For discussion of the larger cultural context in which notions of codes, information, downloading, and the like have taken shape, see the recent book by Hayles, 1999.

[8] As an example, see Kavka, 1994.

[9] For the history of earlier eugenic, contagionist, and "environmentalist" discourses in medicine and society, see works by Kevles, 1985; Paul, 1998, 1995; Bynum, 1994; Delaporte, 1986; and Schiebinger, 1993.

[10] For connections between the tropes of writing, inscription, etc. and those of the book of life and book of nature in molecular biology, see the works noted earlier by Kay. Note as well Lewontin's critique (Lewontin, 1993) of the tropes found in molecular biology. For the historical significance of these tropes and their roots in medieval and Renaissance theories of nature, language, and hermeneutics, their connections to Biblical narratives of Adam, Babel, and the Pentecost, and changing meanings of the Book of Nature, see Bono, 1995.

[11] For this story, see my book (Bono, 1995).

[12] The Western desire for unity and universality is discussed in Bono, 1995 where its connections to language theory and to theories of nature through the trope of the "Word of God" are stressed. I also discuss the various ways in which the study of "variety," "diversity," and "difference" were accommodated to such desires and ultimately legitimated by new religious narratives informing seventeenth-century natural history and natural philosophy. See also my discussion of Newton and the eclipse of the emblematic world view (1999). For a fascinating extension of these concerns into contemporary electronic technologies and cyberspace, see Gunkel, 1999.

[13] See Kay 1995, 1997, 1998, forthcoming.

[14] See the related discussions in Greely, 1992; Keller, 1994, 1992; Lloyd, 1994; and Limoges, 1994. For the concepts of the "normal" and the "abnormal," see the seminal work by Georges Canguilhem (1978, 1988).

[15] See, for example, Joan Scott 1988, 1996; Donna Haraway 1991, 1997; and Drucilla Cornell 1993.

[16] Rheinberger, 1997, p. 107. Rheinberger treats the significance of difference for science in a chapter entitled, "Reproduction and Difference." See also Rheinberger, 1995. I discuss the importance of metaphor, narratives, repetition, and difference for science and the history of science in a number of published and forthcoming works (Bono, 1990, forthcoming). H. Tristram Engelhardt, Jr. notes, in this volume, the effects of difference that the bioethicist must recognize. I do not mean to suggest by my comments that we are fated to an irreducible heterogeneity based upon either biology or culture that precludes shared discourses and understanding and attempts to define common grounds. The problems of interdiscursive exchanges, shared practices, and the possibilities of "translation" are central, to my mind, to the future of science studies and biomedical ethics. I shall address these issues in my new book (Bono, in press)

REFERENCES

Bacon, Sir Francis: 1980, 'The great instauration,' in J. Weinberger (ed.) *Sir Francis Bacon. The Great Instauration and New Atlantis*, Harlan Davidson, Arlington Heights, Illinois.

Bono, J.J.: 1990, 'Science, discourse, and literature: The role/rule of metaphor in science,' in S. Peterfreund (ed.), *Literature and Science: Theory and Practice*, Northeastern University Press, Boston, pp. 59-89.

Bono, J.J.: 1995, *The Word of God and the Languages of Man: Interpreting Nature in Early Modern Science and Medicine. Volume 1, Ficino to Descartes*, University of Wisconsin Press, Madison.

Bono, J.J.: 1999, 'From Paracelsus to Newton: The word of God, the book of nature, and the eclipse of the emblematic world view,' in J. Force and R.H. Popkin (eds), *Newton and Religion: Context, Nature, and Influence*, Kluwer, Dordrecht, pp. 45-76.

Bono, J.J.: *Figuring Science: Metaphor, Narrative, and the Cultural Location of Scientific Revolutions,* Stanford University Press, Stanford, in progress.

Bynum, W.F.: 1994, *Science and the Practice of Medicine in the Nineteenth Century*. Cambridge University Press, Cambridge.

Canguilhem, G.: 1978, *On the Normal and the Pathological*, Carolyn R. Fawcett (trans.), Reidel, Dordrecht/Boston.

Canguilhem, G.: 1988, 'The question of normality in the history of biological thought,' in *Ideology and Rationality in the History of the Life Sciences*, A. Goldhammer (trans.), MIT Press, Cambridge, Massachusetts, pp. 125-145.

de Certeau, M.: 1988, 'Ethno-graphy, speech, or the space of the other: Jean de Léry,' in *The Writing of History*, T. Conley (trans.), Columbia University Press, New York, pp. 209-243.

Cornell, D.: 1993, *Transformations: Recollective Imagination and Sexual Difference*, Routledge, New York.

Cranor, C.F. (ed.): 1994, *Are Genes Us? The Social Consequences of the New Genetics*, Rutgers University Press, New Brunswick, New Jersey.

Delaporte, F.: 1986, *Disease and Civilization: The Cholera in Paris, 1832*, (A. Goldhammer, trans.), MIT Press, Cambridge, Massachusetts.

Doyle, R.: 1994, 'Vital language,' in Cranor, C.F. (ed.), *Are Genes Us? The Social Consequences of the New Genetics*, Rutgers University Press, New Brunswick, New Jersey, pp. 52-68.

Doyle, R.: 1997, *On Beyond Living: Rhetorical Transformations of the Life Sciences,* Stanford University Press, Stanford.

Greely, H.T.: 1992, 'Health insurance, employment discrimination, and the genetics revolution,' in Kevles, D.J. and Hood, L. (eds.), *The Code of Codes: Scientific and Social Issues in the Human Genome Project*, Harvard University Press, Cambridge, Massachusetts, pp. 264-280.

Gunkel, D.J.: 1999, 'Lingua ex machina: Computer-mediated communication and the Tower of Babel,' *Configurations* 7, 61-89.

Haraway, D.J.: 1991, *Simians, Cyborgs, and Women: The Reinvention of Nature*, Routledge, New York.

Haraway, D.J.: 1997, *Modest_Witness@Second_Millenium.FemaleMan©_Meets_OncoMouse™ – Feminism and Technoscience*, Routledge, New York.

Hayles, N.K.: 1999, *How We Became Posthuman: Virtual Bodies in Cybernetics, Literature, and Informatics,* University of Chicago Press, Chicago.

Judson, H.F.: 1992, 'A history of the science and technology behind gene mapping and sequencing,' in Kevles, D.J. and Hood, L. (eds.), *The Code of Codes: Scientific and Social Issues in the Human Genome Project*, Harvard University Press, Cambridge, Massachusetts, pp. 37-80.

Judson, H.F.: 1979, *The Eighth Day of Creation: Makers of the Revolution in Biology*, Simon and Schuster, New York.

Kavka, G.S.: 1994, 'Upside risks: Social consequences of beneficial biotechnology,' in Cranor, C.F. (ed.), *Are Genes Us? The Social Consequences of the New Genetics*, Rutgers University Press, New Brunswick, New Jersey, pp. 155-179.

Kay, L.E.: 1993, *The Molecular Vision of Life: CalTech, The Rockefeller Foundation, and the Rise of the New Biology*, Oxford University Press, New York and Oxford.

Kay, L.E.: 1995, 'Who wrote the book of life? Information and the transformation of molecular biology,' *Science in Context* 8, 609-634.

Kay, L.E.: 1997, 'Cybernetics, information, life: The emergence of scriptural representations of heredity,' *Configurations* 5, 23-91.

Kay, L.E.: 1998, 'A book of life? How the genome became an information system and DNA a language,' *Perspectives in Biology and Medicine* 41, 504-528.

Kay, L.E.: 2000, *Who Wrote the Book of Life? A History of the Genetic Code*, Stanford University Press, Stanford (forthcoming).

Keller, E.F.: 1995, *Refiguring Life: Metaphors of Twentieth-Century Biology*, Columbia University Press, New York.

Keller, E.F.: 1994, 'Master molecules,' in Cranor, C.F. (ed.), *Are Genes Us? The Social Consequences of the New Genetics*, Rutgers University Press, New Brunswick, New Jersey, pp. 89-98.

Keller, E.F.: 1992, 'Nature, nuture, and the Human Genome Project,' in Kevles, D.J. and Hood, L. (eds.), *The Code of Codes: Scientific and Social Issues in the Human Genome Project*, Harvard University Press, Cambridge, Massachusetts, pp. 281-299.

Kevles, D.J.: 1992, 'Out of eugenics: The historical politics of the human genome,' in Kevles, D.J. and Hood, L. (eds.), *The Code of Codes: Scientific and Social Issues in the Human Genome Project*, Harvard University Press, Cambridge, Massachusetts, pp. 3-36.

Kevles, D.J.: 1985, *In the Name of Eugenics: Genetics and the Uses of Human Heredity*, Knopf, New York.

Kevles, D.J. and Hood, L. (eds.): 1992, *The Code of Codes: Scientific and Social Issues in the Human Genome Project*, Harvard University Press, Cambridge, Massachusetts.

Lewontin, R.C.: 1993, *Biology as Ideology: The Doctrine of DNA*, HarperCollins, New York.

Limoges, C.: 1994, 'Errare humanum est: Do genetic errors have a future?' in Cranor, C.F. (ed.), *Are Genes Us? The Social Consequences of the New Genetics*, Rutgers University Press, New Brunswick, New Jersey, pp. 113-124.

Lloyd, E.A.: 1994, 'Normality and variation: The human genome project and the ideal human type,' in Cranor, C.F. (ed.), *Are Genes Us? The Social Consequences of the New Genetics*, Rutgers University Press, New Brunswick, New Jersey, pp. 99-112.

Pagden, A.: 1993, *European Encounters with the New World*, Yale University Press, New Haven.

Paul, D.: 1995, *Controlling Human Heredity, 1865 to the Present*, Humanities Press, Atlantic Highlands, New Jersey.

Paul, D.: 1998, *The Politics of Heredity: Essays on Eugenics, Biomedicine, and the Nature-Nuture Debate*, SUNY Press, Albany.

Rabinow, P.: 1996, *Making PCR: A Story of Biotechnology*, University of Chicago Press, Chicago.

Rheinberger, H.: 1997, *Toward a History of Epistemic Things: Synthesizing Proteins in the Test Tube*, Stanford University Press, Stanford.

Rheinberger, H.: 1995, 'Beyond nature and culture: A note on medicine in the age of molecular biology,' *Science in Context* 8, 249-263.

Schiebinger, L.: 1993, *Nature's Body: Gender in the Making of Modern Science*, Beacon Press, Boston.

Scott, J. W.: 1988, *Gender and the Politics of History*. Columbia University Press, New York..

Scott, J.W.: 1996, *Only Paradoxes to Offer: French Feminists and the Rights of Man*, Harvard University Press, Cambridge, Massachusetts.

Webster, C.: 1975, *The Great Instauration: Science, Medicine, and Reform, 1626-1660*, Duckworth, London.

ERIC T. JUENGST

CONCEPTS OF DISEASE AFTER
THE HUMAN GENOME PROJECT

In an article written in 1990 at the dawn of the U.S. Human Genome
Project, James Watson and Robert Cook-Deegan wrote something
guaranteed to catch the eye of a philosopher of medicine. They said that
"The major impact of the genome project will be a slow but steady
conceptual evolution – a change in the way that we think about disease
and normal physiology" (Watson and Cook-Deegan, p. 3322). The
concepts that frame the way we think about disease and health, of course,
are central to the philosophical task of attempting to clarify and explain
the dynamics of medical reasoning. They also play important roles in
quite practical matters of health policy and medical practice: significant
changes in our disease concepts will affect both what we count as
legitimate medical problems and how we assign the social responsibilities
and roles that attend such problems. For example, recall the way that the
germ theory helped highlight the problem of the infected, but
symptomless, "carrier" of disease for turn of the century medicine, and
the social authority that problem gave to public health initiatives designed
to control the behavior of those cast in the "Typhoid Mary" role (Brandt,
1987). If the major effect of the Human Genome Project is going to be a
change of this sort, it will be important for both medical epistemology
and medical ethics to understand that change as clearly as possible, since
both the logic and norms of medicine could show the consequences.
    Watson and Cook-Deegan go on to suggest the direction in which they
think the Human Genome Project will take our thinking about disease:

    A century ago, a revolution in medicine was in full stride following the
    discovery of infectious organisms and the dawn of bacteriology. Over
    the course of the century, the conceptual base of medicine has
    broadened from gross anatomy of organs to cellular biology to
    dissection of biochemical pathways. The next step is to study the most
    fundamental elements in biology – Mendel's hereditary factors, now
    known as "genes." This will not replace population biology,
    organismal biology, cellular physiology, or biochemistry, but will
    supplement them with a new and powerful foundation of knowledge ....
    Once this foundation is solid, the next stage will be to use the masses

Stephen Wear, James J. Bono, Gerald Logue and Adrianne McEvoy (eds.), Ethical Issues in
Health Care on the Frontiers of the Twenty-First Century, 127–154.
© 2000 Kluwer Academic Publishers. Printed in Great Britain.

of information and new analytical techniques to understand disease and normal biology (Watson and Cooke-Deegan, 1990, pp. 3322, 3324).

This is, in fact, a reasonable thumbnail sketch of how many genome scientists see their place in history: as laying the groundwork for the final flowering of the search for "specific causes" of disease that revolutionized medicine a century ago, by focusing our understanding of disease on genes and their dynamics.[1] Having made that claim, however, Watson and Cook-Deegan go on to describe the Human Genome Project itself in more detail, and leave the conceptual and social implications of their prediction unaddressed. This poses a challenge for the philosophy of medicine: What might it mean to use our knowledge of genes as a new foundation for understanding health and disease?

My goal in this essay is to take one first step in addressing this challenge, by exploring three sets of shorter-range implications of the conceptual sea-change that Watson and Cook-Deegan forecast for medicine that one can already detect in the medical genetic literature. The first set are the implications of what I call "genetic imperialism:" the view that, since genes are foundational to understanding disease, all diseases are best conceptualized as "genetic diseases." The second set are the implications of what I call "genetic contagionism:" the view that conceptualizing health problems as genetic diseases means understanding the genes themselves as the specific causes of the problem, much as the germ theory isolated microbes as the pathogens for infectious disease. The third set are the implications of what I call "genetic humoralism:" the view that the frank health problems caused by genes are also just extreme examples of their influence, which really ranges across the spectrum of human traits from pathological to healthy.

To look ahead, my conclusion in examining these three sets of implications will be that unless we can use genomics to effect a genuine conceptual *revolution* in the way we think about disease, the principal effect of the conceptual *evolution* that Watson and Cook-Deegan anticipate will be an increase in the incidence of genetic health problems in our society. Specifically, I argue that unless we can finally replace two strains of nineteenth-century thinking that still run strong in modern genetic medicine, the doctrine of specific causation and the tradition of constitutional pathology, the principal effect of the Human Genome Project may well be that more health problems will be understood as genetic health problems, more people will be identified as suffering from those genetic health problems, and more genetic differences will be reinterpreted as genetic health problems. In the next three sections I try to explain each of these overlapping claims in turn, and to point out some of the practical consequences that each kind of conceptual change would produce.

I. GENETIC IMPERIALISM:
ARE ALL DISEASES REALLY "GENETIC DISEASES?"

It is becoming commonplace for proponents of genome research to point out that, to the extent that all our physiological responses to the environment and its insults are products of our genes, "all disease is genetic disease," and to back that up with lists of recent findings that suggest that "current research in human genetics is providing, at almost a weekly pace, more and more examples of the central role that genes play in human health" (Guyer and Collins, 1993, p. 1145). But it is one thing to acknowledge that genes constitute part of the background conditions against which health problems occur, and another to give them the central role in the process. The former claim is true by definition, and does little to threaten medicine's existing nosological taxonomy. The boundaries between "infectious diseases," "cancers," and "genetic diseases" can still be protected by the doctrine that amongst all the co-factors involved in any disease there are identifiable "specific causes" that are more definitive than the rest. In the context of this same doctrine, however, giving genes "the central role" in health and disease makes quite a different claim. It suggests that the genetic influences on the expression of disease should be counted as their specific causes, and that every disease that can be shown to be so influenced should be subsumed into the category of genetic diseases (Hesslow, 1984). The result of these reclassifications, of course, is a remarkable increase in the number of diagnoses from this domain, and a corresponding rise in the population incidence of "genetic disease."

It is easy to find the signs of this nosological expansion in the biomedical literature over the last decade. It was forecast in the early discussions of the Human Genome Project by those who argued that the project's main accomplishment would be:

to markedly increase the number of human diseases that we recognize to have major genetic components. We already understand that genetic diseases are not rare medical curiosities with negligible societal impact, but rather constitute a wide spectrum of both rare and extremely common diseases responsible for an immense amount of suffering in all human societies. The characterization of the human genome will lead to the identification of genetic factors in many more human diseases, even those that now seem to be multifactoral or polygenic for ready understanding (Friedman, 1990, p. 413).

At the same time, such "genetic imperialism" was repudiated by those biomedical scientists afraid that the Human Genome Project would divert needed funding from their (non-genomic) research programs (Cook-Deegan, 1994), and by philosophers who argued that the genome research simply amounted to "blind reductionism gone too far" (Tauber and Sarkar, 1992). By setting out to explicate the ways in which particular genes condition the natural histories of medicine's whole suite of diseases, the critics argued, human genome research would inevitably encourage us to discuss those diseases in genetic terms, regardless of the genes' actual etiologic significance and at the expense of more manipulable causes (Hubbard and Wald, 1993; Strohman, 1993).

Of course, those exploiting the genetic strategy are quick to admit that the etiologies of most diseases are more complex than simple Mendelian "single gene" disorders; but there is plenty of room within the traditional nosology of medical genetics for "multi-factoral" and "polygenic" disorders as well. More important to successfully annexing new diseases into genetic medicine is to promise the practical benefits of the new orientation for clinical medicine: e.g., to identify the new therapeutic and prophylactic interventions it makes possible. Thus, proponents of genome research argued that, because "the benefits of public health and good nutrition in Western societies have led to a remarkable decrease in those diseases with a primarily external cause," it is time to give attention to the "internal causes" of disease:

> As a society we need to change our view of disease as an outside enemy and find a new way of thinking about illness. ... This new way of thinking about what determines health or disease will be fueled by our increasing knowledge in genetics. This is giving us a new concept of cause and pathogenesis of disease and helping to explain the variations seen in particular diagnostic categories by showing interaction of nature and nurture for common diseases. Genetics will increasingly allow us to interfere earlier in the cascade of events

leading to overt disease and clinical manifestation (Baird, 1990, p. 205).

Thus, most who hail the "new era of molecular medicine" promised by genome research explain their enthusiasm in terms of the clinical purchase that genes might give us on medicine's remaining challenges, simply as a newly visible element of a complicated process. They write that "the ability to detect individuals at risk for a disease prior to any pathologic evidence for the disease theoretically offers medicine a new strategy – anticipation of disease and pre-emptive therapy" (Caskey, 1993, p. 48).

Clearly, the clinical benefits of geneticizing disease categories are still mostly promissory. However, that promise is already enough to begin to shift our understanding of diseases like cancer or Alzheimer Disease to the point that testing for their (quite marginal) genetic factors is already being called both "presymptomatic" and "diagnostic," as if their presence were indicative of disease (Caskey, 1993; Brandt, J., 1989; Malkin, 1990). In fact, the increasing genetic reorientation of biomedical research all across the nosological spectrum has already encouraged the Director of the NIH's genome research effort to declare a conceptual victory by claiming that:

> The Human Genome Project is arguably the single most important organized research project in the history of biomedicine. Through this international research effort, we will obtain the source book for biomedical research in the 21st century and beyond ... . We will gain unprecedented insight into the manifold ways in which organismal dysfunction can arise and how this can result in disease, and we will elucidate many new ways of altering such situations. From this information will come new therapies and, perhaps more important, new strategies for prevention based on understanding individual risk and how to avoid illness (Guyer and Collins, 1993, p. 1145).

The apparent continuing success of genetic medicine's conceptual imperialism is a testament to the power that Collingwood's "manipulability theory" of causation exerts in medical reasoning (Collingwood, 1974; Juengst, 1993). In this case, just the increasingly visibility of genes as causal factors, in the wake of genome research, and the sheer promise that their contributions will become the most susceptible (of all the possible causal factors in complex diseases) to human control, seems enough to promote them to the status of today's "specific causes" of disease. However, the promotion of genes to the status of pathognomic specific causes does have important epistemic and ethical implications.

First of all, casting genes in the role of specific causes for disease requires disregarding a good bit of what genome research is itself teaching us about genes and their dynamics. In fact, it requires holding onto a strikingly old-fashioned form of genetic thinking which today is disparaged as "bean bag genetics." "Beanbag genetics" is the epithet that the evolutionary biologist, Ernst Mayr, employed to criticize his predecessors' interpretation of the Mendelian principle that genes (and traits) sort themselves independently between generations. Early Mendelians assumed that this meant that particular genes (or "unit characters") were necessary and sufficient causes for their trademark traits: "specific causes," in medicine's terms. Mayr pointed out, however, that unlike colored beans drawn from a bag, genes rarely have only one phenotypic effect, and are never entirely disconnected from each other (Mayr, 1963, p. 263). Since then, the phenomena to which he referred – pleiotropy, heterogeneity, and linkage – have become foundational for much of modern genetics, undergirding both the gene-hunting of the Human Genome Project and the functional analysis of genes to which so much of basic biomedical research has turned.

Despite Mayr's best efforts, however, beanbag genetics has shown stubborn persistence in medicine. "Genetic diseases" are still usually interpreted as self-contained causal associations between a particular allele of a specific gene and a single (though not necessarily simple) clinical syndrome, complicated at most by the magic of mysterious environmental factors (Hull, 1978). As genomic research extends Mayr's arguments into human biology, of course, this simple "beanbag" model for medical genetics will have to break down. Almost all the DNA-based descriptions of new "disease genes" have underscored the fact that multiple alleles at a given locus can produce the same clinical phenotype, and in most cases multiple loci are implicated as well. The effect of this heterogeneity is to mute the predictive power of any particular genetic approach to risk assessment for these conditions, and the expressions of caution that dominate most current genetic testing policy statements are a reflection of this uncertainty (Benjamin, et al., 1994; Holtzman, 1989). It also should not be surprising that mutations at the genomic level should ramify through the body's systems in more than one direction, and end up causing widely different types of health problems. As medicine re-learns human physiology from the genome up, genetic multipotency is likely to become the norm rather than the exception: every genotypic change probably has multiple phenotypic effects, just as any particular effect is likely to have multiple genotypic causes.[2] As a result, genetic risk assessment tests for most multifactoral health problems will have modest predictive power. In fact, they will not usually be able to predict the aspects of illness that will be most important for patients: the time of onset, the severity, the duration, or the treatabilty of their disease experience (Juengst, 1995).

What this suggests is that as an evolutionary stage in medicine's thinking about disease, the "genetic paradigm" carries its own inborn limits. If genomic science can bring medicine all the way through the simplistic "bean-bag" thinking of Mendelian medical genetics and out the other side, the allure of promoting genes as the specific causes of complex diseases should simply evaporate.

Moreover, the re-classification of diseases as "genetic" has another important set of consequences: it adds to the social meaning of these diseases, and social burdens that the afflicted have to bear. When a disease becomes reinterpreted as a genetic disease, it acquires all the peculiar cultural baggage that has traditionally been associated with genetic explanations of health problems. Three elements of this baggage seem particularly important:

1. *Determinism.* Since relatively few genetic mutations create easily discernible patterns of disease inheritance within a family, the clearest examples of genetic disorders (like Huntington Disease) have often been both predictable and intractable: they have appeared in highly penetrant Mendelian patterns in the unfortunate families that inherit them, and unfold inexorably in their individual victims. This history still affects the way many people think about genetic risk information, by leading them to assume that genetic diagnostics of any kind have more predictive power than other kinds of health risk assessments, and that all genetic health problems inevitably unfold in the lock-step fashion of our traditional models. This assumption is corroborated by popular (and academic) accounts giving genetic tests occult powers to expose individual's "future diaries" (Annas, 1994). Unfortunately, overly deterministic under-standings of genes can inhibit, rather than facilitate, the recipient's ability to anticipate and prepare for future illness, by encouraging unnecessary

fatalism and exposing the recipient to discrimination by those who can make exclusionary use of such predictions.

2. *Reductionism.* Because of the causal power genetic risk factors are often (mistakenly) given, they also tend to play an disproportionate role in the social identification of those who carry them, reducing their identities to their carrier status. Genetic information can identify health risks we inherit from (and often share with) our families, and explain those risks at what seems to be a very basic biological level. Together, these facts make it easy to interpret genetic health risks as a reflection on the recipient's basic identity as a person, and to label people accordingly (Fox-Keller, 1991; Brock, 1992), as we do when we refer, for example, to "Down's babies," "sicklers," or "phenylketonurics." To the extent that these genotypic labels cast a (socially disvalued) health problem as a person's defining feature, this reductionistic understanding of genetic test results simply exacerbates any stigmatization that the target disease may carry (Marteau and Drake, 1995; Billings, *et al.*, 1992; Saxton, 1998).

3. *Familial Implications.* Of course, overly deterministic and reductionistic interpretations of genetic health risks are cultural perceptions which the facts of genetics do not demand. However, there is one feature of genetic risk information that is relevant to almost all genetic testing. When genetic information reveals an individual's risk for disease, it immediately suggests the possibility that family members are also at risk. Of course, to the extent that this suggestion alerts relatives to remediable health risks, it is one of the virtues of taking the genetic approach. However, it can also challenge families in a number of ways.

First of all, clinically useful genetic information about individuals often requires knowing the background against which the individual's genome presents itself: the pattern of inheritance of the traits and markers in question within the larger family. This means that in order for a genetic test to be useful to an individual family member, other members of the family have to be willing to provide that background, and, in the process, discover their own status within the pattern. Moreover, like most medical interventions, genetic testing is usually motivated by a crisis – someone diagnosed with breast cancer or genetic disease – which creates a sense of urgency to "get the family in for testing." For those other family members, the decision to participate in a testing program raises a basic moral question: what are the demands of my loyalty-based obligation to help my kin learn their genetic risks? In particular, must I sacrifice my own "right not to know" in order to help my relative enjoy the "right to know," and join him or her in braving the psychosocial risks of having that personal information known about me? When family members decide to protect their own interests and decline to participate, the same question is passed "downstream" to their children and grandchildren: if those downstream kin should decide to be tested, the status of the declining member could be revealed as a simple matter of deduction. What interests must they sacrifice, then, in order to give the decliner the filial respect that they deserve?

Finally, if a decliner's kin do become interested in learning their own genetic risks, but cannot do so without involving the reluctant relative, to what lengths may the family go to persuade the unwilling to do their familial duty? Split decisions about genetic testing have already been observed by genetic counselors to lead to familial discord in some cases, while unanimous decisions in other cases have raised suspicions of undue familial pressure to participate.

Of course, genetic information is as much about our differences as it is about our shared traits, and illuminating those differences is another way in which genetics can challenge the familial virtue of mutual loyalty. As we are able to sort out which lineages within families, and which individuals within a lineage, carry a family's risk-conferring mutations, tension will be created between the divergent interests of the two groups. Whatever their commitment to family solidarity, the family will have to face the fact that it will be in its "normal" members' interests to reveal their non-carrier status in some circumstances and in the interests of the carriers to conceal theirs. Even if no clinical evidence of the family's condition exists in some family members, if they are not proven to be non-carriers, they can suffer from what sociologists call a "courtesy stigma," simply by virtue of their relationship to an individual who has such a condition (Malkin, *et al.*, 1990). Family members free of the mutations in question, for example, will find it in their interests to use that information to counter their family history of a disease in applying for insurance. In doing so, however, they will inevitably raise questions about their kin who do not volunteer their test results in turn. Should families be expected to stick together "in sickness and in health" as we ask of married couples, or do the "limited sympathies" of human nature give us leave to concentrate on the welfare of our own threads within the familial patterns of inheritance?

Since human families will always weave together a combination of different genetic threads, new abilities to identify those differences will continue to expose families to this kind of external pressure, as long as we live in a society that uses such differences to allocate its opportunities. These three features of genetic explanations – their deterministic reading, their reductionistic application, and their familial implications – serve to raise the stakes and animate the discussions of genetic imperialism, because together they give genetic diseases a social meaning that other health problems do not share. Moreover, they are the same social meanings that have fueled some of the worst political abuses of biomedical science over the course of the last century: the U.S. eugenic sterilization and immigration policies (Reilly, 1991) and the European racial hygiene programs. Now, as then, to the extent that genes are understood to implicate patients' futures, their identities, and their closest relationships, they become centrally important to the patients' lives. In fact, as I will show below, most of the ethical, legal and social issues currently being discussed as the 'downstream' risks of the Human Genome Project – i.e., the issues of genetic privacy, discrimination, and education – are direct creations of this old-fashioned "new way of thinking" about our health problems.

If the conceptual evolution that the Human Genome Project promises for medicine merely takes us to the next iteration of the research program that has guided medicine for the last century, then one of the outcomes may be the increased incidence of patients facing genetic diagnoses and their attendant social burdens. But in the process of doing that, it could have other consequences as well: as I show below, it could lead us to relabel more previously healthy people as sick, and it could lead us to interpret more human traits as pathological. The extent to which those consequences occur depends, I will argue, on the extent to which the public and the genetics community continue to rely on two other nineteenth-century pathological concepts: the notions of disease agency and constitutional predisposition.

II. GENETIC CONTAGIONISM

One of the major conceptual effects of the doctrine of specific causation in nineteenth-century medicine was the ontological reification of diseases in terms of their causal pathogenic agents. Both in Pasteur's germ theory and in Virchow's cellular pathology diseases were understood to be reducible to real things in the world: the pathogens or lesions which could provide necessary and sufficient targets for intervention. Under this view, diseases are separable from the patients that suffer them; they are understood best as predators attacking the patient, either as invading germs or as devouring wounds (Rather, 1959; Richmond, 1954). Diseases like schistosomiasis and herpes fit clearly into this scheme: they are diseases identified with the invading entities that cause their clinical signs and symptoms. Explaining a set of clinical problems as a "cancer" – an abnormal body part, consuming other normal body parts – is also to use this model, as is a diagnosis of "spina bifida" – a localizable lesion in the body. On this model, the proper target for therapeutics is not the epiphenomenal clinical symptoms of the disease, but whatever the disease "agent" does to cause those symptoms: the infection, the metastasis, or the break. The great successes of the public health movement in combating infectious disease in the early twentieth-century, and the reorientation of psychiatry to look for the "organic" bases of mental illness during the same period owe much to this interpretation of disease, as does the common correlative view that health is largely a matter of being "clean" and "whole."

One of the important corollaries of this ontologically robust view of disease is that it becomes possible for diseases to be "carried" by organisms who, while unaffected themselves, serve to transmit disease to potential hosts. As historians note:

The simplistic interpretation of the germ theory, one which many physicians embraced at first, was that a pathogenic bacteria in a human host equaled a disease. Before long it became clear that some individuals could harbor large numbers of dangerous bacteria and suffer no effects. The most famous of these was Mary Mallon, whose gallbladder teemed with typhoid bacilli, while she enjoyed perfect health... . The carrier state is now recognized as extremely common in many diseases (Hudson, 1987, p. 164).

Moreover, the lesson of Typhoid Mary was that the "carrier state" is also a crucial target for intervention in any attempt to forestall the spread of a disease of this sort. From the point of view of preventive medicine, carriers do not enjoy perfect health at all: they are infected with disease which could either eventually blossom to harm them or spread to those around them. This made possible the concept of screening otherwise healthy people to detect their hidden diseases, both for the purposes of providing them with "pre-emptive therapy" and providing others with protection from the danger they represented (Brandt, 1987).

It is easy to interpret genetic diseases using this model, and to think of genes as germs or "transmissible lesions." Indeed, the "bean-bag" Mendelian genetics of the 1920s and 30s made the wholesale application of this model irresistible to those who were interested in extending the success of public health methods to solve the social problems of rising health care costs, "feeble-mindedness," crime and immigration (Paul, 1984; Reilly, 1991). Even after immigration restrictions and surgical sterilization were rejected as legitimate means of preventing genetic infection, eugenicists looked forward to the day when

> some biochemical means is discovered for detecting the existence within the organism of the recessive genes responsible for the emergence of defective characters. It would then be a simple matter to apply tests to all individuals and discover those who possess the recessive factors for defective traits. Preventing undesirables from reproducing and continuing this generation after generation leads to the elimination of a considerable number of such defectives from the population, with the result that the perpetuation of the race is left to those individuals that seem to possess normal traits (Fasten, 1935, p. 353).

Unfortunately, it is not so easy to escape the conceptual legacy of this way of thinking about genetic health problems. The view that genetic diseases are reducible to the genetic alleles which they express is a widespread element of genetic medicine's paradigm. In fact, medical genetics seems to have finally succeeded in combining the germ theory and cellular pathology in its view of genetic disease. Like localized lesions, we see genetic diseases as defects at the molecular level, and whose pathogenic effects are, more often than not, the progressive, degenerative, breakdown of the body. This view suggests "surgical" interventions to repair the defects through recombinant DNA gene therapies. Thus, "when geneticists discuss one of the diseases, they are now talking about an anatomic derangement in the same concrete sense that the urologists talk about a kidney stone, or cardiologists talk about a stenotic mitral valve" (Stanbury, et al., 1983, p. 3). Furthermore, they claim that "conceptualization of disease at the level of the gene strengthens the drive to 'cure' the disease at the same level. There is something aesthetically compelling about cutting to the heart of the problem, by treating the disease at the molecular level, where it originates" (Roblin, 1979, p. 111).

On the other hand, like infectious diseases, we think that genetic defects can lie dormant in their "carriers," and be transmitted to their offspring, long before they are clinically manifest. This view suggests preventive measures to forestall the clinical manifestation of the disease in those already infected, and, once again, public health programs to contain the spread of disease genes within the population. Again, appealing to the fact that "the major successes of medicine – for example, antibiotics, immunization, vaccination, prevention of Rh disease, endocrine replacement therapy – have been where the intervention or treatment has been directed early on at the underlying disease mechanisms" (Baird, 1990, p.208), geneticists continue to endorse population screening programs designed to detect carriers of recessive genetic disease in order to "reduce the incidence" of the disease (Caskey, 1993; Palomaki, 1994).[3]

Like the disease agency theories that preceded it, in other words, the separation of disease and patient effected by this robustly ontological view of genetic disease allows the creation of a new class of people with genetic health problems: the "carriers" of genetic disease. Just as the emergence of localized plaques and tangles in the brain tissue can justify a diagnosis of "organic" Alzheimer's disease even before symptoms occur, and just as the presence of HIV virus makes one a reportable "carrier" of HIV disease, so the identification of the "Huntington" gene is understood to secure a diagnosis of Huntington disease, and "carrying" one allele of a recessive disease gene makes one a legitimate client for genetic services.

Unfortunately, if genes do not serve well as "specific causes," they are even less well suited to play the roles of either germs or wounds. Again, as the number and variety of different specific mutations that can all cause the same disease increases, so does the challenge of detecting and correcting them all in a patient. Worse yet, the causal complexity works in both ways: even the paradigmatic examples of clean Mendelian "single gene" disorders, like "recessive" cystic fibrosis and "dominant" Huntington's disease are turning out to be multifactoral enough that carrying one of their (multiple) pathognomic genotypes no longer guarantees that one will experience a problematic clinical syndrome (Benjamin *et al.*, 1994; Strohman, 1993). Since most prophylactic interventions are conceptually committed to a deterministic etiology of specific causation, applying this model on the basis of genetic information risks making (and acting on) both false negative and false positive prognoses. This means that preventive interventions also risk intervening unnecessarily in cases that the "environmental" forces of expression and penetrance would have naturally mitigated (Strohman, 1993). Even the view of genetic defects as molecular lesions is quickly becoming outdated. Many of the "genetic factors" now being associated with disease risks are not the structural breaks, deletions, rearrangements or malformations of DNA that have preoccupied medical geneticists until now. With genome research, the "defects" identified are more often structurally sound variations of the gene, part of a range of differentially benign mutant forms that make up the "polymorphisms" of the gene. One interesting example of the strain this fact puts on our conceptual scheme is the growing practice of using the term "pre-mutations" to describe the benign products of a nucleotide repetition problem, reserving "mutations" for the disease-causing versions: it echoes neatly the manner in which oncologists discuss "pre-cancerous lesions" in attempting to use their own ontological disease language to talk about human variation (Roussea, 1991).

Moreover, applying the disease agency interpretation of disease to genetic disease gives the new classes of patients some additional burdensome social roles. Two kinds of problems are important.

First, just as the germ theory was used to stigmatize "carriers" of infectious disease like Typhoid Mary, gene carriers also suffer stigmatization as tainted sources of disease (Kenen and Schmidt, 1978; Markel, 1992; Marteau and Drake, 1995). This stigmatization can manifest itself in subtle social labeling (as in the characterization of particular families of patients as "cancer families" (Malkin, et al., 1990)), or in frank discrimination (as in the claim that carrying particular genes amounts to a "pre-existing medical condition" for the purposes of insurance underwriting (Billings, et al., 1992)). Here, it is not diagnosis of a disease that triggers the reductionistic genetic labeling: it is merely the identification of an individual or family as a "carrier" of the disease gene. Ironically, this perception that gene carriers are already themselves "diseased" can be exacerbated even in attempts to combat the discrimination that perception can produce: for example, the only way that the Americans with Disablities Act can be interpreted to protect gene carriers from employment discrimination is if, in fact, being a gene carrier is considered to be a physical disability (Rothstein, 1992).

Moreover, disease "carriers" acquire strong social obligations not to spread their diseases, even where those duties are not enforced by the public health authorities. In genetic contexts, where the disease is "vertically transmitted" from parents to offspring, these perceived social obligations can conflict with the reproductive plans and liberties of prospective parents (Faden, 1994; Charo and Rothenberg, 1994). To the extent that genetic service programs are evaluated for funding in public health terms – i.e., in terms of their success in reducing the incidence of particular genes – genetic service providers will have a stake in seeing that their clients make the "correct" reproductive decisions: decisions not to bear children at risk for genetic disease (Clarke, 1990). While responsible couples can and do take genetic considerations into account in making their reproductive decisions, our historical experience with the excesses of the eugenics movement shows the danger of pressuring them to do so: reproductive decisions are wrapped tightly enough with a wide enough diversity of fundamental beliefs and values in our culture that, within wide limits, almost any use of public health authorities to attempt to control them will be perceived as unjustifiably coercive.[4]

III. GENETIC HUMORALISM

The late nineteenth-century alternative to finding a specific causal explanation for a clinical illness was to appeal to the doctrine of "constitutional pathology." Unlike specific causation, the constitution concept provided a view which allowed diseases to be related to their sufferers intrinsically as well as accidentally. The elements composing the body – tissues, fluids and forces – were taken to be maintained in characteristic proportions by the (formal) constraints of the particular definitive nature of that individual. Well into the nineteenth-century the constitution was interpreted as the relative balance of the four humors, placing the concept in a tradition that can be traced almost continuously back to Galen and Hippocrates (Ciocco, 1932; Haller, 1981). The resulting way an individual was composed was called his or her "physical make up" or "constitution" (Ciocco, 1932). An individual's constitution was reflected in emotional states and behavior (as the person's mental "temperament"), and in his or her physical appearance (through the characteristic "habitus" of the body and "facies" of the face) (Stephenson, 1888). More importantly, however, one's constitution also affected one's health. Depending on the relative balance struck between the elements in one's constitution, it could display "diatheses" or special susceptibilities to certain diseases (Rosenberg, 1976). These constitutional diseases were called "dyscrasias," because they represented an intrinsic imbalance within one's make-up. The first edition of *Black's Medical Dictionary* includes this definition, summarizing the doctrine:

> *Constitution*, or diathesis, means the general condition of the body, especially with reference to its liability to certain diseases. A sound constitution is one in which the structure and functions of the various parts and organs are so evenly maintained that there is no apparent liability to any disease. The term "constitutional" is sometimes vaguely applied to diseases which present knowledge does not permit of our attributing to any definite organ or system. A constitution such as the gouty constitution may be inherited, or it may develop as the result of improper food, habits and environment; or, on the other hand, a hereditary predisposition towards some disease may be gradually eliminated by a careful and regular life.

This definition captures a number of the important features of the concept of a constitution. First, it is a concept that was traditionally used in explanations of clinical phenomena at the level of the whole organism. It describes the "general condition of the body" as a whole, and the specific pathological predispositions that condition produces. Lower-level or more specific clinical complaints or symptoms would be explained as a consequence of the individual's intrinsic susceptibilities: i.e., in terms of the relations between the individual's form and functioning. As the "even maintenance" of a "sound constitution" suggests, the concept rests on a "physiological" model of disease-as-disequilibrium (Rather, 1959).

As pathological explanation increasingly moved to the levels of tissues and cells, constitutional thought lost its theoretical foundations in the humoral doctrines. As a result, its explanations became vague, drawing more heavily from contemporary chemical notions like the "dispositions" of compounds (Mann, 1964). Moreover, since many of the old "constitutional" diseases could be reinterpreted ("ontologically") as histological lesions or infectious diseases, the explanatory scope of constitutional theory was limited to illnesses for which no more concrete account could be given: it became a "pathology of the gaps." This meant that in internal medicine, constitutional terms were often used as synonyms for "idiopathic," to indicate that the cause of the problem was actually unknown. At the same time, this focused constitutional thought more squarely on the (mysterious) domain of hereditary diseases, and provided a form of explanation for the patterns of transmission that were observed. If diseases that "run in families" were to be explained as inherited traits, they had to be explained constitutionally as the result of tendencies of the hereditary "stock" or "blood" to function or dysfunction in particular ways (Adams, 1814). This model provided a useful pre-Mendelian source of explanation for the observed variety in the expression of hereditary disease. A constitutional imbalance only made one susceptible to the attack of a disease, rather than directly causing it. Moreover, the imbalance itself could be rectified (or exacerbated) by environmental factors. With the rediscovery of Mendel's work at the turn of the century, constitutionalists quickly adapted Mendelian genetic explanations to constitutional pathology. By 1929, the situation could be summarized this way:

> With the World War, the doctrine of the constitution took a sudden leap forward, and was further helped out by the development of Mendelian reasoning (genetics) and endocrinology. Consideration of the soldier as a whole, and of vast outdoor clinics of men en masse tended to revive the general pathology of Hippocrates just as the pathological lesion and the bacillus forwarded special (local) pathology and specific therapy. The constitution came to be seen as the summation of inherited traits which are basic to disease. The constitution became assimilated to the genotype of Johanssen; the physical habitus and facies to the sometimes illusory phenotypes (Garrison, 1929, p. 678).

Finally, as the definition above suggests, because of the explanatory flexibility of the constitutional doctrine, and the varied hereditary patterns, constitutional pathology retained from its humoral roots a commitment to the notion of the inheritance of acquired characteristics. Like the balance of humors, constitutions, and their pathological tendencies, could be influenced by the activities of the individuals that displayed them and the environments in which they were displayed. Since this seemed to fit the facts of many "constitutional" diseases, linked them to the localized pathologies of the day, and also offered a possibility for their treatment, medicine clung to this view long after the other life sciences had given it up (Churchill, 1976).

In fact, while it is rarely acknowledged as such, constitutional thinking continues to flourish in genetic medicine. As Robert Murray writes, for geneticists, "There is only one context in which genetic health can be unambiguously discussed (if one discusses it at all) and that is from the standpoint of the individual with a genetic constitution that clearly produces a detectable disturbance of the body's equilibrium with its environment or accelerates the process of aging of the body or of particular organs or tissues of the body" (Murray, 1974). There are two important conceptual reasons for this, both of which have been capitalized upon by the Human Genome Project and its proponents.

First, while the inheritance of acquired characteristics is no longer part of medical genetic thought (at least not until the discovery of the expanding nucleotide repeat mutations! (Roussea, 1991)), the constitutionalist's hope that inherited characteristics might be mitigated by behavioral interventions is still at the core of its approach. Genomicists write that "ultimately, the results of the HGP ... will profoundly alter our approach to medical care, from treating disease that is already advanced to a preventative mode focused on identification of individual risk. This should permit early initiation of changes in lifestyle and medical surveillance, preventing individuals from becoming ill in the first place" (Guyer and Collins, 1993, p. 1151).

The disease agency model that is so helpful in justifying reproductive case control measures (like population carrier or prenatal screening programs) is too deterministic to be very helpful in explaining how genetic knowledge can prevent disease in people who have already inherited the relevant "patho-genes." In order to promise preventive interventions that go beyond reproductive interventions (with their uncomfortable eugenic associations), genetic medicine needs the flexibility of constitutional concepts like "predisposition," "susceptibility," and "genetic loads" and "thresholds." Thus, they write that

> We have known for a long time that many common diseases such as atherosclerosis, hypertension, schizophrenia and so on are familial, but the genetic aspects have been ill defined. From the examples given, it is clear that we are learning that most common diseases are genetically heterogeneous, but susceptibility is due to major genes in many cases. Genotypes relatively unusual in the population may come to make up a large proportion of those with common diseases. An essential point is that individuals at risk can be identified for intervention, and there may be a long period to intervene... . Rather than ignore the internal genetic component to disease cause, we should evaluate the genetic input and

then attempt to tailor preventive or therapeutic programs to take it into account. We will then be able to focus our prevention of disease where it will have most effect – to those who are predisposed, and before they start down the pathogenic pathway. ... Our opportunities for preventing expression of predisposition, although limited now, are rapidly increasing. Technological developments are likely in the future to identify at risk individuals at relatively low cost, for example, looking for 80-100 different disease-predisposing genes in one sample from an individual at once... .We need to see our own genetic individuality as a potential origin of disease. We are all different – we are all genetically unique – which means our risk for disease is different one from another. Progress depends on realizing this and applying the knowledge to prevention (Baird, 1990, pp. 207-208).

As this passage suggests, the second reason "constitutional" thinking has been so tenacious in medical genetics is its utility in explaining health problems as genetic diseases *without* the need to show the specific necessary and sufficient genes involved. Any unusual phenotypic variation, in this case, warrants a claim of genotypic "imbalance" or "diathesis" which can be appropriated as the "internal genetic component" of the disease cause. For the complicated conditions that we continue to have to relegate to the "polygenic" or "multifactoral" miscellaneous bin of the genetic disease category, this ability is crucial, since it allows us to posit predispositions or susceptibilities where we cannot yet tell convincing specific causal stories for the mutations they find. Conceptually, this offers a catch-all category that, like most over-inclusive explanatory hypotheses, is open to, and often criticized for, overuse and abuse (Edlin, 1987; Wachbroit, 1994).

Moreover, one way in which this overuse manifests itself is in the creation of entirely new forms of human pathology. Since any deviation from a biological norm should have a constitutional pathological explanation (as a constitutional imbalance, a diathesis, in one direction or another), any deviation can be turned into a new genetic disease. Thus, for those who think about disease in constitutional terms, "as more and more is learned about the genetic underpinnings of various human traits, abilities and physical characteristics, some conditions, which we now regard as 'normal' variations, may come to be viewed as maladies" (Gert, *et al.*, p. 157). In other words, just as deep thought was pathologized as a symptom of the "melancholic" constitution, and heavy drinking has been pathologized as an expression of the "alcoholic" disposition, so can the identification of genetic influences on stature, aggressiveness, or risk-taking be used to pathologize shortness, "criminality," and risk aversion as newly discovered forms of "genetic disease" (Duster, 1989). Troy Duster and Dorothy Nelkin's work shows the pervasive ways in which this tendency to "medicalize" (or, in this case, "geneticize" (Edlin, 1987)) socially problematic differences between people in order to justify efforts at social control has already crept into culture (Duster, 1989; Nelkin and Tancredi, 1989). In this way, to the extent that our enthusiasm for medicine's new "preventive mode" encourages more people's "genetic individuality" to be understood as "sources of disease," the genetic paradigm will have succeeded in increasing our society's perception of its overall genetic morbidity.

The "constitutional" catch-all interpretation of genetic deviations also has its share of special social connotations. Like "carriers" of disease agents, those that are known to be constitutionally prone to pathology face a special kind of stigmatization that accidental victims of disease avoid. However, while those labeled as "carriers" can be stigmatized as irresponsible and unclean vectors of disease, the "predisposed" are more likely to be labeled as weaklings in the face of disease. For the Victorians, for example, while the asymptomatic syphilitic was to be chastised as negligent and profligate, the consumptive, the alcoholic, and the "latent schizophrenic" had to be protected from their own inability to resist disease. Today, this same "vulnerability stigma" can be seen to be animating the way we think about those at genetic risk of disease, from the grounding of airmen with sickle cell trait in the 1970s (Duster, 1989, pp. 24-28), to the recent concern over the "psychological vulnerability" of children at genetic risk for colon cancer (American Studies of Human Genetics, 1995).

Moreover, as many have pointed out, the constitutional way of thinking about genetic disease shifts the burden of responsibility to response to the disease. If particular individuals are understood to be unusually vulnerable outliers of the normal population, people whose genetic health problems are primarily rooted in "their own genetic individuality," it becomes natural to assign them the sole responsibility for discerning and avoiding the illnesses to which they are prone. In settings like the industrial workplace, where the relevant "external causes" of illness are controllable environmental mutagens and toxins, this way of thinking may relieve institutions from what would otherwise be their obligations to insure the general safety of their employees (Draper, 1992; Duster, 1989).

In summary, then, the "slow but steady conceptual evolution" forecast by Watson and Cook-Deegan needs to be watched carefully. If it unfolds against the array of nineteenth-century concepts we currently use to understand genetic health problems, it risks exacerbating the overall morbidity associated with genetic health problems in three ways: 1. We risk adding to the burden of existing disease (by geneticising them and thereby giving them the fatalism and "courtesy stigma" that accompanies genetic diseases); 2. We risk making more people sick with those diseases (by identifying carriers and thereby giving them the "infection stigma" and carrier obligations that accompany that ontological conception of disease), and: 3. We risk creating more diseases (by medicalizing differences as diatheses and thereby giving people the "vulnerability stigma" that accompanies constitutional thinking about disease).

IV. CONCLUSION

How can we avoid these consequences? Through its investment in research on the ethical, legal, and social implications of genome research, the U.S. Human Genome Project has been a pioneer in experimenting with one way to deal with the implications of Watson and Cook-Deegan's evolution. The goal of that effort is to try to reduce the social burden of the (perceived) increases in genetic morbidity that genome research will produce, by establishing policies that can help keep personal problems – like reproductive choices – appropriately personal, and social problems – like workplace safety – securely on the shoulders of society. The possibilities and limitations of that effort have been discussed extensively elsewhere (American Society of Human Genetics, 1995; Billings, *et al.*, 1992; Brock, 1992; Duster, 1989; Holtzman, 1989; Juengst, 1993; Nelkin and Tancredi, 1989; Nolan and Swenson, 1988). But efforts at making the social world safe for genomics may not be the only approach that is possible. To take a phrase from the advocates of gene therapy, would it not be more aesthetically compelling to "cut to the heart of the problem" and address it at its conceptual roots, "where it originates?" Could we, in other words, actually use genetics and the Human Genome Project to foment a revolution in the ways we think about disease, beyond simply substituting "genes" for both "germs" and "humors?"

In the glare of medicine's nineteenth-century paradigms, it is hard to see what that revolution might produce. In broad outline, however, given what we are learning about the dynamics of the human genome, it would be likely to encourage the shift to a multidimensional understanding of disease causation, with three consequences. First, we should begin to see a new conceptual separation between diseases and their causes. The etiological reductionism that allows one to "diagnose" the active presence of a genetic disease upon finding its pathognomic molecular defect is crumbling under the weight of counter-examples, as genomic research reveals the many ways that specific clinical syndromes and molecular mutations can be mis-matched. As mutation-based nosologies begin to fray beyond clinical utility, a return to clinically described entities seems likely. This, in turn, should mean a return to taking the patient's complaints as the source of meaning for medicine, as "mere symptoms" regain their status as organizational tools for medical science, and the dangers of focusing too heavily on either populational or molecular concerns become clearer. Finally, both trends would promote the acceptance of multiple approaches to health problems, via multiple causal "handles," as opposed to the determinism with which genetic explanations are now invested. The fact that the National Institute of Environmental Health Sciences has recently announced a collaboration with the Human Genome Project to promote the study of gene-environment interactions in health and disease is a sign of this progress: far from eclipsing environmental approaches to disease, the Human Genome Project may in the end force biomedical research in that direction, by exposing the limits of the genetic paradigm that currently dominates the community. Developing these hints is a project for the philosophy of medicine that goes well beyond the aims of this paper. However, consider one example of the sort of conceptual clarification I have in mind.

One small step in the direction of this revolution would be to take more care with the terms we use in genetic medicine to describe its domain and its tools. One way in which the old models are perpetuated is through the generic use of deterministic adjectives, such as presymptomatic or diagnostic, to describe the entire range of risk assessments that genetics makes possible or reductionistic labels like "disease genes" and "gene carriers."

As a starting point for such a discussion, and the conclusion to this one, consider the following suggestions for a new genetic lexicon: the only genes that should be considered "diagnostic" of genetic disease should be those that can be used in confirming the diagnosis of an active genetic disease process, such as the use of mutation analysis to diagnose fragile X syndrome in developmentally delayed children (Roussea, *et al.*, 1991). By contrast, "prognostic" genes are capable of being used to forecast the emergence of a clinical health problem with a high degree of certainty, such as mutation analysis for Huntington disease (Benjamin, *et al.*, 1994). Such testing is only "presymptomatic" if one concedes that to carry the mutations is to have Huntington disease in its earliest stages.

Similarly, "predictive" mutations should be those that can be said to identify a true genetic predisposition to a clinical health problem: that is, a tendency or inclination to go wrong in a particular way, if not inhibited by other genetic or environmental checks. In these cases, a positive test result would allow us to predict that, unless the predisposition is controlled, the clinical problem will result. Newborn testing for phenylketonuria (PKU) fits this model of genetic testing. By contrast, genes that only confer a genetic susceptibility to disease – that is, a vulnerability to a particular environmental insult – might be called "prophylactic" genes or "contingency" genes (Nolan and Swenson, 1988). For example, Alpha-1-antitripsin deficiency creates such a susceptibility: in the absence of tobacco smoke it does no harm; but in those who do smoke it represents a serious liability (Stokinger and Scheel, 1973). By calling it prophylactic, we would simply be emphasizing that there are interventions to be made; that the problem is not internal or inevitable.

"Probabilistic" genetic testing would be a less determined category of genetic risk assessment. Here, one thinks of a test like the test for P53 mutations in Li-Fraumeni family members (Malkin, *et al.*, 1990); a test which can serve to alert the clients that they are at a statistically higher risk than the population for a particular kind of health problem, but cannot make stronger claims about the specific course the future will take. By contrast, "Genetic Profiling" would be the category of tests that simply identify a loose empirical association between a particular mutation and an increased incidence of a given health problem. An illustration here would be the putative association between deletions in the gene for angiotensin converting enzyme and the risk of myocardial infarction (Cambien, 1992).[5]

Medicine has always been devoted to interpreting signs in order to help patients plan their futures. To that extent, clinicians share an occupational hazard with weather forecasters and fortune-tellers: people set great store by the predictions they make, even when they are notoriously inaccurate. Much of that inaccuracy just reflects the limits of technique: lab test results, barometer readings and palm line lengths are not always precise and reliable indicators of things to come. But even sure signs can yield false predictions when their meanings are misinterpreted. And since false predictions carry the appearance of certainty, they can be dangerous for both professionals and their clients.

How professionals should interpret predictive signs for their clients depends upon the nature of their services. Meteorologists predict the circumstances their clients will face as they go about their lives, but not how their clients will experience that environment. By contrast, fortune-tellers are expected to predict the course of their clients' future life experiences. Admittedly, the meteorological approach to genetic diagnostics strains against our modern inclination to invest genes with occult powers to determine the fate of individuals. However, if genetic explanations of disease can be reinterpreted as barometer readings rather than palm lines, their forecasts can strengthen their carriers, by giving them the opportunity to prepare for the environmental pressures that they will face. If the purpose of medical prediction is to enhance rather than constrain personal autonomy, the weather-person's perspective may be worth remembering in trying to sort through what it means to "have the genes" for the health problem that concern us.

Scientific critics of the Human Genome Project's attempts to anticipate and address the ethical, legal and social implications of new advances in human genetics have sometimes queried why it should be the responsibility of the genome research community to become the "lightening rod of human genetics" or moral and political purposes, given the fact that the social problems of genetic privacy, stigmatization and discrimination seem so far out of their control. In fact, however, the scientific community is crucial to this effort, because it is their evolving understanding of genetic disease that animates most of our issues. If they can teach the rest of us to think about disease in a way that avoids the old-fashioned errors of genetic imperialism, genetic contagionism, and genetic humoralism, they will have succeeded in giving us both new wineskins and new wine with which to toast the new millennium.

Case Western Reserve University School of Medicine
Cleveland, Ohio

NOTES

[1] There is no shortage of evidence that genome scientists see themselves as poised at the beginning of the "end of history" for medical science. For example, Walter Gilbert writes that:

> The genome project is not just an isolated effort on the part of molecular biologists. It is a natural development of the current themes of biology as a whole. ... The information carried on the DNA, that genetic information passed down from our parents, is the most fundamental property of the body. To work out our DNA sequence is to achieve a historic step forward in knowledge. Even after we have made that step we will still need to refer back to the sequence, to try to unravel its secrets more and more completely. But there is no more basic or more fundamental information that could be available (Gilbert, 1992, p. 83).

See also Guyer and Collins, 1993; Hood, 1988 for similarly eschatological claims for genome research.

[2] For example, the recent claims of association between the APOE4 allele and Alzheimer's disease have quite complicated implications for clinical practice because of the already established uses of APOE4 testing for coronary artery disease risk and head injury prognosis (Mayeux, et al., 1995).

[3] Thus, it is not too surprising to find direct echoes of the old eugenics' rhetoric in the current literature: for example, the recent editorial suggesting that, while gene therapy has little hope of reaching "the theoretical ideal of 'purifying' the human gene pool," "a broader approach, based on systematic screening of the whole population for carriers and the elimination of new carriers among their offspring, would in principle be effective" toward that preventive goal (Davis, 1992, p. 361).

[4] Ruth Faden makes this point nicely in the context of prenatal HIV screening:

> Eliminating an incident of disease or disability by "preventing" the person who would have the disease or disability from being born is not an instance of prevention – not in the sense in which it is ordinarily meant and not as the term ought to be used. ... It suggests that the lives of some persons with a disability or illness are not worth living, that such persons are to be understood only as social or economic drains and never as sources of either independent value or enrichment for the lives of others (Faden, 1994, p. 92).

[5] The point of such a lexicon is not to create water-tight categories to which particular genes and genetic tests would be permanently assigned. The categories are clearly overlapping, and one test could fall into several depending on the clinical situation. For example, a HD mutation test could be "diagnostic" if it were used to rule out a diagnosis of HD in a neurologically impaired patient. Rather, the point is to develop some conventions for describing tests in ways that give the public a more nuanced understanding of their epistemic power and practical significance.

REFERENCES

Adams, J.: 1814, *A Treatise on the Supposed Hereditary Properties of Diseases*, J. Callow Publishers, London.

American Society of Human Genetics, American College of Medical Genetics.: 1995, 'Points to consider: Ethical, legal and psychosocial implications of genetic testing in children and adolescents,' *American Journal of Human Genetics* 57, 1233-1241.

Annas G.: 1994, 'Rules for "gene banks:" Protecting privacy in the genetics age,' in T. Murphy and M. Lappe, (eds.), *Justice and the Human Genome Project*, University of California Press, Berkeley, pp. 75-91.

Baird, P.A.: 1990, 'Genetics and health care: A paradigm shift,' *Perspectives in Biology and Medicine* 33, 203-213.

Bauer, J.: 1942, *Constitution and Disease*, Grune and Stratton, New York.

Benjamin, C.M., Adam, S., Wiggins, S., Theilmann, J.L., Copley, T.T., Bloch, M., *et al*.: 1994, 'Proceed with care: Direct predictive testing for Huntington Disease,' *American Journal of Human Genetics* 55, 606-617.

Billings, P., Kohn, M., De Cuevas, M., Beckwith, J., Alper, J.S., and Natowicz, M.R.: 1992, 'Discrimination as a consequence of genetic screening,' *American Journal of Human Genetics* 50, 476-482.

Brandt, A.: 1987, *No Magic Bullet: A Social History of Venereal Disease in the United States Since 1880*, Oxford University Press, New York.

Brandt J.: 1989, 'Presymptomatic diagnosis of delayed onset diseases with linked DNA markers: The experience of HD,' *Journal of the American Medical Association* 261, 3108-3114.

Brock D.: 1992, 'The Human Genome Project and human identity,' *Houston Law Review* 29, 19-21.

Cambien, F.: 1992, 'Deletion polymorphism in the gene for angiotensin converting enzyme a potent risk factor for myocardial infarction,' *Nature* 359, 5641-5644.

Caskey, C.T.: 1993, 'Presymptomatic diagnosis: A first step toward genetic health care,' *Science* 262, 48-49.

Charo, A., Rothenberg, K.: 1994, '"The Good Mother:" The limits of reproductive accountability and genetic choice,' in K. Rothenberg and E. Thomson (eds.), *Women and Prenatal Testing: Facing the Challenges of Genetic Technology*, Ohio State University Press, Columbus, Ohio, pp. 105-131.

Churchill, F.: 1976, 'Rudolf Vichow and the pathologist's criteria for the inheritance of acquired characteristics,' *Journal of the History of Medicine* 31, 117-148.

Ciocco, A.: 1932, 'The historical background of the modern study of constitution,' *Bulletin of the History of Medicine* 4, 23-28.

Clarke, A.: 1990, 'Genetics, ethics, and audit,' *The Lancet* 335, 1145-1147.

Collingwood, R.G.: 1974, 'Three senses of the word "cause,"' in T. Beauchamp (ed.) *Philosophical Problems of Causation*, Dickenson Publishing Co., Encino, California, 118-126.

Comry, J., (ed.): 1906, *Black's Medical Dictionary*, 1st Edition, A. and C. Black, Ltd., London.

Cook-Deegan, R.: 1994, *The Gene Wars: Science, Politics and the Human Genome*, W.W. Norton, New York.

Davis, B.: 1992, 'Germ-line gene therapy: Evolutionary and moral considerations,' *Human Gene Therapy* 3, 361-365.

Draper, E.: 1992, *Risky Business: Genetic Testing and Exclusionary Practices in the Hazardous Workplace*, Cambridge University Press, New York.

Duster, T.: 1989, *Backdoor to Eugenics*, Routledge Publishing Co., New York.

Edlin, J.G.: 1987, 'Inappropriate use of genetic terminology in medical research: A public health issue,' *Perspectives in Biology and Medicine* 31, 47-56.

Engelhardt, H.T.: 1984, 'Clinical problems and the concept of disease,' in L. Nordenfelt and B. Lindahl, (eds.), *Health, Disease and Causal Explanation in Medicine*, D. Reidel Publishers, Boston, pp. 27-41.

Faden, R.: 1994, 'Reproductive genetic testing, prevention and the ethics of mothering,' in E. Thomson and K. Rothenberg (eds.), *Women and Prenatal Testing: Facing the Challenges of Genetic Technology*, Ohio State U. Press, Columbus, Ohio, pp. 88-98.

Fasten, N.: 1935, *Principles of Genetics and Eugenics*, Ginn and Co., New York.

Fox-Keller, E.: 1991, 'Genetics, reductionism and normative uses of biological information,' *Southern California Law Review* 65, 285-291.

Friedman, T.: 1990, 'The Human Genome Project – Some implications of extensive "reverse genetic" medicine,' *American Journal of Human Genetics* 46, 407-414.

Garrison, F.: 1929, *An Introduction to the History of Medicine*, W.B. Saunders, Philadelphia.

Gert, B., Berger, E., Cahill, G., *et al.*: 1996, *Morality and the New Genetics*, Jones and Bartlett Publishing Co., Boston.

Gilbert, W.: 1992, 'A vision of the Grail,' in D. Kevles and L. Hood (eds.), *The Code of Codes: Scientific and Social Issues in the Human Genome Project*, Harvard University Press, Boston, p. 83-98.

Guyer, M. and Collins F.C.: 1993, 'The Human Genome Project and the future of medicine,' *American Journal of Diseases of Children* 147, 1145-1152.

Harding, A.E.: 1992, 'Growing old: The most common mitochondrial disease of all?' *Nature Genetics* 2, 51-252.

Haller, J.: 1981, *American Medicine in Transition: 1840-1910*, University of Illinois Press, Chicago.

Hesslow, G.: 1984, 'What is a genetic disease?' in L. Nordenfelt and B. Lindahl (eds.), *Health, Disease and Causal Explanation in Medicine*, D. Reidel Publishers, Boston, pp. 183-193.

Holtzman, N.: 1989, *Proceed with Caution: Predicting Genetic Risks in the Recombinant DNA Era*, Johns Hopkins University Press, Baltimore, Maryland.

Hood, L.: 1988, 'Biotechnology and medicine of the future,' *Journal of the American Medical Association* 259, 1837-1844.

Hubbard, R., Wald, I.: 1993, *Exploding the Gene Myth*, Colophon Books, Boston.

Hudson, R.: 1987, *Disease and Its Control: The Shaping of Modern Thought*, Praeger Press, New York.

Hull, R.: 1978, 'On getting "genetic" out of "genetic disease,"' in J. Davis (ed.), *Contemporary Issues in Biomedical Ethics*, Humana Press, Clifton, New Jersey, pp. 71-87.

Juengst, E.: 1993, 'Causation and the conceptual scheme of medical knowledge,' in C. Delkeskamp-Hayes and M.A.G. Cutter (eds.), *Science, Technology and the Art of Medicine*, Kluwer, Dordrecht, pp. 127-152.

Juengst, E.: 1995, 'The ethics of prediction: Genetic risk and the physician-patient relationship,' *Genome Science and Technology* 1, 21-36.

Kenen, R.H., Schmidt, R.M.: 1978, 'Stigmatization of carrier status: Social implications of heterozygote screening,' *American Journal of Public Health* 49, 116-120.

Malkin, D., Li, F.P., Strong L.C., *et al.*: 1990, 'Germ-line p53 mutations in a familial syndrome of breast cancer, sarcomas and other neoplasms,' *Science* 250, 1233-1238.

Mann, G.: 1964, 'The concept of predisposition,' *Journal of Environmental Health* 8, 840-845.

Markel, H.: 1992, 'The stigma of disease: The implications of genetic screening,' *American Journal of Medicine* 93, 209-215.

Marteau, T.M., Drake, H.: 1995, 'Attributions for disability: The influence of genetic screening,' *Social Science and Medicine* 40, 1127-1132.

154   ERIC T. JUENGST

Mayeux, R., Ottoman, R., Maestre, G., *et al.*: 1995, 'Synergistic effects of traumatic head injury and apolipoprotein E4 in patients with Alzheimer's disease,' *Neurology* 45, pp. 555-557.

Mayr, E.: 1963, *Animal Species and Evolution*, Harvard University Press, Boston.

Murray, R.: 1974, 'Genetic disease and human health: A clinical perspective,' *Hastings Center Report*, 4-7.

Nelkin, D., Tancredi, L.: 1989, *Dangerous Diagnostics: The Social Power of Biological Information*, Basic Books, New York.

Nolan, K., Swenson, S.: 1988, 'New tools, new dilemmas: Genetic frontiers,' *Hastings Center Report*, 40-46.

Palomaki, G.E.: 1994, 'Population-based prenatal screening for the fragile X syndrome,' *Journal of Medical Screening* 1, 65-72.

Paul, D.: 1984, 'Eugenics and the left,' *Journal of the History of Ideas* 45, 567-590

Rather, L.J.: 1959, 'Towards a philosophical study of the idea of disease,' in C. Brooks and P. Cranefield (eds.), *The Historical Development of Physiological Thought*, Hafner Publishing Co., New York, pp. 351-375.

Reilly, P.: 1991, *The Surgical Solution: A History of Involuntary Sterilization in the U.S*, Johns Hopkins University Press, Baltimore.

Richmond, P.A.: 1954, 'American attitudes towards the germ theory of disease, 1860-1880,' *Journal of the History of Medicine* 9, 428-454.

Roblin, R.: 1979, 'Human genetic therapy: Outlook and apprehensions,' in G. Chacko (ed.), *Health Handbook*, Amsterdam, North Holland Publishing Co, 104-114.

Rosenberg, C.: 1976, *No Other Gods: On Science and American Social Thought*, John Hopkins University Press, Baltimore.

Roussea, F. *et al.*: 1991, 'Direct diagnosis by DNA analysis of the Fragile X syndrome of mental retardation,' *New England Journal of Medicine* 325, 1673-1681.

Rothstein, M.: 1992, 'Genetic discrimination in employment and the Americans with Disabilities Act,' *Houston Law Review* 29, 23-85.

Saxton, M.: 1988, 'Prenatal screening and discriminatory attitudes about disability,' in E. Baruch, A. D'Adamo, J. Seager (eds.), *Embryos, Ethics and Women's Rights*, Haworth Press, New York.

Stanbury, J., *et al.*: 1983, 'Inborn errors of metabolism in the 1980's,' in J. Stanbury, *et al.*, (eds.)., *The Metabolic Basis of Inherited Diseases*, McGraw Hill, Inc., New York.

Stephenson, F.: 1888, 'Temperament and diathesis in disease,' *Medical Record* 34, 362.

Stokinger, H.D., Scheel, L.D.: 1973, 'Hypersusceptibility and genetic problems in occupational medicine: A consensus report,' *Journal of Occupational Medicine* 15, 564-573.

Strohman, R.C.: 1993, 'Ancient genomes, wise bodies, unhealthy people: Limits of a genetic paradigm in biology and medicine,' *Perspectives in Biology and Medicine* 37, 112-145.

Tauber, A.I. and Sarkar, S.: 1992, 'The Human Genome Project: Has blind reductionism gone too far?' *Perspectives in Biology and Medicine* 35, 220-235.

Wachbroit, R.: 1994, 'Distinguishing genetic disease and genetic susceptibility,' *American Journal of Medical Genetics* 53, 236-240.

Watson, J., Cook-Deegan, R.: 1990, 'The Human Genome Project and international health,' *Journal of the American Medical Association* 263, 3322-3324.

DOROTHY NELKIN

FROM PROMISES OF PROGRESS TO PORTENTS OF
PERIL: PUBLIC RESPONSES TO GENETIC
ENGINEERING

In 1993, scientists at a George Washington University laboratory conducted a genetic engineering experiment that "twinned" a nonviable human embryo. The purpose was to find a way to create additional embryos for In Vitro Fertilization, but major newspapers, popular magazines and talk shows covered the experiment as if it had actually yielded a cloning technology for the mass production of human beings. The responses were remarkably polarized. The *Los Angeles Times* announced the glorious news that "infertility, virginity and menopause are no longer bars to pregnancy" (October 27, 1993). But also envisioned were embryo and selective breeding factories, cloning on consumer demand, breeding of children as organ donors, a cloning industry for selling multiples of human beings, and even a "freezer section of the biomarket."[1] *Time* wrote of the "Brave New World of cooky cutter humans" (November 8, 1993). And scientists were repeatedly accused of "playing God."

Five years later, the media covered Ian Wilmut's creation of Dolly, the cloned sheep, and then Dr Richard Seed's plan to clone babies, with similar extravagance. Again the responses were polarized, ranging from promises of progress to portents of peril, from images of miracles to visions of apocalypse. At the cutting edge of a highly publicized science and the focus of an aggressive and competitive industry, experiments in genetic engineering attract media attention. While such hype is encouraging public expectations about instant cures and futuristic visions about controlling disease, it is also evoking Frankenstein fears, concerns about risks, and sometimes cynicism about a science that is driven by commercial interests. Indeed, the hype that surrounds genetic research these days may be unfortunate; for the gap between the extravagant promises of gene therapy and actual clinical realities could result in a significant public backlash.

In *The DNA Mystique: The Gene as a Cultural Icon*, historian Susan Lindee and I explored the ubiquitous presence of the gene in popular culture, analyzing the deterministic powers attributed to this biological

Stephen Wear, James J. Bono, Gerald Logue and Adrianne McEvoy (eds.), Ethical Issues in
Health Care on the Frontiers of the Twenty-First Century, 155–170.
© 2000 Kluwer Academic Publishers. Printed in Great Britain.

entity in popular stories and media representations. In this essay, I extend
this analysis to explore the public response to gene therapy and genetic
engineering. Research in these areas has moved rapidly from the
laboratory to mass culture, from professional journals to the television
screen. I will describe the impressions conveyed to the public through the
images and stories about gene therapy that are appearing in a wide range
of media. For stories of genetic engineering appear not only in news
reports, but also in advertisements, soaps, comic books, childcare books,
prime time television, and mass culture magazines – all media that reach
a very large lay audience.

   Media messages matter. As a window on broader social attitudes,
mass culture helps to create the unarticulated assumptions and
fundamental beliefs underlying personal decisions, social policies, and
institutional practices. Widely disseminated images of gene technologies
will bear on public expectations, influencing consumer demands for
medical procedures. They shape the way we think about new technologies
and develop ways to control them. And they influence how people
respond to genetic information. In her numerous interviews with women
in genetic counseling clinics, anthropologist Rayna Rapp found their
choices influenced mainly by the world of medicalized melodramas:
"Dallas," "St. Elsewhere," and the Jerry Lewis Telethon (Rapp, 1988). In
our image society, media representations are likely to be increasingly
important in shaping public opinion about science and technology in
future years.

   The public response to biotechnology has been ambivalent from the
very early days of recombinant DNA research in the 1970s. In 1975,
scientists organized the Asilomar Conference, intended to evaluate the
risks of this emerging technology. The discussion at this conference was
constrained: speakers were instructed to deal with technical issues of risk
to human health and to avoid questions about the philosophical
implications of creating life, or the social implications of genetic
manipulation. Reporters, however, worried about "Frankenstein mon-
sters," "biological holocaust," and the dangers of "warping the genetic
endowment of the human race." The media interest went well beyond the
possible risks to health to include the dangers of genetic engineering. The
message conveyed to the public in this early debate over genetics was
runaway science needs to be controlled.[2]

   Only a few years later, media messages changed as attention turned to
the   potential   economic   benefits   and   medical   applications   of

biotechnology. Attracted by the development of a new venture capital industry, reporters during the 1980s covered biotechnology as the next economic miracle. "The bankers are in hot pursuit." Techniques of gene splicing, once represented as dangerous, became "a mundane tool," and headlines began to tout the potential clinical applications of the research as "miracles."[3] The change, over just a few years, was remarkable. In 1977, *Time* magazine had run a cover story called "The DNA Furor: Tinkering with Life." Three years later its cover story was called "DNA: New Miracle." In 1976, the *New York Times* magazine section published a skeptical article called "New Strains of Life or Death?" In 1980, an article in the same section was called "Gene Splicing: The Race towards Better Human Health." In recent years, the media have continued to welcome every new discovery of a marker or a gene for a disease with extravagant headlines and promotional hype, but ambivalence about this technology persists, focused especially on gene manipulation.

The media convey – uncritically – a wild array of futuristic scenarios presented by enthusiastic scientists who declare that genetics will "unlock the secrets of life" and allow the prediction and control of disease. It is not just the media that lack restraint in their extravagant declarations. In the future, said a geneticist to *Discover*, "present methods of treating depression will seem as crude as former pneumonia treatments seem now" (Wingerson, 1982, p. 60). In the future, said another scientist to a reporter, food companies will sell infirmity-related breakfast cereals targeted to aid those with genetic predispositions to particular diseases. "Computer models in the home will provide consumers with a diet customized to fit their genetic individuality, which will have been predetermined by simple diagnostic tests" (Clydesdale, 1989). In the future, editorialized Daniel Koshland, former editor of *Science*, the field of behavioral genetics promises a means of crime control: "When we can accurately predict future [criminal] behavior, we may be able to prevent the damage" (Koshland, 1992, p. 777).

The recent development of pharmaceutical products and the proliferation of clinical trials on new therapeutic procedures have encouraged technological optimism in the media. For example, the introduction of therapeutic molecules, especially TPA (tissue plasminogen activator) for dissolving clots, and the use of human growth hormone to treat dwarfism were newsworthy issues. The first official gene therapy experiment in 1990 – the injection of cells containing ADA genes in a child with an immune system malfunction – became a major news event. "The long awaited era of genetic therapy has at last arrived," said a writer in *The Sciences* (Culver, 1990, p. 18). Quoting geneticist French Anderson, *Time* reported, "Physicians will simply treat patients by injecting a snippet of DNA and send them home cured" (Dewitt, 1994, p. 46). *Discover* called gene therapy "The Ultimate Medicine." Writing on gene transfer techniques, the reporter proclaimed, "Genetic surgeons can now go into your cell and fix those genes with an unlikely scalpel: a virus." Interviewed for this *Discover* article, molecular biologist Richard Mulligan declared: "We can use gene transfer to make a cell do whatever we want .... We can play God in that cell" (Montgomery, 1990, p. 60). Similarly, *US News and World Report* told its readers that gene therapy is the medicine of the future. "No disease has given up more of its secrets to genetic sleuths than cancer" (Brownlee and Silbemer, 1991). Genetics, promised the writer, will allow doctors to "do something" about disease. The isolation of the colon cancer gene in 1993 prompted an enthusiastic scientist to tell a *New York Times* reporter of its implications: "Deaths are entirely preventable" (Angier, 1993, A1).

Seeking the publicity that will attract funds for research, scientists and the press offices of their institutions often turn tentative experimental findings into magic bullets. For example, a 1994 handbook for physicians on gene therapy was advertised as offering practical information on treatment options as if clinically effective gene therapies were available (Culver, 1994). In early 1995, the Scripps Research Institute promised a cure for cancer through a small injection of a protein that would cut off the blood supply from tumors and cause them to shrink, while leaving normal tissue intact. This press release was about an observation on laboratory animals; there had been no experimental trials testing the relevance of the observation to human pathology.

The biotechnology industry has made a major commitment to gene therapy, expecting that this will be the basis of future medicine. Several new journals are entirely devoted to developments in gene therapy research. The promises of gene therapy have led to an important expansion of the biotechnology industry as firms anticipate a "mega-market" for gene products. More than twenty companies are specializing in gene therapy despite economic and technical hurdles, and over 60% of gene therapy studies are directly financed by industry in anticipation of a profitable future. They believe that media coverage will help create a market for their products.

In our entrepreneurial culture, commercial interests are a source of public information. Corporate advertisements announce "A great leap in the treatment of disease," or promise "a healthy future one gene at a time." One ad says it all: "Bad Genetics? Use Opti-genetics – The first genetic optimizer."[4] The Vivigen Genetic Repository advertises its services as a gene bank: "Here's a healthy gift that really counts: DNA for your loved ones for future genetic analysis."[5] The idea is to store genetic data so that children will benefit from the eventual availability of diagnoses and cures. A pharmaceutical company ad, appearing right after the discovery of the mutation in the BRCA-1 breast cancer gene, announced its progress in finding "breakthrough ... new treatments and ultimately the cure for breast cancer."

When they report on complex scientific issues, journalists, usually untrained in scientific methods, rely heavily on press releases from scientists and biotechnology firms. They respond with enthusiasm to promotional hype that will appeal to their readers. "Genetic research leaves Doctors hopeful for Cures," "New Hope for Victims of Disease," "Genetics, the war on aging ... [is] the medical story of the century .... Genetic technologies will dramatically curtail heart disease, aging, and much more" (Rosenfeld, 1992, p. 42). In a story called "the Age of Genes," *U.S. News* reported that "advances bring closer the day when parents can endow children not only with health but also with genes for height, good balance, or lofty intelligence" (Brownlee and Silberner, 1991, p. 60). A popular 1997 film, *Gattaca*, is about a society in the not-so-distant future where parents select and engineer the traits of their newborns. Such media reports and popular stories foster public hopes and expectations.

So too do parental advice books and magazines. And they too convey the promise of genetic therapy. Anxious parents are advised to discover their childrens' predispositions. They should monitor their children's problems and inborn traits to try to decipher the "script" imposed by the genes. A medical journalist advised parents to make a genogram with all the information they can find about all their relatives.

Such messages, coming to the public from diverse sources, and often using scientific authority to advance commercial goals, are reaching a vulnerable public. They share a ubiquitous theme: genetic research will resolve the devastating uncertainties about disease, providing a scientific means of prediction through genetic testing, and then, through gene therapy or engineering, an effective means of control. But there is a counter side to promotional hyperbole. Exaggerated promises that dovetail with people's hopes and expectations open the way to cynicism should promises falter or fail. Indeed, the death of a patient, Jessie Gelsinger, during a gene therapy experiment at the University of Pennsylvania in September 1999, evoked a cynical and angry response from journalists. Just as the Challenger accident evoked cynicism about the hype in the space program, so this accident set back media confidence in gene therapy. In particular, journalists blamed it on the need for quick success in a corporate driven science.

In 1996, a National Institutes of Health advisory committee had warned about the "exaggerated and oversold expectations" created by scientists working on gene therapy. They noted that present tools and techniques are inadequate for therapeutic gene transfer, and recommended greater emphasis on basic research that would better illuminate disease mechanisms. Describing this committee report, the journal *Nature Medicine* observed the "upside-down history" of gene therapy – "one in which conceptual advance has become widely accepted and firmly established as medical principle before even a single clinical instance of clinical efficacy has been demonstrated" (Friedman, 1996, p. 145).

The remarkable advances in genetic diagnostics and the growing scientific understanding of cellular processes will probably lead eventually to therapies for genetic defects, but they are not likely to be very widely available in the near future. Meanwhile, the growing realization of the practical difficulties of extending laboratory studies to clinical applications is puncturing inflated expectations, and this is increasingly reflected in media reports. Describing the research on the unusual frequency of a mutation in the BRCA1 gene among Ashkenazy Jewish women, reporters commented again and again on the absence of effective therapies. "Does it make sense to screen healthy women for the defect given that there is no good therapy to offer those in whom it is found?" (Kolata, 1997, p. 9).

The problems of clinical application follow in part from the complexity of genetic disease. Some diseases are caused by the absence of the activity of a gene product; others by altered proteins that disrupt cellular function; and still others by alterations of chromosomal structure. The efforts to find a therapy for Cystic Fibrosis were complicated by the fact that this disease has many more mutations than scientists had originally anticipated.

There are also problems in finding safe vectors capable of transporting genes into targeted cells. Most gene therapies use viruses as the carrier mechanism, for this is the most feasible way of targeting appropriate cells. But a Cystic Fibrosis therapy that used inactivated cold viruses as a vector failed to benefit patients. So too, research on a therapy for Duchenne Muscular Dystrophy had no effect. And there are the unanticipated risks of adverse side effects. Inserting a gene in the wrong place along a strand of DNA could cause an undesirable mutation. Also, the immune system may attack cells treated with gene therapy, responding to them as foreign or infected, especially when the genes are transferred by a virus. Gene transfer experiments on monkeys were found to cause malignant T-cell Lymphoma. Clinical trials of Genentech's promising drug called Pulmozyme, developed to treat Cystic Fibrosis, were halted when they found significant mortality rates among treated patients.

Discovering risks and side effects is, of course, the purpose of animal research and clinical trials. But in an area that is hyped as "the medicine of the future," failures become more than routine science; they also become a newspeg for journalists who seldom convey the fact that a failed clinical trial can in fact yield very useful information about appropriate or risky products.

Aside from the technical difficulties in the application of gene therapies, the very high cost of developing products for diseases that affect relatively few people may block their widespread availability. Biotechnology companies have benefited from the Orphan Drug Act which was designed to provide financial incentives to find cures for diseases affecting fewer than 100,000 people. As a reward for developing a drug for only a limited market, the company that produces the drug is protected for seven years from competition by an identical product made by another company. In effect, it is given a monopoly on the drug in order to recover its investment. However, the monopoly allowed by the Orphan Drug Act may also drastically increased the cost of a treatment.

This is the situation with an enzyme replacement theory that has been developed to treat the genetic disorder called Gaucher's disease. This devastating disease, typically expressed in enlarged livers and spleens and bone lesions, affects an estimated 20,000 Americans who, of course, welcome the possibility of relief. But the cost of treatment is so prohibitive, that economists suggest it would not only use up the insurance of affected individuals, but significantly change the national health care budget. "The cost of treating 1200 to 1400 Gaucher's patients worldwide will soon approach one billion dollars .... It would pay for 50,000 patient years of renal dialysis" (Beutler, 1996, p. 523).

Similarly, the much-publicized experiment on gene therapy for children who lack an immune system enzyme appeared to be a technical success, but its very high cost as an ex-vivo treatment has limited its use beyond an experimental procedure. Indeed, this problem concerns all rare disorders; prohibitive costs are precluding their wide practical use.

The distance between developments in genetic diagnostics and the availability of therapy – between the ability to find the genetic sources of a disease and the possibility of either preventing or curing it – is an old and familiar story. The development of diagnostic techniques has always outpaced the ability of medical practitioners to provide appropriate treatment. Until the twentieth century, clinicians had focused most of their attention on the art of diagnosis because there were few treatments available for the many existing diseases. Lacking therapeutic options, they emphasized the importance of describing their patients as accurately and scientifically as possible. The purpose of diagnosis was to understand the bio-physical system for purposes of prognosis and prediction. The process of discovery became an objective in itself, and the value of a test rested less on therapeutic implications than on its contribution to the refinement of medical evaluation.

Today, however, with greater knowledge about bodily processes and promises about "our genetic future," the public expects results. But understanding genetic diseases does not necessarily lead to benefits beyond the ability to predict. Indeed, the growing access to genetic information has left us in a state of what historian Robert Proctor has called "enlightened impotence" (Proctor, 1995). And the value of enlightenment is increasingly tempered by awareness that predictive information about genetic status may expose individuals identified as "predisposed" to stigmatization and discrimination.

Media reports about the discovery of new genes often include stories about families who have lost their employment or insurance as a result of genetic testing. In the absence of therapeutic interventions, many people who know from their family history that they are vulnerable to a disease choose not to be tested even when tests are available as in the case of Huntington's (Wexler, 1992). And when they are tested, they worry about the confidentiality of the information. Personal accounts reveal the concerns of people whose "lives are lived in the shadow of genetic disease" (Marteau and Richards, 1996). Two sisters who were diagnosed as having the mutation in the BRCA-1 gene that increased their susceptibility to breast cancer chose to keep their genetic status secret. Insisting on anonymity, they spoke as Jackie and Emma at a workshop on insurance discrimination. They had chosen to be tested in the hope that the knowledge might lead to a therapeutic intervention, but, they said, to protect their families from discrimination, they were often forced to lie about the results (National Action Plan on Breast Cancer, 1995).

Practical concerns about the social implications of genetic research are often compounded by moral reservations about genetic manipulation. While often dismissed as a marginal response from people with religious agendas, these concerns are important in considering public attitudes. For moral concerns about genetic manipulation – even the idea of injecting DNA into human cells with the goal of replacing damaged genes – are often expressed in mass culture media. The power of these concerns reflects in part the iconic importance of the gene in American culture where DNA is often treated as an essential entity, the "secret of life," a way to define the essence of personhood and identity. Scientists have encouraged such images by calling the genome the Bible, the Book of Man or the Holy Grail, and by conveying an image of this molecular structure as more than a biological entity. "DNA is what makes us human," said James Watson, the first director of the Human Genome project. In this context, any research that involves genetic manipulation can be viewed as taboo.

Body parts – the pineal gland, the heart, the blood, the brain – have all at various times in human history been believed to contain the essential human self. Today it is the genes that appear to be the essence of humanity. DNA, in many media stories, is a sacred territory, a taboo arena that should never be manipulated. These concerns are most explicit in religious publications, where genes appear as "life's smallest components" and the "core" of humanness. Articles in religious magazines are critical of research involving "tinkering" or "tampering" with genes. In 1989, a writer in the evangelical journal *Christianity Today* asked: "Is it permissible to alter humanness at its core, to tamper with our essential humanity? Genes are a core that should not be monkeyed with." An essay in the Jehovah's Witness publication *Plain Truth* questioned the hubris of contemporary genetics, which has made "man himself the new God" (Kroll, 1990, p. 8). To manipulate genes is to move them to the profane realm of engineering and technology. This, it is feared, will compromise their sacred status.

Concerns about genetic manipulation appear frequently in religious publications, but they also extend beyond such groups. For example, in his 1995 Morgenthau Memorial lecture on "The New Dimensions of Human Rights," Zbignieuw Brzezinski defined what he believes to be one of the most critical problems of human rights today: "the rapidly growing potential for the actual alteration of human individuality and for the inequitable social exploitation of that potential." This "momentous challenge" posed by science, he said threatens "dehumanization and destruction" (Brzeskinsky, 1995).

The fear of tampering with genes is also a favorite subject in films about mutants, science fiction novels, and numerous revivals of the Frankenstein myth. In these stories, DNA is sacred or forbidden territory, to be transgressed at a very high cost. Many are traditional narratives of divine retribution for violating the sanctity of human life. But since the 1970s, they have appropriated the language of contemporary genetics. And they focus on the potential abuses of genetic manipulation. Take, for example, a comic book series called "DNAgent" in which a company, Matrix Inc., exploits techniques of genetic engineering to create special agents who will help expand the company's control. The company sends these DNAgents on missions. However, the agents cannot really be controlled because they contain human DNA and therefore have an irrepressible human essence. The message of the series is that "Science has made them but no man owns DNAgents." In another series, "Extinction Agenda," scientists from a country called Genosha genetically engineer mutants who will serve as slaves. Genetic engineering in all of these fantasies is technology out of control.

The "perils" in "uncontrolled tampering" is also a common theme in news reports. "Lurking behind every genetic dream come true is a possible Brave New World nightmare," says a *Time* reporter. "To unlock the secrets hidden in the chromosomes is to open up the question of who should play God with man's genes." An accompanying image portrayed scientists balancing on a tightrope of coiled DNA (Dewitt, 1989, p. 70). And an illustration for a *New York Times* article on gene therapy featured a drawing imitative of the famous Edvard Munch painting, "The Scream." A figure stands, horrified, mouth ajar, eyes wide open, its hair a mass of coiled DNA (Goldenberg, 1990).

The power of these moral reservations must be considered in light of the strongly embedded fundamentalist tradition in American society where moral and religious agendas have extraordinary importance. Gallup polls suggest that some 90% of Americans profess their belief in God and 70% belong to a church. Every U.S. President has invoked God in his inaugural address, and surveys find that 63% of adult Americans would not vote for a President who did not believe in God (Gibbs, 1991, p. 60). Indeed, as we know from the growing ascendance of the religious right, religious values seem to be increasingly important in shaping American attitudes, and in particular, attitudes toward many aspects of science. Recall, for example, the moral opposition to fetal research that succeeded in stopping federal funds for this medical research for over 12 years, and has now succeeded in blocking research on the in vitro embryo. Note the persistent religious resistance to the teaching of evolution that generated the creation controversy, and is once again influencing the public school biology curriculum. The religious impulse also underlies the anti-instrumental values that drive the animal rights movement and the wide support for its opposition to animal experimentation (Jasper and Nelkin, 1994). In America, any technology that threatens to tamper with nature is bound to confront organized opposition.

Today these ubiquitous concerns have come to focus on genetic engineering. Religious forces have organized against research in genetics and biotechnology since 1980. In the wake of the 1980 U.S. Supreme Court decision authorizing the patenting of new life forms, the National Council of Churches mobilized religious leaders to demand a government review of genetic engineering. "New life forms may have dramatic potential for improving human life," they said, but "the cure may be worse than the original problem." Again, in June 1983, a resolution signed by 50 religious leaders opposed "efforts to engineer specific genetic traits into the germline of the human species." And in 1995, The United Methodist Church, along with biotechnology critic Jeremy Rifkin, organized eighty religious leaders to protest against the patenting of genes and gene products as a "commodification of life."

Moral reservations about genetic engineering also follow from the historical relationship of the science of genetics to eugenic practices. Until the mid 1990s, therapeutic developments mainly focused on somatic cell therapy, the replacement in an individual of a problematic gene. But scientific interest lies increasingly in germ line therapy that would alter the reproductive lineage. Those who advocate these practices point out their beneficial implications for people with severe genetic conditions who want to have normal children. And they point out that there may be a long term cost efficiency in using a technology that would solve a medical problem for multiple generations. But some perceive such techniques in religious terms, as interfering with "God's Will," and others worry about infringing on the "integrity" of the genome. Still others see that famous geological catastrophe – the slippery slope – the use of the technology not for purposes of therapy, but for enhancing normal traits. This concern was reinforced when a March of Dimes Survey in 1992 found that 43% of Americans approve of using gene therapy to enhance the physical and mental or personality traits of their children as well as for resolving the problems of genetic disease.

Critics anticipate that commercial interests, hoping to market their gene products and recoup their investments, will lead to many problems. Therapeutic agents have often been used in treatments for which they were not originally approved. These so-called "off-approval" uses sometimes result in therapeutic discoveries. But they may also result in abuse. The human growth hormone, for example, was originally developed to benefit children with a specific form of hereditary dwarfism. But to expand its market for the drug, one company supported programs for screening normal school children to detect those who were expected to be of less than average height. And doctors were encouraged to prescribe the hormone more widely to young boys of short stature.

Critics also worry that gene enhancement techniques will change the way society defines the "normal," and that social pressures will encourage the expansion of their use. Yet, access to such techniques will remain limited. Questions of equity are thus another theme in popular culture narratives. In a story called, "Pursuit of Excellence," Rena Yount portrayed a future society in which wealthy parents bought genetic engineering services so that their children would possess spectacular physical beauty and intelligence. The bio-engineered – the children of the wealthy – ran the society. They held all the top positions, made the most money, and solved the important scientific and technological problems. Yount's story focused on the anguish of a mother who could not afford genetic services for her first child, but who struggled to pull together enough money to buy intelligence and beauty for her second. In the end, her struggles tore the family apart. Yount (1994) presented a critical perspective on the idea of genetic responsibility in an age of genetic engineering: "Make better people ... and they will put the rest of the world in order." And like many such narratives, it builds on the fear that genetic engineering will be used to extend the relative advantages of those with resources to use it, thus widening existing gaps in social opportunity.

In science fiction, manipulating DNA leads to the creation of immoral or amoral human beings. The scientist in Robin Cook's novel *Mutation* produces a monster when he injects genes for intelligence into his own IVF-conceived baby son (Cook, 1989). Michael Stewart's novel, *Prodigy,* featured a geneticist who injected his IVF-conceived child with an extra intelligence gene. His wife objected, saying "it is just a short step from Mendel to Mengele." But the geneticist proceeded and their daughter became an intellectual prodigy, but also a "living nightmare" with "evil built into her genes." The moral? "No man has the right to tamper with the building blocks of human life" (Stewart, 1991).

Beyond science fiction, genetic engineering has real life implications for a range of special interest groups. A spokesman from the Gay and Lesbian Task Force worried that genetic thinking would give rise to the idea that "by tweaking or zapping our chromosomes and rearranging our cells, presto, we'd no longer be gay" (Wilson, 1991). Disability groups, such as the National Association of the Mentally Ill (NAMI), have also expressed concerns about the consequences of genetic engineering. NAMI had long welcomed studies of the genetic basis of mental disorders because defining them in biological terms would ease the burden of blame that had afflicted families of the mentally ill. But they also worry that advances in gene therapy may bear on public definitions of normality and result in devaluing their lives.

The availability of gene therapies also raises legal and policy questions. Could the desire to eradicate disease lead to policies mandating gene therapy, for example as a condition for obtaining a marriage license? In a political climate where cost-containment is a major priority, will genetic conditions be treated like communicable diseases, requiring controls that will limit their transmission?

Finally, the ties between the science of genetics and its commercial applications – the conflicts of interest when profits and ethics collide – have invited considerable cynicism. The media have called attention to the non-medical interests, to the investments and profits in an intensely competitive field, that are driving many aspects of this science and its clinical applications. Corporate-driven science attracts the skepticism of those reporters who still hold an ideal of science as a pure and unsullied profession. These days, according to some disillusioned journalists, scientists are "greedy entrepreneurs," "gene merchants," or "molecular millionaires." They are driven by economic interests that threaten their objectivity and override concerns about abuse. Cartoon images of geneticists include corporate graphs and dollar signs. Scientists are portrayed in less than flattering terms as bumbling, naive, and unaware of the social implications of their discoveries. This image of the scientist as tainted by the relentless drive for profit is the theme of some of the most popular films about science such as "Gattaca," "The Fugitive," and "Jurassic Park."

In this paper, I have described some themes that have emerged from the convergence of science hype, commercial interests, and popular hopes and fears about gene therapy and genetic engineering. These themes appear repeatedly in a wide range of media sources and they are critical in shaping public understanding of genetics. They are setting a framework for public opinion that will effect clinical practices, individual decisions, and institutional policies. The messages are polarized – projecting both futuristic fantasies and moral reservations, batting readers back and forth from inflated optimism to skepticism and doubt. They reflect the perpetual hope, fostered by science hype, of finding technological solutions to dread disease. But they also reflect a pervasive cynicism that is increasingly part of the broader public image of science and technology in the 1990s.

At the end of the twentieth century, American science is increasingly becoming a focus of public scrutiny. Scientists are no longer seen as the "new priesthood," and corporate ties to science – especially the field of genetics – have undermined the popular view of science as an objective, disinterested profession. In this context, there may be a high cost to the hype that surrounds the promise of genetics. A 1994 cover of *Time* depicts a figure on a pedestal, his arms extended in a Christ-like pose, his torso inscribed with a double helix. The caption is: "Genetics – The Future is Now." The image is the Ascension. But the public interpretation of this genetic future is far less benign. Genetics is not quite perceived as the route to heaven.

New York University
New York, New York

NOTES

[1]   Some of these scenarios are from *The Newsletter of the Center for Biotechnology Policy and Ethics of Texas A&M University*, which reviewed the media coverage of human embryo cloning on January 1, 1994.

[2]   For coverage of the conference, see Krimsky, 1982, and Rogers, 1977. For a review of news coverage, see Altimore, 1982.

[3]   See, for example, the use of the language of miracles in a featured article, "The Miracle of Spliced Genes," *Newsweek*, March 17, 1980, and in *U.S. News and World Report*, December 12, 1979.

[4]   Opti-Genetics, Metabolism Nutrition Inc., Miami, Florida

[5]   Vivigen Genetic Repository brochure, 1991.

REFERENCES

Altimore, M.: 1982, 'The social construction of a scientific controversy,' *Science, Technology and Human Values*, 24-31.

Angier, N.: 1993, 'Scientists isolate gene that causes cancer of colon,' *New York Times*, December 3, A1

Beutler, E.: 1996, 'The cost of treating Gaucher's disease,' *Nature Medicine* 2(5), 523.

Brownlee, S., Silberner, J.: 1991, 'The age of genes,' *US News and World Report*, November 4, 60.

Brzeszinsky, Z.: 1995, 'The new dimension of human rights,' Morgenthau Memorial Lecture.

Clydesdale, F.: 1989, 'Present and future food science and technology in industrialized countries,' *Food Technology* (September), 134-146.

Cook, R.: 1989, *Mutation*, Putnam, New York.

Culver, K.W.: 1990, 'The splice of life: Gene therapy comes of age,' *The Sciences* 1(7), 18-24.

Culver, K.W.: 1994, *Gene Therapy: A Handbook for Physicians*, Mary Ann Liebert, New York.

Dewitt, P. E.: 1994, 'The genetic revolution,' *Time*, January 17, 46-47.

Dewitt, P. E.: 1989, 'The perils of treading on heredity,' *Time*, March 20, 70.

Friedman, T.: 1996, 'Human gene therapy: An immature genie, but one that is certainly out of the bottle,' *Nature Medicine* 2(2), 145.

Goldenberg, S.: 1990, *New York Times*, Cartoon, September 16.

Jasper, J. and Nelkin, D.: 1994, *The Animal Rights Crusade*, The Free Press, New York.

Kolata, G.: 1997, 'Tests to assess risks for cancer raising questions,' *New York Times*, March 27, A9.

Koshland, D.: 1992, 'Elephants, monstrosities and the law,' *Science* 225, 777.

Krimsky, S.: 1982, *Genetic Alchemy*, Massachusetts Institute of Technology Press, Cambridge.

Kroll, P.: 1990, 'The gene healers: Curing inherited diseases,' *The Plain Truth: A Magazine of Understanding* 55(8), 8.

Marteau, T. and Richards, M., (eds.): 1996, *The Troubled Helix*, Cambridge University Press, Cambridge.

Montgomerey, G.: 1990, 'Ultimate medicine,' *Discover* 11(March), 60.

National Action Plan on Breast Cancer: 1995, 'One family's experience,' Workshop on Genetic Discrimination, sponsored by the National Action Plan and the ELSI Working Group of the NIH Human Genome Project, Bethesda, Maryland, July 11.

Nelkin, D. and Lindee, S.: 1995, *The DNA Mystique: The Gene as a Cultural Icon,* W.H. Freeman, New York.

Proctor, R.: 1995, *Cancer Wars,* Basic Books, New York.

Rapp, R.: 1988, 'Chromosomes and communication,' *Medical Anthropology Quarterly* 2, 143-157.

Rogers, M: 1977, *Biohazard,* Knopf, New York.

Rosenfeld, A.: 1992, 'The medical story of the century,' *Longevity* (May), 42-53.

Stewart, M.: 1991, *Prodigy,* Harper Collins, New York.

Vivigen Genetic Repository brochure, Sante Fe, New Mexico, June 15.

Wexler, N.: 1992, 'Clairvoyance and caution,' in D.J. Kevles and L. Hood (eds.), *The Code of Codes,* Harvard University Press, Cambridge, Massachussetts.

Wilson, R.: 1991, 'Study raises issue of biological basis for homosexuality,' *Wall Street Journal,* August 30.

Wingerson, L.: 1982, 'Searching for depression gene,' *Discover,* February 4, 60-64.

Yount, R.: 1994, 'Pursuit of excellence' in D. Knight (ed.), *The Clarion Awards,* Doubleday, New York.

DIANE B. PAUL

PKU AND PROCREATIVE LIBERTY:
HISTORICAL AND ETHICAL CONSIDERATIONS

I. INTRODUCTION: PKU IN THE HISTORY OF EUGENICS

In 1946, Lionel Penrose was appointed Galton Professor of Eugenics in University College London. At least in one respect, this biochemical geneticist was a very odd choice for a professorship dedicated to the science of improving the human race through selective breeding. No geneticist in Britain, and perhaps in all Europe and North America, was more deeply and consistently hostile to that enterprise. While Penrose's peers may have criticized one or another aspect of the eugenics movement (such as its race or class bias), his own rejection was nearly complete. Indeed, Penrose employed his inaugural lecture as Galton Professor, "Phenylketonuria: a problem in eugenics" (Penrose, 1946), to demonstrate the futility of eugenical selection, and the need for a fundamentally new approach to human genetics.

Penrose was particularly concerned to show that sterilization laws, which had been adopted by over thirty American states, all the Scandinavian countries, and Germany, were based on a scientific mistake. In the early part of the century, eugenicists thought that if "defectives" could be prevented from breeding, their numbers in the next generation would be vastly reduced. For example, a committee of the American Breeders Association concluded that two generations of segregation and sterilization would largely "eliminate from the race the source of supply of the great anti-social human varieties" (Laughlin, 1914, p. 60), while Charles B. Davenport (1912, p. 286), one of America's leading geneticists in the 1910s-'30s, predicted even more optimistically that "the crop of defectives will be reduced to practically nothing" as the result of a single generation of segregation.[1]

At the time, mental defect was generally thought to be a recessive trait (where symptoms are expressed only if the responsible gene is inherited both from the mother and father). But according to the Hardy-Weinberg principle, which allows one to calculate the frequency of heterozygote carriers when the frequency of the gene is known, rare recessive genes are maintained in the population almost entirely by the reproduction of symptomless heterozygotes. Individuals who are actually affected by the

Stephen Wear, James J. Bono, Gerald Logue and Adrianne McEvoy (eds.), Ethical Issues in Health Care on the Frontiers of the Twenty-First Century, 171–190.
© 2000 *Kluwer Academic Publishers. Printed in Great Britain.*

condition represent but the tip of the iceberg. Preventing these individuals from breeding will thus do little to reduce the incidence of the undesirable trait.

To exemplify this and other problems with eugenics, Penrose chose a recessively-inherited disease that he and biochemist Juda Quastel had named "phenylketonuria" (or PKU) in 1935, a year after it was first described by Norwegian physician/biochemist Ivar Asjborn Følling. Sterilization of individuals with PKU is pointless, Penrose argued, since the disease is associated with such severe mental retardation and other abnormalities that those affected rarely reproduce. Moreover, since it is both recessive and rare, the carriers would vastly outnumber those actually affected. Assuming an incidence of 1 in 50,000 births, Penrose estimated that, to eradicate the gene, it would be necessary to sterilize one percent of the normal population, and concluded that "only a lunatic would advocate such a procedure to prevent the occurrence of a handful of harmless imbeciles."[2] (The incidence of the disease in Britain, and most of Western Europe, is actually about 1 in 10,000-15,000 births, so the carrier frequency is over 2 percent). He further noted that since there are many individually rare recessive disorders, it is likely that most people would be carriers of some serious defect. Thus eugenical selection against PKU was not a practical policy.

What then, if anything, might be done to reduce its incidence? Penrose did think that consanguineous matings in affected families should be avoided and, ultimately, also the mating of two carriers for the same defective gene. Indeed, in the 1930s, he had tried to develop methods to detect PKU carriers, who could then be discouraged from mating with each other (a policy that was expected to produce a slight increase in the number of heterozygotes, but whose immediate effect would be a reduction in cases of the disease). This path eventually led to contemporary carrier testing programs for diseases such as Tay-Sachs, sickle-cell anemia, and the thalassemias. Penrose also suggested that the course of hereditary biochemical illnesses might be influenced by deliberately altering the body's metabolism, a path that led in a radically different direction – to treatment.

At the time, the view that genetic disease might be treatable contradicted conventional wisdom. It was generally assumed that when traits were hereditary, the only way to increase or decrease their frequency was by selective breeding. "Since we are dealing with a defect due to heredity," Penrose noted, "the problem of controlling the incidence apparently ceases to be a medical question and becomes a matter of pure eugenics." But he considered that view mistaken. Although inborn, there might well be ways to alleviate the symptoms of PKU "in a manner analogous to the way in which a child with club-feet may be helped to walk, or a child with congenital cataract enabled to see." In particular, he thought it might be possible to treat this and other forms of mental illness through changes in diet.

The symptoms associated with PKU result from an excess of phenylalanine, an essential amino acid that can not be synthesized by humans but that is necessary for normal growth and development. A deficiency in a liver enzyme produces defective hydroxylation of phenylananine to another amino acid, tyrosine; as a result, the phenylalanine accumulates in the blood and tissues. In the 1930s, both Penrose and the American biochemist George Jervis proposed that affected babies be fed a diet from which the phenylalanine had been removed. Jervis' proposal aroused little interest at the time, since specialists in mental retardation considered it unfeasible. Penrose had actually experimented with dietary treatment, but was discouraged from continuing by Frederick Gowland Hopkins (then one of Britain's leading biochemists) on grounds that it would cost £1,000 a week to feed one patient on the synthetic diet (Kevles, 1985, pp. 177-178).

However, in 1951 English biochemists Louis Woolf and David Vulliamy tried again, and found some improvement in the three small children on whom the foods were tested. In the early 1950s, other groups in England and U.S. also reported some success with a low-phenylalanine diet. Although not all experiments were successful, the positive reports generated great excitement. Mental retardation was then considered hopelessly untreatable. Now it seemed that an understanding of biochemistry might actually lead to a cure – and as Penrose had speculated, not just for PKU. However, it soon became clear that to be effective, treatment had to begin shortly after birth. The only available test, involving urine analysis, was not reliable until the age of six to eight weeks, by which point the infant had already left the hospital and possibly suffered some degree of irreversible brain damage. Thus the urine test was unsuitable for purposes of mass screening (Paul and Edelson, 1998; Edelson, 1994).

In 1960 Robert Guthrie, a physician and microbiologist at Children's Hospital in Buffalo, developed a simple and sensitive blood test that could be administered a few days after birth. In late 1961, the Children's

Bureau began a massive field trial of the Guthrie test in order to determine its suitability for use in a national screening program for all newborns. Even before the trial had ended, newspapers and magazines hailed the test as a major discovery, while the American Medical Association included it in its 1962 year-end report on several major medical breakthroughs, including the cracking of the genetic code.

Given the rarity of the disease, which was the cause of retardation in less than one percent of the institutionalized mentally-retarded, why did the Guthrie test generate such excitement? At the time, it was widely assumed that early dietary treatment would prevent many forms of mental retardation and/or mental illness; thus the PKU control program was seen as a model for the prevention of other metabolic disorders. A story appearing in the *Family Weekly* typifies contemporary expectations. According to the author, PKU "strikes only one child in 20,000. But circumventing this disease has opened a way toward eradicating the blight of mental retardation which, in the United States alone, afflicts 5,500,000 persons" (Rosenberg, 1962, p. 6).

PKU and the "Therapeutic Gap"

While hopes that dietary therapy would solve the problem of mental retardation were disappointed, screening and treatment did succeed in virtually eliminating PKU as a cause of mental retardation – thereby confirming Penrose's point that "genetic" need not be equated with "fixed." During the 1970s, when the IQ and sociobiology controversies were at their height, the PKU control program became a favorite illustration of the mistakes of genetic determinists. Although PKU is an inborn error of metabolism, a knowledge of its biochemistry enables us to limit the supply of the damaging substrate and thereby avoid or mitigate the symptoms of the disease. It thus demonstrates that biology is not destiny (see Paul, 1998).

The frequency with which PKU was (and is) invoked to illustrate this point is explained by the dearth of other examples of successful interventions for hereditary disorders. For the same reason, it has become an exemplar of the value of genetic medicine. In the 1970s, it was assumed that advances in molecular genetics would lead quickly to cures. Indeed, some Protestant theologians embraced the concept of genetic engineering of humans on the grounds that it would provide an alternative to abortion (see van Dijck, 1998, pp. 45-49). However, progress in genetic knowledge has not been matched by corresponding progress in treatment. A 1985 study that attempted to measure the efficacy of treatment in 351 single-gene diseases reported that, in general, the response to treatment was slight; there was little improvement in lifespan, reproductive capability, or social adaptation (Hayes *et al.*, 1985, p. 246).

The study's authors poignantly remarked that a paper on "Treatment of Genetic Disease" presented by one of their number at the 3rd International Congress of Human Genetics in 1967 "evoked an enthusiastic response in the audience (and in the media), probably because the report emphasized those disorders with some promising responses to treatment." But disappointment followed, since "eighteen years later it is evident that success in treatment has not kept pace either with the discovery of 'new' diseases or with earlier expectations. Although there is good news for patients with some inborn errors . . ., we can do little with about half of them and treatment for the rest leaves much to be wished. And as for Mendelian disorders not designated as inborn errors, the record is even worse" (Hayes, *et al.*, 1985, p. 251). Since 1985, whole new classes of disease-causing genes have been discovered, with hundreds of such genes cloned and thousands of markers identified (Epstein, 1997). But this increase in medical knowledge has hardly closed the "therapeutic gap." In 1995, a study comparing responses to treatment for the same sixty-five inborn errors of metabolism included in the 1985 study reported that 57% were at least partially ameliorated by treatment (compared to 40% previously). But in only 12% were all the manifestations of the disease eliminated, a "cure" rate that had not changed in the intervening decade (Treacy, Childs, and Scriver, 1995).

Advocates of course remain optimistic, arguing that lags between scientific discovery and practical application are only to be expected.[3] In their view, knowledge of the ways in which genes are implicated in disease will eventually result in effective interventions. Eric Juengst (this volume) notes that proponents of genome research promise great benefits for clinical medicine. He cites P. A. Baird's claim that "genetics will increasingly allow us to interfere earlier in the cascade of events leading to overt disease and clinical manifestation" (Baird, 1990, p. 205). With even greater optimism, Leroy Hood predicts that for every defective gene there will be a corresponding therapeutic regime "that will circumvent the limitations of the defective gene." Medicine will move from a reactive to a preventive mode, enabling "most individuals to live a normal, healthy, and intellectually alert life without disease" (Hood, 1992, pp. 157-158).

Perhaps these predictions will eventually prove true. But as Juengst remarks, at present the benefits are still mostly promissory. An Institute of Medicine report notes that, "the explosion in our knowledge of the structure and function of human genes" has led to a plethora of tests but few measures to cure, to prevent, or even to mitigate the symptoms of disease. As a result, genetic services today largely consist of diagnosis and counseling (Andrews *et al.*, 1994).

Moreover, the therapies on offer, such as bone marrow transplants and blood transfusions for sickle-cell anemia or kidney transplants for polycystic kidney disease, are often drastic or dangerous or both. The options facing women who test positive for disease-predisposing mutations in the BRCA1/2 genes are illustrative. Women with these mutations are at high (but still indefinite) risk of developing breast and ovarian cancer. Their limited choices include adherence to an intensive surveillance regime, which will not affect their chances of developing cancer, and prophylactic surgery (radical bilateral mastectomy and/or oophorectomy), which reduces risk to some uncertain degree. As Glenn McGee has recently noted, "the proactive medicine we have the power to do with genetic diagnosis is not the bucolic prophylaxis practiced by Marcus Welby. It is draconian stuff" (McGee, 1997, p. 22).

Given this harsh reality, it is not surprising that PKU is so often invoked by those who favor the expansion of genetic testing. While the ability to thwart mental retardation in PKU is not a product of specifically genetic knowledge,[4] it does represent a clear case of an effective intervention in the course of a genetic disease. Infants started on the diet early and maintained with reasonable control in infancy and early childhood are not retarded. Indeed, screening and treatment have allowed many persons with PKU to attend college, hold jobs, and raise families.

II. THE REALITY OF TREATMENT

However, treatment is neither as effective or easy as was once thought, or as those who invoke PKU as an exemplar (either of the errors of biological determinism or the value of genetic testing) now imply. When mass screening began in the early 1960s, it was commonly asserted that only infants and young children would require treatment; that when gross brain development was complete (around the age of five), children could resume a normal diet. But studies revealed that IQ scores declined and behavioral problems increased after the diet was terminated; as a consequence, dietary recommendations became progressively more conservative. The American Public Health Association and the American Academy of Pediatrics now recommend that all young women with PKU remain on-diet through at least their child-bearing years (because phenylalanine is a potent teratogen, damaging to the developing fetus). Most treatment centers in the U.S. and Canada (87%) as well as the Medical Research Council in Britain are even more conservative, recommending "diet for life" for PKU patients of both sexes (Fisch et al., 1997). However, strict adherence to the diet is difficult to achieve.

There are a number of reasons why. Both the special foods and the formula are boring and burdensome to prepare. The PKU diet involves substitutes for most natural protein foods, including bread, cake, meat, fish, eggs, and dairy products, supplemented by a phenylalanine-free formula that contains extra tyrosine and other amino acids, vitamins, and minerals. It is perhaps the most restrictive of all medical diets, involving foods that must be eaten as well as foods that must be avoided. The formula is unpalatable and it is necessary to drink quite a lot of it (typically three 8-ounce servings a day and about 25% more during pregnancy). Since most individuals with PKU do not feel sick, there is little incentive to comply. (In general, promises of long-term improvement provide much less incentive for compliance than do immediate benefits.)

The formula and special low-protein foods are also costly. A recent study found that the mean weekly cost of the formula is $99.51, and the cost of food (excluding formula) $77. By school age, it costs three times as much to feed an affected than unaffected child (California Department of Health Services, 1997, p. 47). The formula was originally classified as a drug, which could only be obtained by prescription, and was thus reimbursable for those with health insurance. When it lost this status in 1972, many insurers came to treat it as a food and refused to provide reimbursement. While some states have passed laws requiring insurance companies to do so, these laws do not cover self-insurers (who now provide most employee coverage). Many individuals with PKU do not have insurance and rely instead on a patchwork of federal and state programs and/or their own resources to finance the diet (Office of Technology Assessment 1988, pp. 112-114). One New York study found that most patients who had health insurance or Medicaid coverage were unable to obtain reimbursement for the formula or low-protein foods (Millner, 1993).

But even when financing is not an issue, it is hard to obtain strict dietary control, especially with older children and young adults. As one of the pioneers of screening notes, "the expectation that otherwise normal children will willingly adhere closely to a restricted diet of any kind is misplaced, and the lack of short-term consequences of dietary cheating provides no reinforcement" (O'Flynn, 1992, pp. 161-162). Adolescents in particular tend to be insecure and highly susceptible to advertising and peer pressure; they want to eat what their friends eat. The literature on susceptibility genes seems particularly naive on the link between knowledge and behavior. (For a broader critique of what he terms "genetic humoralism," see Juengst's essay in this volume.) Many commentators assume that knowledge of environmental factors that trigger harmful responses will automatically prompt changes in life-style. "It's easier to change the environment than to change your genes" writes one researcher (Kahn, 1996, p. 497). She may be right, but changing environments and life-styles is hardly a simple matter. Food habits are particularly resistant to change. Eating is not simply a matter of nutrition; like sex, it has a social as well as biological component. Food is integral to religious and ethnic identify (which is why immigrants' food habits are the last to change); eating the same foods is one way of showing that we belong to a group. Most important, meals express friendship and are used to establish intimacy (Douglas, 1975). They are loaded with social and symbolic meaning.

Moreover, while nutritional therapy in infancy and childhood does prevent mental retardation, it does not necessarily eliminate significant cognitive deficits. Thus a recent study found that 27% of young adults with the disease had IQs over two standard deviations below estimated population norms (compared to an expected two percent), a degree of impairment that would likely limit their employment prospects (Beasley, Costello, and Smith, 1994).

Significant impairments usually occur in cases of poor dietary control. But even early and well-treated individuals with PKU often experience more subtle cognitive impairments, such as lower than expected IQs, difficulties in problem-solving, reduced attention spans, and impaired fine motor ability (Burgard, *et al.*, 1996; Diamond, 1994; Faust, Libon, and Pueschel, 1986-87; Levy, 1991; Mabry, 1991; Pennington, *et al.*, 1985; Ris, *et al.*, 1994; Smith, Beasley, and Ades, 1990, 1991; Waisbren, *et al.*, 1994; Weglage, *et al.*, 1993, 1995; 1996a; Welsh, *et al.*, 1990). Behavioral and emotional problems are also common (Kalverboer, *et al.*,

1994; Waisbren and Levy, 1991; Weglage, *et al.*, 1996b). (But see also Koch, *et al.*, 1985; Mazzocco, *et al.*, 1994; and Schmidt, *et al.*, 1996 for more positive findings). Thus individuals with PKU generally require long-term medical, social, and rehabilitative services.

The most serious problem associated with treatment is "maternal PKU" (MPKU). If women do not resume the diet prior to conception and maintain it throughout pregnancy, the effects on their offspring can be catastrophic. (Metabolic control needs to be established before conception given the time it takes to bring phenylalanine levels down to an acceptable level.) Over 90% of infants born to untreated mothers suffer from mental retardation and microencephaly and 12-15% from congenital birth defects. Many also suffer from emotional disturbances (Levy and Ghavami, 1996, p. 177).

The affected infants do not themselves usually have PKU. However, the phenylalanine circulating at high levels in the maternal blood is teratogenic and it easily crosses the placental barrier and concentrates in the fetus at a level about 1.5 times that in the mother. Thus levels of the amino acid that are safe for the mother may be damaging to the fetus (Koch, 1993, p. 1224). To avoid these risks, women are advised to maintain very strict dietary control (at or below 6 mg/dl). Maintaining this control is especially difficult during pregnancy, most notably in the first trimester when tolerance for the formula is lowest and nausea and vomiting common. Conversely, attempts to maintain very strict control may sometimes produce deficiencies of phenylalanine and other amino acids. This overtreatment may also damage the fetus (Brenton and Lilburn, 1966, S180; Levy and Ghavami, 1996, p. 182).

Given the earlier assumptions about when it was safe to terminate treatment, young women generally discontinued the diet during childhood and were not followed for many years. Many have not been located and may be unaware of the risk. Most women with PKU seek medical attention after they became pregnant, when some damage to the fetus has quite possibly already occurred (Koch, *et al.*, 1993, 1229). The Maternal PKU Collaborative Study, which began in 1984, identified 402 pregnancies; researchers found that few of the young women were on diet (and 101 had IQs lower than 80 (Koch, *et al.*, 1994, p. 113)). Before the advent of screening, women with PKU were severely retarded and bore very few children. Their fertility is now nearly normal, with the incidence of the maternal PKU syndrome 1 in 30,000-40,000 pregnancies (U.S. Preventive Services Task Force, 1996, p. 495). That situation has prompted concern that all the benefits of screening may be neutralized by the birth of retarded children to women who have ended the diet (Kirkman, 1982). (There were so few pre-conceptually treated and well-controlled pregnancies when the study began that researchers could not determine whether the current diet provides for adequate fetal growth and development (Koch, *et al.*, 1994, pp. 112, 118).) But it is even harder to resume than to stay on the diet (see O'Flynn, 1992, p. 163).

III. PKU AND THE ISSUE OF REPRODUCTIVE AUTONOMY

Lionel Penrose made PKU the paradigmatic example of what was wrong with eugenics. But at the time he wrote, treatment for PKU – and the problem of MPKU – did not yet exist. By the 1980s, when the problem was finally acknowledged, attitudes towards what was once called reproductive responsibility had been transformed.

In Penrose's day, nearly everyone agreed that individuals at high risk of transmitting a serious disease (genetic or otherwise) should refrain from childbearing. As noted earlier, Penrose himself assumed that carriers of the same defective gene should not marry each other. Even in the 1950s and '60s, the principle of reproductive responsibility was generally taken for granted. The view of Sheldon Reed, who coined the expression "genetic counseling," was typical: "No couple has the right to produce a child with a 100 per cent chance of having PKU, and it is doubtful whether a couple has the right deliberately to take a 50 per cent chance of producing such a serious defect" (Reed, 1964, p. 85). That attitude was shared even by individuals who are associated with opposition to biological determinism and eugenics. For example, according to Ashley Montagu, "there can be no question that infantile amaurotic family idiocy [Tay-Sachs disease] is a disorder that no one has a right to visit upon a small infant. Persons carrying this gene, if they marry, should never have children, and should, if they desire children, adopt them" (Montagu, 1959, pp. 305-306).[5] Similarly, in his anti-eugenics classic *Mankind Evolving*, Theodosius Dobzhansky asserted that, "persons known to carry serious hereditary defects ought to be educated to realize the significance of this fact, if they are likely to be persuaded to refrain from reproducing their kind. Or, if they are not mentally competent to reach a decision, their segregation or sterilization is justified. We need not accept a Brave New World to introduce this much of eugenics" (Dobzhansky, 1962, p. 333).

Today, few bioethicists – the people to whom governments and the media today tend to turn for advice and commentary on genetics-related issues – would agree. Indeed, for a complex of reasons, including the rise of the women's and patients' rights movements, the concept of reproductive responsibility is now decidedly out of fashion (see Paul, 1992). On the reigning view, genetic counseling should be value-neutral, and clinicians should scrupulously avoid giving any opinion regarding their clients' reproductive decisions. That perspective is exemplified by the recent Institute of Medicine report, whose authors assert that, "reproductive genetic services should not be used to pursue eugenic goals, but should be aimed at increasing individual control over reproductive options," and that "the goal of reducing the incidence of genetic conditions is not acceptable, since this aim is explicitly eugenic; professionals should not present any reproductive decisions as 'correct' or advantageous for a person or society" (Andrews, et al., 1994, pp. 8, 15).

However, bioethics is not monolithic, and some scholars, while abjuring coercion, have argued that individuals at high risk of transmitting a serious disease should not reproduce (e.g., Steinbock and McClamrock, 1994; Purdy, 1995). Moreover, there is evidence that the autonomy-orientation of most bioethicists and genetics professionals is not necessarily shared by other clinicians or by the public at large. The plurality of moral views that Tris Engelhardt describes in his introductory essay is evident in this sphere as well. For example, according to a survey conducted by Dorothy Wertz and her colleagues (Wertz, 1997b), only 5% of geneticists in the U.S. thought that governments should require carrier tests for common genetic disorders before marriage, whereas 36% of primary care physicians, and 31% of patients thought they should. Asked whether they thought that people at high risk should not have children unless they use prenatal diagnosis and selective abortion, only 10% of geneticists but 78% of physicians and 81% of patients agreed they should not (Wertz, 1997b). In general, physicians and the public seem to think it is unfair to the child, to sibs, and to society to knowingly run a significant risk of having a child with a serious genetic disorder (Wertz, 1997a, p. 339).

Conflicting professional perspectives are evident when we compare genetic counseling with common practices in the counseling of HIV-infected women. According to M. Gregg Bloche, during the late middle and late 1980s, articles in leading medical journals and pronouncements by public health and professional authorities maintained that HIV-infected women should be counseled not to get pregnant and if pregnant, to abort, and he notes that, "at the clinical level, the counseling of reproductive abstinence has been widely recommended and is probably an established practice" (Bloche, 1996, pp. 258, 260).[6]

The tensions that might otherwise be produced by conflicting principles about reproduction are avoided, or at least muted, by the fact that the clinicians who deal with HIV/AIDS are not those who deal with genetic diseases such as Tay-Sachs or Huntington's disease. Maternal PKU represents an exception to this rule of professional compartmentalization. PKU itself is a genetic disease, while MPKU is not. The same metabolic clinics attend to both.

It might be said that the cases of HIV/AIDS and MPKU raise radically different issues than do cases of genetic disease. There is at least one relevant distinction. To avoid the risk of transmitting a genetic disease, prospective parents must be willing either to forgo having biological children or to abort an affected fetus. Since raising children is often viewed as a crucial element of human flourishing, and abortion may not be an acceptable option, some parents face very harsh choices. In the case of HIV/AIDS or MPKU, however, risk can be greatly mitigated by compliance with a therapeutic regime.

For some philosophers, there exists an additional distinction between genetics and non-genetics cases. According to one philosophical line of argument (deriving primarily from the work of Derek Parfit), there is an important moral difference between bearing a damaged child who could have been born healthy (a situation that arises in the non-genetics cases, where harm can be avoided if the mother complies with treatment) and bearing such a child who otherwise would not have been born (a situation that arises in genetics cases, where harm is avoided only if life is). In the latter cases, the child could only be harmed if it could be shown that damaged life is a worse fate than no life. From this philosophical standpoint, there are no grounds to counsel parents to delay child-bearing, since even a severely disabled child has not been harmed by the parents' decision. In the non-genetics cases, the parent would be morally obligated to take the steps necessary to avoid harm to the future child.

But even if this line of argument were compelling (and it has been strenuously contested by other philosophers (e.g., Powers, 1996; Green, 1997)), it does not address a problem that also arises in respect to the first distinction. There is little dispute over the pregnant woman's moral responsibility to avoid cocaine or be treated with AZT or maintain good metabolic control, or even the clinicians' duty to urge them to take these actions. The thorny issue is what to do when those efforts fail or can be predicted to fail, with significant resulting damage to the fetus. In the case of MPKU, the risk of damage is very high – much higher than in the case of HIV/AIDS, where even in the absence of therapy, the rate of parent-to-offspring transmission is 25%. If it is legitimate for clinicians to advise against childbearing until the mother agrees to treatment in the HIV/AIDS case, it is surely so in the case of MPKU. Conversely, if it is always wrong to give advice regarding reproductive decision-making, it is irrelevant whether or not the condition that provokes it happens to be genetic.

Indeed, the increasing penetration of the genetics model – according to which the clinicians' role is simply to help the client/patient achieve their own goals – reflects the implicit understanding that there are more similarities than differences between the genetics and non-genetics cases. Genetics has been treated as a special case because of its link to reproduction. If we believe that laissez-faire is the only acceptable policy in respect to reproduction, why distinguish MPKU from PKU or HIV/AIDS from Tay-Sachs disease? In fact, we less and less frequently make that distinction.

The important issue is not whether the state should intervene to prevent prenatal harm, although this is how the question of preventing prenatal harm is often framed. For a variety of reasons, both moral and prudential, the exercise of state power in this area is undesirable (see Clayton, 1997; Mathieu, 1996, pp. 70-101; Engelhardt, this volume). It is a common mistake to conflate judgments about morality and public policy, for we often have good reasons not to compel people to do what we think is right (Murray 1996, p. 109). In any case, at least in the U.S. and most of Europe, the use of state power in this sphere is extremely unlikely.

The eugenicists' primary tools were segregation and sterilization of affected individuals. But their overriding concern was with the inheritance of mental traits, and they conceived a sharp difference between themselves and the "feeble-minded," criminals, paupers, and despised ethnic groups who constituted their targets. When, in the post-World War II period, geneticists' focus shifted to disease, eugenics of that sort quickly lost its appeal. After all, middle-class people from favored ethnic groups also get sick. (In his inaugural address, Penrose noted that no individuals with phenyketonuria had been discovered among either Jews or American blacks.) And U.S. state governments are hardly likely to require women to have abortions; they are much more prone to try to discourage or even ban them. Thus, the real question is what attitude should be adopted by professionals. What should they say, if anything, when women who are not on-diet, and often have a difficult time functioning themselves, want children – who will be at high risk for mental retardation? Should clinicians attempt to discourage them? Or should they view their role as helping clients fulfill their own wishes, whatever those wishes may be? Dorothy Wertz found that 59% of geneticists and 39% of primary-care physicians would support a parental decision to take their chances given the following situation: "Woman with PKU, uncontrolled phenylalanine level in 3rd trimester, 100% chance of birth defects" (Wertz, 1997a). That is a striking illustration of the extent to which a laissez-faire attitude toward reproduction has replaced the view that reproduction is a social concern. But is support for such decisions reasonable?

An affirmative response might be grounded in the claim that reproductive autonomy is an absolute right. But even if this usually undefended claim were compelling, giving advice does not necessarily infringe it. In medicine generally, clinicians are expected to make recommendations; no one claims that simply by doing so they are undermining the autonomy of their patients. As long as the client's decision-making role is acknowledged, voicing an opinion per se is not coercive, whether or not reproduction is involved (Bloche, 1996, p. 260).

In some philosophical perspectives, autonomy may be undermined by pressures arising from unfavorable social circumstances. On any perspective, it is undermined by slanting facts to produce a desired outcome, a practice that was once common in the U.S. (Paul, 1997) and apparently often remains so in genetic counseling outside of North America (Wertz, 1997a, pp. 326-327, 336). As Dorothy Wertz notes, "clients given false 'facts' have no opportunity to resist" (Wertz, 1997a, p. 326). But unless one assumes the existence (and desirability) of the unfettered individual assumed by classical liberal theory, advice-giving in and of itself does not thwart autonomy.

However, those who would support, for example, the decision of a woman with an uncontrolled phenylalanine level to become pregnant might also be moved by recognition that since it is the mother who carries the child-to-be, it is almost always women rather than men who are asked to sacrifice their desires (Murray, 1996, p. 101). Moreover, these women are disproportionately poor, and often minority. Thus, asking for reproductive restraint places an additional burden on those whom society already disadvantages. They might also note that the suffering associated with genetic disease is sometimes exaggerated, and that many of the real problems could be relieved were society willing to provide greater social support and more accommodations for people with disabilities (Hubbard and Wald, 1993). These points have real force. As the previous discussion suggests, financing the low-phenylalanine diet is often a struggle. While the Women, Infants, and Children (WIC) program and other federal and some state programs cover the cost of formula for women during pregnancy, they do not usually do so prior to conception.

Even more important, these young women require long-term social and psychological support. Indeed, in MPKU, the amount of social support better predicts compliance with the diet than does IQ or knowledge (Waisbren, et al., 1995). However, the American socio-medical system is geared primarily toward curing individuals with acute illnesses, rather than helping those with chronic conditions to live as well as they can (Knox, 1996; see also Hoffman, et al., 1996). Thus, it is much more difficult to obtain help with bathing, cooking, shopping, and other activities of daily living than to obtain access to even quite expensive diagnostic tests. But assistance with ordinary activities is what individuals with PKU most need. Compliance with the diet would thus be fostered by a broader and deeper web of social support. However, even greatly improved services are likely to mitigate, rather than eliminate, the problem of MPKU.

It is easy to understand why the concept of reproductive responsibility has gone out of fashion, associated as it is with the oppression of women, paternalism toward patients, and the abuses of the eugenics movement. But reproductive decisions do have social consequences, at a bare minimum, for another person. To admit this is not to justify state intervention, the slanting of facts to achieve a desired outcome, or even strong directivity in counseling. In the end, we should respect prospective parents' right to make poor choices (even if we could do otherwise). Nor is the interest of the child-to-be or the larger society all that morally counts. It is surely legitimate for women to consider their own welfare, and not only that of the child-to-be. As Thomas Murray writes, pregnancy should not become "a kind of moral trump card, carrying the trick no matter how strong the other cards on the table" (1996, 105). Thus, the nature and extent of the sacrifice asked of the parents matter, as do the degree of risk and the kind of harm to the child-to-be. In MPKU, the harm is severe and near-certain when phenylalanine levels are uncontrolled. In such a case, even Lionel Penrose would surely agree that the decision to bear a child is unwise. As long as they acknowledge that the final decision belongs to the woman, and do not misrepresent the facts, clinicians who express that opinion do their clients no wrong.

ACKNOWLEDGMENTS

Research for this essay was supported by the National Science Foundation under Grant No. SBR-9511909. I am also grateful to the members of the PKU CORPS at Children's Hospital, who labor mightily to help young women (and men) cope with PKU, to Bruno Sasser, for helpful comments on a draft of the essay, and to the editors of this volume for their patience.

University of Massachusetts,
Boston, Massachusetts

NOTES

[1] See also Paul and Spencer, 1995

[2] This is a somewhat misleading illustration of the futility of eugenics since its advocates were not concerned with rare diseases like PKU but with mental defect, which they thought was frighteningly common. See Paul and Spencer, 1995.

[3] To skeptics the gap between knowledge and treatment reflects a faulty framework for thinking about the relation of genes to disease. They note that medical measures in general are only weakly related to declines in mortality (see Kass, 1971; McKeown, 1976; McKinlay and McKinlay, 1977; Mann, 1997; Magnus, 1996). Genetic research is considered particularly unlikely to result in major benefits to human health since the involvement of genes in most common disorders, such as coronary heart disease, diabetes, rheumatoid arthritis, cancer, or stroke, is modest at best, and they believe that, in respect to the many but rare single-gene disorders, such as Duchenne's muscular dystrophy, cystic fibrosis, or Huntington disease, a knowledge of DNA sequences provides little causal information (Lewontin, 1992; Hubbard and Wald, 1993; Golub, 1997, 205-219; Juengst, this volume).

[4] It has been known since the 1930s that the brain damage associated with PKU is somehow related to high levels of blood phenylalanine. Dietary therapy based on this understanding began in the 1950s – more than thirty years before the gene was cloned and in the absence of any detailed knowledge of the pathogenesis of the disease. Even today, it is unclear whether the damage associated with PKU results from phenylalanine acting directly or indirectly (by inhibiting the transport of tyrosine and other essential amino acids), what exactly is damaged (e.g., the nerve cells or the myelin sheath) or even whether phenylalanine or one of its metabolites is the primary biochemical culprit (Levy and Ghavami, 1996, pp. 179-80).

⁵ It is notable that none of these scientists is concerned with effects on the gene pool. In the
    cases discussed by Reed and Montagu, the fitness of the recessive homozygote is nearly zero.
⁶ On physicians' view of directivity, see also Geller and Holtzman, 1995, pp. 103-105.

REFERENCES

Andrews, L., *et al.*: 1994, *Assessing Genetic Risks*, National Academy Press, Washington, D.C.
Baird, P.A.: 1990, 'Genetics and health care: A paradigm shift,' *Perspectives in Biology and Medicine* 33, 203-213.
Beasley, M.G., Costello, P.M., and Smith, I.: 1994, 'Intellectual status of young adults with phenylketonuria (PKU),' in J.-P. Farriaux and J.-L. Dhondt (eds.), *New Horizons in Neonatal Screening*, Amsterdam and New York, Elsevier Science, pp. 109-110.
Bloche, M.G.: 1996, 'Clinical counseling and the problem of autonomy-negating influence,' in R.R. Faden and N.E. Kass (eds.), *HIV, AIDS & Childbearing: Public Policy, Private Lives*, Oxford University Press, New York, pp. 257-319.
Brenton, D.P., and Lilburn, M.: 1996, 'Maternal phenylketonuria: A study from the United Kingdom,' *European Journal of Pediatrics* 155 (Suppl 1), S177-S180.
Burgard, P., *et al.*: 1996, 'Intellectual development of the patients of the German collaborative study of children treated for Phenylketonuria,' *European Journal of Pediatrics* 155 (Suppl 1), S33-S38.
California Department of Health Services, Genetic Disease Branch: 1997, *Cost and Availability of Dietary Treatment of Phenylketonuria (PKU): Report of a National Survey*, Berkeley, California.
Clayton, E.W.: 1997, 'Legal and ethical commentary: The dangers of reading duty too broadly,' *Journal of Law, Medicine & Ethics* 25, 19-21.
Coulter, W.E., *et al.*: (1912), *Heredity and Eugenics*, University of Chicago Press, Chicago.
Davenport, C.B.: 1912, 'The inheritance of physical and mental traits of man and their application to eugenics,' in W.E. Coulter, *et al.*, *Heredity and Eugenics*, University of Chicago Press, Chicago, pp. 269-288.
Diamond, A.: 1994, 'Phenylalanine levels of 6-10 mg/dl may not be as benign as once thought,' *Acta Paediatrica* Supplement 407, 89-91.
Dobzhansky, T.: 1962, *Mankind Evolving,* Yale University Press, New Haven.
Douglas, M.: 1975, 'Deciphering a meal,' in *Implicit Meanings: Essays in Anthropology*, Routledge and Kegan Paul, London, pp. 249-275.
Edelson, P.J.: 1994, 'Lessons from the history of genetic screening in the US: Policy, past, present, and future,' unpublished ms.
Epstein, C.J.: 1997, '1996 ASHG Presidential Address. Toward the 21st Century,' *American Journal of Human Genetics* 60, 1-9.
Faust, D., Libon, D., Pueschel, S.: 1986-87, 'Neuropsychological functioning in treated phenylketonuria,' *International Journal of Psychological Medicine* 16, 169-177.
Fisch, R.O., *et al.*: 1997, 'Phenylketonuria: Current dietary treatment practices in the United States and Canada,' *Journal of the American College of Nutrition* 16, 147-151.
Geller, G., Holtzman, N.A.: 1995, 'A qualitative assessment of primary care: Physicians' perceptions about the ethical and social implications of offering genetic testing,' *Qualitative Health Research* 5, 97-116.

Golub, E.S.: 1997, *The Limits of Medicine: How Science Shapes our Hope for the Cure*, University of Chicago Press, Chicago.

Green, R.M.: 1997, 'Parental autonomy and the obligation not to harm one's child genetically,' *Journal of Law, Medicine & Ethics* 25, 5-15.

Hayes, A., Costa, T., Scriver, C.R., Childs, B.: 1985, 'The effect of mendelian disease on human health II: Response to treatment,' *American Journal of Medical Genetics* 2, 243-255.

Hoffman, C., Rice, D., Sung, Hai-Yen: 1996, 'Persons with chronic conditions: Their prevalence and costs,' *Journal of the American Medical Association* 276 (November 13), 1473-1479.

Hood, L.: 1992, 'Biology and medicine in the twenty-first century,' in D.J. Kevles and L. Hood, (eds.), *The Code of Codes: Scientific and Social Issues in the Human Genome Project*, Harvard University Press, Cambridge.

Hubbard, R., Wald, E.: 1993, *Exploding the Gene Myth: How Genetic Information is Produced and Manipulated by Scientists, Physicians, Employers, Insurance Companies, Educators, and Law Enforcers*, Beacon Press, Boston.

Kahn, P.: 1996, 'Coming to grips with genes and risk,' *Science* (October 25), 497.

Kalverboer, A.F.: 1994, 'Social behavior and task orientation in early-treated PKU,' *Acta Paediatrica* supplement 407, 104-105.

Kass, E.H.: 1971, 'Infectious disease and social change,' *Journal of Infectious Diseases* 123, 110-114.

Kevles, D.J.: 1985, *In the Name of Eugenics: Genetics and the Uses of Human Heredity*, Alfred A. Knopf, New York.

Kevles, D.J., Hood, L., (eds.): 1992, *The Code of Codes: Scientific and Social Issues in the Human Genome Project*, Harvard University Press, Cambridge.

Kirkman, H.N.: 1982, 'Projections of a rebound in frequency of mental retardation from phenylketonuria,' *Applied Research in Mental Retardation* 3, 319-328.

Knox, R.: 1996, 'Widespread chronic illness cited,' *Boston Globe*, November 13.

Koch, R., Yusin, M., Fishler, K.: 1985, 'Successful adjustment to society in young adults with phenylketonuria,' *Journal of Inherited Metabolic Disorders* 8, 209-211.

Koch, R., et al.: 1993, 'The North American Collaborative Study of Maternal Phenylketonuria. Status Report 1993,' *American Journal of Diseases of Children* 147, 1224-1230.

Koch, R., et al.: 1994, 'The International Collaborative Study of maternal phenylketonuria: Status Report 1994,' *Acta Paediatrica Supplement* 407, 111-119.

Laughlin, H.H.: 1914, *Report of the Committee to Study and to Report on the Best Practical Means of Cutting Off the Defective Germ-Plasm in the American Population*. *I. The Scope of the Committee's Work*, Bulletin No. 10A, Eugenics Record Office, Cold Spring Harbor, New York.

Levy, H.L.: 1991, 'Nutritional therapy in inborn errors of metabolism,' in R.J. Desnick (ed.), *Treatment of Genetic Diseases*, Churchill and Livingstone, New York.

Levy, H.L. and Ghavami, M.: 1996, 'Maternal phenylketonuria: A metabolic teratogen,' *Teratology* 53, 176-184.

Lewontin, R.C.: 1992, 'The dream of the human genome,' *The New York Review of Books* 39; reprinted in Lewontin, R.C.: 1993, *The Doctrine of DNA*, Penguin Books, New York, pp. 60-83.

Lewontin, R.C.: 1993, *The Doctrine of DNA*, Penguin Books, New York.

Mabry, C.C.: 1991, 'Status report on Phenylketonuria treatment,' *American Journal of Diseases of Childhood* 145, 33.

Magnus, D.: 1996, 'Gene therapy and the concept of genetic disease,' available online at 'Ethics and Genetics: A Global Conversation,' Glenn McGee (ed.), http://www.med.upenn.edu/bioethic/genetics.html.

Mann, J.M.: 1997, 'Medicine and public health, ethics and human rights,' *Hastings Center Report* 27 (May-June), 6-13.

Mathieu, D.: 1996, *Preventing Prenatal Harm: Should the State Intervene?* Georgetown University Press, Washington, D.C.

Mazzocco, M.M., *et al.*: 1994, 'Cognitive development among children with early-treated phenylketonuria,' *Developmental Neuropsychology* 10, 133-151.

McGee, G.: 1997, *The Perfect Baby: A Pragmatic Approach to Genetics*, Rowman and Littlefield, Lanham, MD.

McKeown, T.: 1976, *The Role of Medicine: Dream, Mirage or Nemesis*, Nuffield Provincial Hospitals Trust, London.

McKinlay, J.B., McKinlay, S.M.: 1977, 'The questionable contribution of medical measures to the decline in mortality in the United States in the twentieth century,' *Health and Society* (Summer), 405-428.

Millner, B.N.: 1993, 'Insurance coverage of special foods needed in the treatment of phenylketonuria,' *Public Health Reports* 108 (Jan.-Feb.), 60-65.

Milunsky, A.M., Annas, G.J. (eds.): 1980, *Genetics and the Law II*, Plenum Press, New York.

Montagu, A.: 1959, *Human Heredity*, World Publishing, Cleveland.

Murray, T.H.: 1996, *The Worth of a Child*, University of California Press, Berkeley.

Office of Technology Assessment, U.S. Congress: 1988, 'Newborn screening for congenital disorders,' in *Healthy Children: Investing in the Future*, USGPO, Washington, D.C., pp. 93-116.

O'Flynn, M.E.: 1992, 'Newborn screening for Phenylketonuria: Thirty years of progress,' *Current Problems in Pediatrics* 22 (April), 159-165.

Paul, D.B.: 1992, 'Eugenic anxieties, social realities, and political choices,' *Social Research* 59 (Fall), 663-683.

Paul, D.B.: 1997, 'From eugenics to medical geneticism,' *Journal of Policy History* 9: 96-116.

Paul, D.B.: 1998, 'Competing agendas, converging stories: The case of PKU,' in *The Politics of Heredity: Essays on Eugenics, Biomedicine, and the Nature-Nurture Debate*, SUNY Press, Albany, and M. Fortun and E. Mendelsohn, (eds.), *The Practices of Human Genetics*, Kluwer, Dordrecht (in press).

Paul, D.B., Spencer, H.G.: 1995, 'The hidden science of eugenics,' *Nature* 374 (March 23), 302-304.

Paul, D.B., Edelson, P.J.: 1998, 'The struggle over metabolic screening,' in S. de Chadarevian and H. Kamminga (eds.), *Molecularising Biology and Medicine: New Practices and Alliances, 1930s-1970s*, Harwood Academic Publishers, Reading, pp. 203-220.

Pennington, B.F., *et al.*: 1985, 'Neuropsychological deficits in early-treated phenylketonuric Children,' *American Journal of Mental Deficiency* 5, 467-474.

Penrose, L. S.: 1946, 'Phenylketonuria: A problem in eugenics,' *Lancet* (June 29), 949-953.

Powers, M.: 1996, 'The moral right to have children,' in R.R. Faden and N.E. Kass, (eds.), *HIV, AIDS and Childbearing: Public Policy, Private Lives,* Oxford University Press, New York, pp. 320-344

Purdy, L.: 1995, 'Loving future people,' in J.C. Callahan (ed.), *Reproduction, Ethics, and the Law*, Indiana University Press, Bloomington, pp. 300-327.

Reed, S.: 1964, *Parenthood and Heredity*, John Wiley, New York.

Ris, M.D. *et al.*: 1994, 'Early-treated Phenylketonuria: Adult neuropsychologic outcome,' *Journal of Pediatrics* 124, 388-392.

Rosenberg. S.S.: 1962, 'A new life for Karen,' *Family Weekly*. (December 16), 6-7.

190 DIANE B. PAUL

Schmidt, H., *et al.*: 1996, 'Intelligence and professional career in young adults treated early for phenylketonuria,' *European Journal of Pediatrics* 155 (Suppl 1), S97-S100.

Smith, I.: 1994, 'Treatment of phenylalanine hydroxylase deficiency,' *Acta Paediatrica Supplement* 407, 60-65.

Smith, I., Beasley, M.G., Ades, A.E.: 1990, 'Intelligence and quality of dietary treatment in phenylketonuria,' *Archives of Disease in Childhood* 65, 472-478.

Smith, I., Beasley, M.G., Ades, A.E.: 1991, 'Effect on intelligence of relaxing the low phenylalanine diet in phenylketonuria,' *Archives of Disease in Childhood* 66, 311-316.

Steinbock, B. and McClamrock, R.: 1994, 'When is birth unfair to the child?' *Hastings Center Report* 24, 15-21.

Treacy, E., Childs, B., Scriver, C.R.: 1995, 'Response to treatment in hereditary metabolic disease: 1993 survey and 10-year comparison,' *American Journal of Human Genetics* 56, 359-367.

U.S. Preventive Services Task Force: 1996, *Guide to Clinical Preventive Services*, 2nd ed., Williams and Wilkins, Baltimore.

van Dijck, J.: 1998, *ImagEnation: Popular Images of Genetics*, New York University Press, New York.

Waisbren, S.E., *et al.*: 1995, 'Psychosocial factors in maternal phenylketoneuria: Women's adherence to medical recommendations,' *American Journal of Public Health* 85, 1636-1641.

Waisbren, S.E., *et al.*: 1994, 'Review of neuropsychological functioning in treated phenylketonuria: An information processing approach,' *Acta Paediatrica Supplement* 407, 98-103.

Waisbren, S.E., Levy, H.L.: 1991, 'Agoraphobia in Phenylketonuria,' *Journal of Inherited Metabolic Disorders* 3, 149-153.

Weglage, J., *et al.*: 1993, 'School performance and intellectual outcome in adolescents with Phenylketonuria,' *Acta Paediatrica* 82, 582-586.

Weglage, J., *et al.*: 1995, 'Neurological findings in early treated phenylketonuria,' *Acta Paediatrica* 84, 411-415.

Weglage, J., *et al.*: 1996a, 'Deficits in selective and sustained attention processes in early treated children with Phenylketonuria – result of impaired frontal lobe functions,' *European Journal of Pediatrics* 155, 200-204.

Weglage, J., *et al.*: 1996b, 'Psychosocial aspects in Phenylketonuria,' *European Journal of Pediatrics* 155 [Suppl 1] S101-S104.

Welsh, M.C., *et al.*: 1990, 'Neuropsychology of early treated Phenylketonuria: Specific executive function deficits,' *Child Development* 61, 1697-1713.

Wertz, D.C.: 1997a, 'Society and the not-so-new genetics: What are we afraid of? Some Future Predictions from a social scientist,' *Journal of Contemporary Health Law and Policy* 13, 299-346.

Wertz, D.C.: 1997b, 'Data distributed at the "Workshop on Eugenic Thought and Practice: A Reappraisal towards the End of the Twentieth Century,"' (May 26-29), Van Leer Institute, Jerusalem. (Published in part as Wertz, D.C.: 1998, 'Eugenics is alive and well: A survey of genetics professionals around the world,' *Science in Context* 11, 493-510.)

JONATHAN D. MORENO

EVERYBODY'S GOT SOMETHING

A few years ago a major newspaper ran what used to be called a "human interest" story about a remarkable woman who, alone, had adopted eight children, each of whom had some significant disability. Everyone learned how to help take care of each other, she explained to the interviewer who expressed amazement at her ability to manage it all. And how did she handle it when someone outside their family stared or made fun of them during an outing? "I just tell them that those people have problems, too, even though you might not be able to see it. I just tell them that everybody's got something."

And that might well be the motto of the now-not-so-new genetics. As Eric Juengst notes in "Concepts of disease after the human genome project" (this volume), for eons those without manifestations of disease visible to the naked eye could tell themselves they were disease free, perhaps indulging in the arrogance and moral superiority of the well. Then a few hundred years ago what had once been a somewhat vague notion of uncleanness became transformed, as bacteria became identifiable, into a far more refined concept of "carrier." These people, if not exactly sick, at least were not disease free, either, and when certain diseases became stigmatized it was easy to extend the danger sign in their direction.

What seems to be the reasonable next step is to see genes as the underlying "carriers" of disease, rather than the people who have them. At first glance this should be an important step toward the goal of destigmatizing individuals, who cannot be blamed for having the genetic endowment they ended up with, anymore than they can be blamed for having the parents they had. Distinguishing between the person and the disease is one of the great humanistic achievements of Hippocratic medicine, probably far more important in this respect than the Hippocratic Oath. This inherited attitude allowed Ronald Reagan, who had a much-publicized basal cell carcinoma on his nose, to respond in typically sunny manner to a journalist who asked him after the operation how it felt to have cancer. "I didn't have cancer," he replied, "something on me had cancer and it's gone."

Stephen Wear, James J. Bono, Gerald Logue and Adrianne McEvoy (eds.), Ethical Issues in Health Care on the Frontiers of the Twenty-First Century, 191–196.
© 2000 *Kluwer Academic Publishers. Printed in Great Britain.*

But could even the irrepressible Reagan have said something like that following a gene therapy procedure? "I didn't have cancer, my genes did." The trouble is, people are widely understood to "be" their genes, or at least their identities are more intimately bound up with their genes than with our long-familiar, grosser somatic structures. Sick genes are frequently taken to imply sick people, even if they are only "carriers." Hence the 1970s sickle cell experience in the United States, in which carrier status became so confused with the risk of clinical disease that an ambitious public health program threatened to stigmatize a large portion of the African-American community.

A second sense in which the image of genes themselves as disease carriers is too strong is still more basic. Unlike anatomic lesions, genes are not specific causes of disease, not even the so-called single gene disorders. Some people with the alteration have a clinical problem and some don't.

Where in all this is the much-vaunted "revolution in genetics?" There is nothing new in the charge that, so far at least, the new medical genetics is all show and no go. Like the record-setting opening price of Genentech shares, the science of the future so far looks mostly late-twentieth century hype. Early experiments, such as the well-publicized attempts to treat autoimmune disease, have enjoyed at best modest success and certainly failed to live up to their advance notices. With the significant exception of agricultural products – which have often been met with open hostility, especially in Europe – fans of medical progress can understandably ask, "Where's the beef?"

Diane Paul's account of PKU seems only to make matters worse. During my freshman year in college, my introductory psychology professor lectured us on the importance of PKU screening and of putting affected individuals on the appropriate diet ("PKUees" she called them). She also wore a snazzy button with a PKU testing logo. That was 1969, when enthusiasm for social movements for every cause ran high. Only two years earlier, Paul reminds us, the International Congress of Human Genetics caused a stir about genetic diseases' treatability.

The Hippocratic texts suggest that physicians have long been fairly good at predicting the course of a disease and identifying the crisis period, without being able to do much about it. Unfortunately, not only has it turned out that genetic diseases continue the traditional gap between knowledge and treatment, but PKU is a particularly vigorous example of a further problem: the slack between treatability and effective therapy.

The knowledge-treatment gap can be theoretically resolved by science, but the treatability-efficacy gap is more frustrating because it turns on a grasp of the human sciences, psychology, sociology, and economics. Thus the recommended diet for persons affected with PKU is a bad combination from the standpoint of patient compliance. It is boring, expensive, and with no benefits that are immediately obvious to the sufferer. And of course it bears no resemblance to fancy genetic interventions.

The issues surrounding genetic testing for cystic fibrosis provide another illustration of the promissory character of genetic medical science. In the past ten years a genetic test was developed to identify carriers of the alteration ("mutation," I have been told by an NIH geneticist, sounds judgmental and should be replaced), and in the past twenty years median survival of sufferers has increased from 18 to 30 years. But these two facts have nothing to do with each other. Survival has increased because of more aggressive medical and dietary management of CF complications. There is no CF gene therapy on the horizon. The test itself has limited application. A National Institutes of Health consensus conference concluded that carrier testing is recommended for those with a positive family history who are contemplating pregnancy, but not for the general population. The cost-benefit ratio for general population screening is poor because the sensitivity of the test varies among different groups, and is especially poor in Americans of Hispanic and Asian origin (NIH, 1997).

The empirical limitations of testing and treating genetic disease did not come as a surprise to everyone, though Juengst shows how even more cautious commentators relied on behavioral interventions to modify an individual's "predisposition." This "constitutional" thinking, imported from an earlier model, allowed both an avoidance of the limitations of detailed genetic knowledge (attributing specific ignorance to blanks that can be filled in later), and an avoidance of the ethically messier business of reproductive genetics.

Juengst's agenda is mainly to encourage a radically new way of seeing disease in the era of genetic medicine, especially in terms of causation. Genetic science requires giving up the familiar and comfortable one-to-one correspondence between a lesion site and a disease. Such a shift includes conceptually three rules: separating diseases from their sources, returning to patients' complaints as the source of the meaning of disease, and accepting multiple approaches to managing disease.

Juengst's three rules might well render our concept of disease more compatible with genetics, but what does it have to say about the treatability problem? What does it say especially for those already born with alterations that contribute to (not "cause") a perceived (patient-defined) disease state?

Consider Juengst Rule 3, the one about accepting multiple approaches to managing disease. A most efficient way to manage disease, even in a multifactorial causal universe, is to eliminate a necessary condition for the disease's appearance. The technical apparatus of the new genetics is almost certainly going to be far better at doing that than at coming up with effective interventions after the earliest stages of reproduction, as suggested by the PKU experience. Similarly, the gene test for CF is not recommended for children or newborns because the result wouldn't make any difference in symptom management.

Rather than attempting somatic cell therapy, it is a much more straightforward and efficient matter to insert "cassettes" of genes into newly fertilized eggs. These chromosome-sets or gene-clusters can then confer a disposition for various metabolic processes throughout the cells of the new individual. For instance, a cancer cell-eating process might be set to react to a certain agent that can be injected into the body following a cancer diagnosis.

Germline genetic engineering. Ever since the President's Commission there has been a rough consensus that crossing that moral Rubicon into altered heritable characteristics just isn't worth the trouble. Nearly two decades ago, however, it was also easier to tell ourselves that practicable somatic cell therapies were just around the corner, and even that behavioral interventions had a future. Now as the limitations of what Juengst calls the nineteenth-century disease paradigm become more apparent, and as the ever-increasing costs of health care have become a familiar political problem, germline therapy seems less like forbidden fruit.

Yet the philosophical conundrums that were part of the birth of bioethics remain, especially the disease prevention/species enhancement distinction: Where does the elimination of some of humanity's most detested scourges end and frivolous enhancement of the species begin? And who is to decide which communities, nations, and peoples enjoy early access to having their heirs "improved," and which ones become part of the genetic proletariat? In a market economy such as that dominant in today's health care system little imagination is required to foresee the consequences.

Dorothy Nelkin richly documents the view that the "overselling" of genetics may partly account for a backlash in the public mind. If she is right, then any indication that the wealthy will have more access to "enhanced" offspring – even if the actual results are disappointing – could over the next years stimulate a renewed basis for class antagonisms that have been unfashionable for decades. Not only class but also racial distinctions could be reinforced. That is, if germline interventions are aimed at characteristics perceived to be related to race, the perverse result could be a reinforcement of racial essentialism in the minds of many.

It might be alleged that Nelkin's survey of attitudes traceable to and expressed in pop cultural images represents a reaction to poor information, that well-informed individuals will have a better understanding and a more favorable impression of modern genetics' potential benefits. One might especially expect a warmer reaction among those who have a specific, health-related interest. But as clinicians have recently discovered, even individuals with family histories of breast or colon cancer – people who could theoretically benefit from more specific information about their risk – have hardly rushed to be tested. On the contrary, from a marketing standpoint the first commercially available gene tests have been a bust. Some researchers are also reporting difficulty in enrolling numbers of subjects in genetic prevalence studies, even in populations that are reputed to be at higher risk than others for certain diseases.

The failure to embrace genetic testing, though disappointing to some, is in fact a triumph of informed consent. Consent forms properly warn potential test subjects that attitudes and understanding about the meaning of gene test results is in flux. Accordingly, there can be no guarantee that, once information finds its way onto a medical record, it will not be used as a basis for denial of future insurance coverage. The social risks of genetic knowledge have proved to be a stronger deterrent to testing than was anticipated by proponents of genetic diagnosis.

With even those known to be at risk for certain serious conditions so slow to embrace the new science, one can only wonder how a general patient population will respond when genetic information (the kind traditionally available based on medical history or physical exam) is combined with gene test results to create personal risk profile. One or two large biotech firms are making the initial moves, and, as owners of the gene tests, they would have proprietary rights over individual genetic information that is stored in HMO computers. Ironically, in rushing their gene tests to the market, these companies may have helped stimulate a backlash that could create popular resistance to the introduction of routine genetic profiles, making for a significant drag on their business. Such are the risks of the marketplace in a world where everybody's got something.

University of Virginia
Charlottesville, Virginia

REFERENCE

Office of the Director, National Institute of Health: 1997, 'Genetic testing for cystic fibrosis,' *NIH Consensus Statement* 15(4), April 14-16.

PART III

THE PHYSICIAN/PATIENT RELATIONSHIP

GERALD LOGUE

THE PHYSICIAN/PATIENT RELATIONSHIP

The physician/patient relationship is at the core of discussions regarding modern health care. Volumes of articles have been written and models proposed to describe this interaction. The following major section of this volume is devoted to the further exploration of this relationship as well as attempts to further define, and hopefully better understand, the doctor/patient relationship as it pertains to medical changes that are rapidly occurring as we approach the new millennium.

In this section, the various roles of the physician are discussed in greater detail by multiple authors. Dr. Howard Brody, in his section "Can Relationships Heal – At a Reasonable Cost?" explores the physician as a "healer." Brody identifies the scientific underpinning of this role as the "placebo effect." That is, while there are certainly scientific aspects of medicine, in an on-going physician patient relationship, the role of healer can be separated from that of communicator. Brody points out that numerous different individuals, including primary care doctors, specialists, and also physician assistants and nurse practitioners, can function very effectively in these roles. He then discusses the various financial implications of the mix of these different healers. The role of managed care in altering this healing role is also discussed. As he so effectively points out, the healing role is often overlooked, or worse discouraged, in a managed care paradigm. Brody then goes on to give us more specific details about how the placebo response may be harnessed within the medical care model. The role of the patient's "story" in this placebo effect is emphasized. The same disease in different patients may produce different narrative stories which then proceed to affect the healing function or placebo effect. Thus, in this shared partnership, Brody feels positive healing function can occur, but at greater expense. The role of physician as leader, especially leading other, less expensive, members of the health care team, may help them to participate in this healing function.

In the second paper, Dr. Julie Rothstein Rosenbaum stresses the aspects of continuity of care and trust in the physician/patient relationship. She discusses the issue of trust with two separate entities, one the physician, and the other the institution. As she points out,

Stephen Wear, James J. Bono, Gerald Logue and Adrianne McEvoy (eds.), Ethical Issues in Health Care on the Frontiers of the Twenty-First Century, 199–204.
© 2000 *Kluwer Academic Publishers. Printed in Great Britain.*

continuity of care may be insured primarily through interaction with a physician, or it may be more through the interaction with an institution with different physicians moving in and out of the therapeutic relationship. She feels that trust is the bedrock of this interaction, improving a therapeutic relationship between the patient and the physician. She also points out that trust is important in communicating specific knowledge to the patient. Obviously, the patient is less likely to hear or believe the physician's explanation if she does not trust the physician. This trust may be between individuals or with an institution, but in either case, the level of trust will define the effectiveness of the various roles that the physician plays in the relationship. She then discusses how medicine can enhance the environment to provide a greater degree of trust. Here she stresses the importance of increasing the likelihood of mutually beneficial outcomes both for patients and their physicians. The ability of the patient and the provider to communicate about their individual roles may help to improve the levels of trust. One of the major challenges to managed care is to optimize the potential for benefit to both sides of this equation. Thus, continuity takes on a much more complex meaning in modern medicine. It does not just imply a unique single physician/patient relationship, but rather an environment including enlightened institutions, which produce a perception of continuity in both patients and physicians. Thus the role of the physician as a manager involves not only financial management, but also optimization of the appearance of continuity of care and hopefully continuity of care itself.

In the third offering, Dr. Kathryn Montgomery presents a provocative thesis on the classification of the physician/patient relationship in modern society. She presents an interesting argument that a medicine of friends may not be ideal for the various roles that the physician must play. While most would like a "friendly physician," modeling the relationship on friendship is not correct. In a perverse sense, the physician may come to idealize the friendship over the welfare of the patient. Specifically, recommendations for treatment or providing information may become more difficult for friends than for a somewhat more detached physician model. As she points out, medicine is an inherently unequal relationship while friendship tends to be a more equal partnership. She addresses how illness narratives from various patients point out that there are limitations of a friendship model. In that context she presents a different model, one that she describes as a medicine of neighbors. This relationship carries a need for mutually acceptable goals, but does provide some distance. Neighbors, in contrast to friends, are party to a somewhat accidental relationship but there is the need for duty, self-preservation, and to emphasize performance above and beyond that which occurs for total strangers. A unique type of recognition exists here, and this is well described by Dr. Montgomery in her chapter. She points out that with respect to some aspects, neighborliness can transcend friendship. Neighborliness also requires a continual learning and adaptation of both parties. It is this learning and adaptation which is also essential to the business of medicine. As the parties move through their various roles, including decision-maker, communicator and healer, there is the potential for a continual learning between neighbors that might not occur between friends.

Much of the discussion regarding physician/patient relationship involves the development of overall models in to which to fit the relationship. Another construct is defining the components of the physician/patient relationship. In taking this microethical view, we would define the components of the physician's role in the relationship. These overlapping components are listed in Table 1. This dissection of the physician's role can be applied to the papers mentioned.

In Table 1 the physician's global role is seen as involving three components: The physician as an advocate of the patient, as an advocate for the associated medical care organization, and as an advocate for society at large. While the traditional role of the physician as the patient's advocate is ideal, much literature clearly points out the sometimes conflicting issues of serving a medical care organization as well as society at large. The former, the medical care organization, is often described as arising out of managed care but this issue predates these newer concepts. "Medical Ethics" of the first half of this century often related to recruiting and retaining patients in sufficient numbers to sustain a practice. Without sufficient patients returning to even a solo practice, the solo provider's practice would fail financially. Thus, specialists were expected to return patients to their general practitioner in order to receive more referrals. Such decisions about referrals related more to advocating for the medical care organization than advocacy for the patient. Clearly such issues have been expanded and made much more complex with newer types of health care financing. "Profiling" of physicians by test ordering, use of therapeutic agents or admission to hospitals represent only new twists on older financial issues, namely that medical care is a business operation with major financial considerations.

Table 1: Components of physician's role in the physician/patient relationship

A. Global
  1. Patient Advocate
  2. Medical Care Organization Advocate
  3. Society Advocate
B. Specific Skills
  1. Data Gatherer
  2. Decision Maker
     a. Diagnostician
     b. Treatment Recommendations
  3. Communicator (Interpreter)
  4. Possessor of Specialized Skills (Endoscopy, Catheter, Echo, Anesthesia, etc.)
  5. Team Leader
  6. Customer Service Representative
  7. Representative of "Society"
     a. Compensation/Pension Issues
     b. Drivers license (regulation issues)

Also, a physician's role as an advocate for society is sometimes in conflict with that of a pure patient advocate. Public health issues often become conflicting. Mandatory reporting of certain diseases is in the interest of society as a whole, but often puts the physician in a conflicting situation. Therapeutic issues such as the use of selected antibiotics also place the physician in a conflict between advocating for the patient and advocating for society. Such issues as the development of antibiotic resistant organisms and health care costs drive societies' interests in these decisions. As described below, the physician is also placed in conflict over issues such as medical clearance for driver's licenses, and medical opinions regarding medical disability issues. These two areas often involve interpretation of vague regulations regarding eligibility. If the physician is too strong as a patient advocate, society will suffer.

Table 1 also outlines the physician's specific role in the physician/patient relationship. A traditional, primary role of the physician is as a data gatherer. Teaching of the "history and physical examination" remains one of the strongest aspects of medical education and remains the foundation of most medical practice. Clearly physicians differ in their skills at collecting data, and especially related to aspects of the patient's history, listening skills are essential. Unfortunately, as the business aspect of medicine become more intense, less and less time is available for this role.

A second major role of the physician is as a decision-maker. These decisions are of two types. The first is to attempt to diagnose what the patient's problem is, and the second is to make treatment recommendations regarding the patient's illness. Multiple disciplines including nursing, pharmacy, social work and psychology assist the role of the physician as a decision maker. In contrast, however, the ever-increasing role of "evidence-based medicine" leads to clinical guidelines, pathways and recommendations which tend to drive therapeutic decisions in a more standardized and "scientific" fashion. Also as patients' direct access to Internet databases occur, patients themselves begin to feel more empowered to take over this role as decision-maker.

A third specific role of physicians is as communicators or as interpreters of medical information. That is, not only are they expected to gather data and make decisions, but to be able to communicate the reasons for these decisions to the patients. Also, physicians enter this relationship as the possessors of specialized skills such as endoscopy, anesthesia, cardiac catheterization, or surgery. In this setting, the physician's role is expected to be that of possessing highly technically developed skills which exceed some minimum standard for performance.

In modern medicine the physician's role as a team leader becomes increasingly evident. The model physician as a sole practitioner is becoming less and less frequent and physicians are being leveraged by the addition of "physician extenders," such as nurse practitioners or physicians assistants, as well as other specialized nursing staff, and other professional staff including social work, pharmacy, psychology, physical therapy, etc. This role of a team leader requires different skill sets. The communication skills, for instance, of the other members of the team may supplant the physician's primary role of communicator, and it is clear that in many patient care situations, primary communication occurs via other members of the team rather than the physician.

As modern medicine become more competitive, there is increased emphasis on "customer service." The rapidly and ease of access to care including testing and treatments become an aspect of patient care which is different than that related to traditional medicine. Here the physician's role is in part as customer service representative to attempt to obtain efficient and appropriate care for patients within complex medical care organizations.

Also, physicians play a role in the physician/patient relationship as a representative of society at large. Specifically here, physicians are expected to play a role in compensation or pension issues including disability as mentioned above. Other regulatory issues such as the ability of certain patients to drive, and the ability of patients to return to work, is part of a physician's role in modern health care. Unfortunately, many of these situations require a difficult balancing of priorities, and different skills must be used simultaneously. An overriding issue is how the elements of moral conduct impact on the different aspects of the physician's role in this health care enterprise.

Department of Medicine
University at Buffalo

KATHRYN MONTGOMERY

A MEDICINE OF NEIGHBORS

The title Charles Rosenberg gave his 1987 history of American hospitals struck a nerve: *The Care of Strangers*. Although the book concerns the nineteenth-century origins of an institution that now seems to have been with us always, its title could serve as well for an account of health care in the United States at the turn of the millennium. "The care of strangers" is an apt description of that agglomeration of professional mores, economic practices, laws, customs, and social habit that makes up the non-system we praise and disparage. The patient-physician relationship, already weakened by patients' geographic mobility and physicians' proliferative sub-specialization, has been dealt yet another blow by managed care. Increasingly, the profession that we appeal to in our direst need, is becoming – in hospital and out – a medicine of strangers.

In the city where I live, not long after Rosenberg's book appeared, Mark Siegler and Leon Kass assembled a group of scholars, many of them physicians, to consider the doctor-patient relationship. Each of us was to make a presentation. Mine turned into a long overdue essay on the poems of John Stone that delineate the physician's vulnerability to patients' grief and pain. Troubled by the increasing relevance of Rosenberg's title, Mark Siegler spoke of expanding an earlier essay into a book about "a medicine of friends" (Childress and Siegler, 1984, pp. 17-30). Images are a subtle, powerful force, and although I shared his concern for the way medical education and the organization of practice shape the patient-physician relationship, the phrase "a medicine of friends" disturbed me. It took some time to work out why. Although Siegler has not (yet) written the book, this essay is a dialogue with a hypothetical, an answer to what I imagine it might have said. In this, as in many other ways, I am indebted to my neighbors.

I. A MEDICINE OF FRIENDS

The nature of the patient-physician relationship is a psychological, epistemological, and moral matter about which thoughtful observers of medicine have had much to say. The place of emotion in medical care is

Stephen Wear, James J. Bono, Gerald Logue and Adrianne McEvoy (eds.), Ethical Issues in Health Care on the Frontiers of the Twenty-First Century, 205–219.
© 2000 *Kluwer Academic Publishers. Printed in Great Britain.*

contested, and the imprecise connection between role and self often affects the physician's practice (Callahan, 1988; Connelly, 1998; Rosenbaum, this volume). The question of the right metaphor for the relationship between patient and physician not only engages hopes and expectations on both sides but also – and not incidentally – implies a vision of society. Images for the physician have included teacher, friend, parent, priest, advocate, engineer, carpenter, scientist, detective, plumber, mechanic.[1] In such a list, "friend" seems trustworthy and solid, even if potentially a little too simple.

The image of the physician as friend has a thin but venerable history, and arguments for it are appealing. Pedro Lain Entralgo's (1969, p. 7) call for the exercise of "medical philia" is perhaps the best known. He begins his historical survey of the patient-physician relationship, *Doctor and Patient*, with a long quotation from Seneca:

> Why is it that I owe something more to my physician and my teacher and yet do not complete the payment of what is due to them? Because from being physician and teacher they become friends, and we are under an obligation to them, not because of their skill, which they sell, but because of their kind and friendly goodwill.[2]

The idea goes farther back than Seneca. Lain Entralgo finds its roots in the value Socrates placed on friends and, above all, in Aristotle's *Nicomachean Ethics*, (VIII 3, 1156a20 *et seq.*, pp. 211ff) where friendship, a mutual goodwill motivated by utility, pleasure, or virtue, is examined as an essential part of human happiness. But in adopting the Greek ideal, Lain Entralgo must set aside the class structure of slaveholding Athens, and his readers must imagine themselves rich, free men who deserve the comradeship of their physician. As a psychiatrist he identifies medical friendship with a more nearly universalizable Freudian transference, a bond which facilitates the patient's recovery (pp. 9, 159); yet friendship for him retains its discriminative character. It is a benevolence accorded an individual "because he is the individual he is" (p. 53).

So attractive is the image of the physician as friend – and I do not mean a "friendly physician" – that many writers, without argument or explanation, assume it as the ideal. This may be because, as Robert Bellah and his colleagues have observed, Americans nowadays regard friendship as a variety of the therapeutic (Bellah, *et al.*, 1985).[3] Or it may be that friendship, freely contracted between individuals, has come to replace what we lack in the way of community. Whatever the case, contemporary advocates tend to regard the idea of the physician as friend as an overlooked but self-evident value. Edmund Pellegrino and David Thomasma (1993, pp. 82-83) describe friendship as related to the medical virtue of compassion – a good physician is compassionate, like a friend – but they stop short of adopting friendship as a model for the relatonship. In Rosamond Rhodes's argument for the reconciliation of justice and care (1995), she finds that "a theoretical perspective (call it justice, beneficence, utility, etc.) is inadequate grounding for ethics," and that in the patient-physician relationship, "justice...requires a foundation of loving friendship." M. Therese Lysaught (1992) celebrates friendship as an ethical standard in her response to David Hilfiker's story of Clyde, a once homeless HIV-positive man making probably his last attempt to get off drugs. She remarks particularly on the class difference that often bars friendship between doctor and patient (Lysaught, 1992). The ideal of friendship caps Linda Emanuel and Ezekiel Emanuel's (1992) models of the doctor-patient relationship; their "deliberative model" calls for the physician to engage in values clarification and moral persuasion as a teacher or friend.

The ideal can also be read back into classic works on the doctor-patient relationship. The title of Rhodes' essay is a reminder that Francis Peabody's (1927) often quoted wisdom might be understood as an appeal for friendship: "The secret to the care of the patient is to care for the patient." The image of the physician as friend can be found quite literally in W. Eugene Smith's widely admired 1948 photographic essay, "Country Doctor." His images of the physician have become icons of the way doctoring used to be: they depict the physician out at all hours, no matter the weather, and comfortable among everyday objects in the kitchens and sickrooms of his patients' homes.

The ideal of friendship is not exclusive to medicine. Rather than survey the obvious – advice to the clergy or philosophies of teaching – I will cite only the unlikely: friendship as an ethical ideal for lawyers. In the 1976 *Yale Law Review* Charles Fried (1976) proposes friendship – focused on a goal, of course, but friendship nevertheless – as a model for the relationship between attorney and client.

Such arguments suggest that friendship is a necessary part of the moral life of a professional. Whether motivated by nostalgia for an imagined past or by longing for a rarely attained ideal, these writers encourage us to believe that good physicians should count patients as friends and that friendship between patient and physician ought to be a goal of medical practice.

II. AGAINST FRIENDSHIP

Attractive as the ideal of friendship is, I have misgivings about its value as an ethical goal for medicine. I suspect that "a medicine of friends" is instead something of a rhetorical turn, a vivid rejection of the current state of professional relationships. As a critique of impersonal medical care and the increasing commodification of medicine, the image suggests what the speaker or writer finds wanting in the way things are. But as an ideal, a medicine of friends is in so little danger of being realized that no very detailed descriptions of it exist. Instead, it has the character of a binary response to the alienation of patient and physician, as if "friend" were not only the ideational opposite of "stranger" but, despite the wide range of human relationships, the only possible contrast. The best indication that the ideal of friendship is not a true goal of medicine is the odd fact that it is expressed primarily by those who belong to the profession in question and not by the people they serve.[4] That physicians propose "a medicine of friends" far more often than patients suggests that friendship may represent peak experiences in the lives of practitioners themselves rather than their ordinary procedural assumption. Indeed, William Branch and Anthony Suchman (1990), who asked internists about the most significant occasions in their professional lives, discovered that they are times of crisis when physicians felt emotionally close to their patients and their families. These are the valuable moments of connection in the service of other human beings. To recognize these connections and to maintain the hope of experiencing more moments like them are laudable aims. But they do not authorize a criterion by which such relationships could be judged nor do they establish friendship as a professional goal. It is very much open to question how such an ideal might operate in medicine and whether it could serve as a useful model for the patient-physician relationship.

A few scholars have called the ideal of friendship into question. Michel Foucault does so indirectly when he underlines the romanticized view of the patient-physician relationship by noting the closed off, erotic implications of *le couple malade*, the metaphoric phrase that designates the relationship in French (Foucault, 1975). Anne Hudson Jones and

Edward Erde have described friendship between physician and patient as simply bad medicine (Erde and Jones, 1983). In the novels they examine they, too, find that it is the physician who idealizes friendship and not the patient. Patricia M.L. Illingworth (1988), considering the interactions of AIDS care, has argued that the friendship model violates autonomy of the patient who does not request it and diminishes the autonomy of the patient who is psychologically needy.[5] Ann Folwell Stanford and Nancy M. P. King, although they do not address the concept of friendship, express strong reservations about the related ideal of understanding the "whole patient," particularly when it licenses, for example, an otherwise well respected physician (one who makes house calls!) to snoop in his patients' medicine cabinets (Stanford and King, 1992). Even Arthur Kleinman's more dialogic remedy for a medicine of strangers, the recommendation that physicians elicit their patients' beliefs about their illness, has been criticized by Michael Taussig as open to manipulative use (Taussig, 1980, p. 12).

Medicine is an inherently unequal relationship, one that for its efficacy depends in part on that inequality. In such circumstances, the physician's desire to know the patient may be an admirable change from regarding "those people" as instances of disease, problems to be solved or, worst, mere "teaching material," but it nevertheless can be invasive and even coercive. It was something like this that I felt at the suggestion of "a medicine of friends." A good patient-physician relationship and a high standard of patient care requires that the person who is ill not be coopted nor her story be alienated or appropriated by attempts to achieve something that looks or feels like friendship.

The doctor is not a friend. Or if she is, that is not who she is being when she is being my physician. She is scraping my abraded palm with a small wire brush, tweezing out the fine grit that remains, then applying iodine. She has paid no attention to my grimace or to the fact that I want out of here – which is not an examining room, but the more neutral and non-billable lab in the clinic of our hospital. She ignores my dawning realization that I was wildly mistaken when I said I'd rather not have a pain-killer an hour before my class. This is not friendly! It is doctorly: she is intent on my hand, telling me what she's going to do next, keeping the small Ulysses contract we implicitly made about no analgesic, and in two or three minutes now, if I can just hold still, I will be eternally grateful and we will go back to being friends.

The physician has distanced herself, narrowing her focus to an injured, almost entirely decontextualized hand. This is a necessary distance, an essential decontextualization, and while it may be relatively easy to manage for a minor, accidental injury, it is far more difficult, if not impossible, for good, sustained care. If she were my doctor, could she persuade me to stop smoking? Would she take a sexual history? Would I "bother" her with what I think are trivial symptoms? Or bother her outside the office with things that *are* trivial? These abrogations of good medical care are the reason physicians are advised not to treat family members. Physicians, too, stand to lose a great deal by taking care of their friends.[6] How can they tell them "bad news"?" Or turn to them as friends when their own lives are difficult? Either friendship or the doctor-patient relationship must give way.

III. WHAT DO PATIENTS WANT?

Friendship is not what patients want from their physicians. Certainly they are quick to condemn those who are discourteous and unfriendly. Fiction offers powerful examples of the suffering such physicians cause. Tolstoy's Ivan Ilych encounters in his diagnostician the same bland and uninflected bureaucratic facade he has so valued in his own work as a magistrate. It is a distant, objectifying, above all *professional* regard that prompts the physician to debate the etiology of his patient's condition and licenses him to ignore the patient's burning question: "Was his case serious or not?" (Tolstoy, 1960, p. 161).

Contemporary illness narratives confirm the patient's need. The hundreds of pathographies that have appeared in the last two decades attempt to fill the void created by illness. They are written not only to make sense of baffling experience but to assuage the suffering that is often not merely ignored by physicians but exacerbated by their chill objectivity (Hawkins, 1993; Frank, 1994). Reynolds Price, who is given his dire diagnosis in a public hallway by two physicians who hurry off to their next task, observes: "Surely a doctor should be expected to share – and to offer at all appropriate hours – the skills we expect of a teacher, a fireman, a priest, a cop, the neighborhood milkman or the dog-pound manager" (Price, 1994, p. 145).

Few patient-writers, however, despite their need for courtesy and respect, find it necessary or desirable that the physician be a friend. Franz

Ingelfinger's (1980) tellingly mis-titled 'Arrogance' is illustrative.[7] The essay chronicles his quest for real medical attention, frustrated at first by consulting friends and colleagues about the very sort of cancer he himself studied and treated. He is rescued by his son, who tells him, "You need a doctor." The one he finds is young and far less experienced than he, but a good and attentive physician. Inside this patient-physician relationship, something important, indeed essential, can occur, and Ingelfinger's title suggests that it is not friendship but the more inegalitarian authority that is a part of care.

Even those few patients who seem to see their physicians as friends do not advocate friendship as a goal of their care. Norman Cousins tells of the collegial physician he persuades to release him from a hospital and treat him instead in a hotel where he can have round-the-clock access to film comedies that reduce his need for medication (Cousins, 1979). But it is the physician's trust and willingness to experiment that Cousins values, not his friendship. In *'The Art of Healing: In Memoriam David Protetch, MD,'* W. H. Auden addresses his physician both as a trusted confidant who left the poet's bad habits untouched and as a fellow patient, "yourself a victim." But the poem's leitmotif is the poet's sad surprise that a doctor can die, "not [my] physician,/ that white-coated sage" (Auden, 1991, pp. 168-170). Both Cousins and Auden value the recognition of their predicament by their physicians, and recognition that grows into a kind of partnership certainly can become a friendship over time, especially when the patient has a chronic illness. But surely the friendship, if that is what it is, is an accidental reward and not a precondition of the relationship. This is particularly important since few patients are editors of *Saturday Review* or major twentieth-century poets and likely to receive the same sort of treatment. Nor do these examples suggest that the equality implicit in friendship is finally desirable in the patient-physician relationship. We all may need to exercise control over the course of a long illness or in the face of life-threatening uncertainty, but we need someone to ask about our "minor vices" too.

Rather than friendship, I propose that as a part of competent medical care people who are ill want a certain committed but disinterested attention and that in times of crisis they also need a recognition of their circumstances and their potential meaning. Ivan Ilych finds this not in any of the physicians he consults but in Gerasim, the butler's helper. The young peasant eases the sick man's pain by bearing his legs on his shoulders; he carries out his bedpan; he alone does not lie about the gravity of his master's illness.

Patients do not want physicians to feel their pain or to circumvent the usual stark procedures lest they be incapacitated, make mistakes, or miss important signs. Anatole Broyard spoke for most patients at their most desperate when he wrote of his uncommunicative surgeon, "I see no reason or need for my doctor to love me – nor would I expect him to suffer with me. I wouldn't demand a lot of my doctor's time: I just wish he would *brood* on my situation for perhaps five minutes, that he would give me his whole mind just once, be *bonded* with me for a brief space, survey my soul as well as my flesh, to get at my illness, for each man is ill in his own way" (Broyard, 1992, p. 44) Five minutes is a long time to see another human being steadily and whole. It is rare indeed and a lot like love, but it is not friendship.

Shall we then discard the ideal of a "medicine of friends?" I doubt that we can, and I suspect that even if we could, we should proceed with caution. Besides the obvious part ideals play in shaping thought and guiding action, they have more subterranean uses: they may act as a counterweight to forces that, less than ideal, are nevertheless inescapable. For example, medicine's aspiration to be an anecdote-free exercise of objective reasoning runs counter to its reliance on case narrative that, especially in academic medical centers, makes it among the most anecdotal of human activities (Montgomery Hunter, 1991). The contempt for anecdotes works to keep the flood of stories under control. The profession's working assumption that drugs are targeted agents entirely distinct from placebos is similar. (Brody, 1980). It contradicts the bone-deep understanding of illness and therapy of most experienced physicians, yet the placebo effect is seldom discussed or acknowledged. These two examples are only a patch on the profession's related, overarching claim: that medicine is itself a science – and in the grand nineteenth-century Newtonian sense of that word. Well researched, rational medicine is a radically experimental enterprise: the application of scientific principles and technological know-how to individual sick people in need of help. Physicians ignore the situational maxims that guide their practice and claim their work is a science in order to function with a modicum of intellectual and existential security in a field that at its best and most professional is inalienably uncertain (Montgomery Hunter, 1996).

Like the ideal of science, "a medicine of friends" works as counterweight to a necessary but somehow suspect attitude or practice. The ideal of friendship is an effort to redress medicine's necessary decontextualization of the patient. Physicians long not just to exercise their skills but also to have a safe way to be in relationship with their patients. The problem lies with the paradox of intimacy and distance that is central to the patient-physician relationship. It exists for other professions, of course, but the license physicians have to touch the body and their familiarity with matters of life and death make the balance both especially difficult and especially important. Advocating a medicine of friends is at once a reminder of the rewards of service and an antidote to a mindset and habits that physicians find inevitable but impoverishing in their practice. Friendship prompts an ideational balance. To invoke the ideal is to remember that the patient, recontextualized, is after all a person. Moreover, friendship does not go too far – as another translation of *philia* might – because a medicine of friends, even in that never-never land where ideals are realized, stops well short of what contemporary Westerners categorize as love. Friendship guards against, even forbids, sexual intimacy.

IV. R$_X$: A MEDICINE OF NEIGHBORS

If the image of friendship serves an ideational need in medicine and can be neither recommended nor discarded, we nevertheless are not doomed to a medicine of strangers. A medicine of neighbors offers a promising alternative. Neighbors are party to an accidental, almost gratuitous relationship, but it is no less full of possibility for all that. Neighborliness is a duty, especially in time of need, but a limited duty that leaves considerable room for both self-preservation and performance above and beyond its call. Fulfillment is judged by acts rather than by motives or emotions. Distinct from love and liking, being a neighbor requires only the fundamental respect involved in one human being's recognition of another.[8] Above all, in its randomness it is a relationship open to time, chance, difference, surprise.

As a model, a medicine of neighbors expresses much of what is valuable in the ethos of medicine, particularly its goal of disinterested service. Equally important, as managed care threatens the trust patients place in physicians, the implication of a medicine of neighbors extends beyond the patient-physician dyad. Medicine is distinctive, even in a democracy, for its attention to all comers. With the decline in public education, it now may be – outside the military – the most nearly egalitarian American institution. If their need is dire – and with the neglect of primary care it is often is – people who are poor have access to excellent medical care. Moreover, the same emergency room treats villains as well as their victims just as (once the battle is over) military physicians are expected treat enemy soldiers. Prisoners are well cared for; indeed, for guaranteed access to medical care in the United States, a citizen must either join the military or go to jail. There are exceptions to this openness, of course, but they are sources of shame: Nazi doctors, for example, or the South African physician who certified that Steven Biko could withstand more torture. Where medicine's openness does not extend to the poor, it is because institutional policies and appointment clerks exclude them. If the poor and uninsured can somehow reach the examination room, physicians by and large will treat them.

As a part of this disinterestedness, physicians are understood to be non-judgmental. Even the old-fashioned, avuncular family doctor is imagined to be capable of attending with relative equanimity to the consequences of our weaknesses of flesh and will. Contemporary medical students learn to ask, "Are you sexually active? With men or women or both?" Physicians see this ameliorating acceptance as desirable not just for its own sake but for the contribution to good patient care made by the information it yields. A model of the good patient-physician relationship must capture such non-judgmental attention and its rewards for both parties. The physician as neighbor is such an image. It neither requires a violation of the physician's boundaries nor licenses a trespass of the patient's; it guarantees these limits even as it enables proximity.

A medicine of neighbors is a theme of William Carlos Williams's short stories, accounts of (among many other things) a physician-poet's decision to stay in small town, industrial New Jersey, tending to immigrant factory workers. Stories like "Jean Beicke," "The Use of Force," and "A Face of Stone," are ethical morbidity-and-mortality conferences in which the narrator calls himself to account for various errors and mistakes of judgment. He fails to diagnose a child's meningitis, calls children in his hospital "brats," loses his patience and his temper, suffers bouts of xenophobia and anti-Semitism, and is never, ever unaware of class. But his patients are his neighbors and he learns from them.

Far more celebratory is John Berger's third-person account of *A Fortunate Man*, another exemplar of a medicine of neighbors (Berger, 1967). Describing his relation to his patients, the general practitioner in the north of England says he is the "requested clerk of their records," for he witnesses the births, rites of passage, marriages, losses, death that affect his townspeople.[9] He knows and keeps their secrets, and just when he thinks he cannot be surprised, he is called to treat an elderly farm wife whose complaint "down there" is unrelated (like the rest of her life) to the penis revealed on physical examination. Likewise, the physician as neighbor is celebrated in John Stone's "He Makes a House Call." Visiting the patient who first taught him cardiac catheterization in his fellowship year, the poet speaks of his intimate knowledge of her body (an aortic valve "that still pops and clicks/ inside like a ping-pong ball") and acknowledges his limited knowledge of her life: "someone named Bill I'm supposed to know." In her garden he accepts his place – "here you are in charge/ of figs, beans, tomatoes, life" – and he is rewarded with an vision of his work in the larger scheme of human endeavor: "health is whatever works/ and for as long" (Stone, 1981, p. 4-5).

The best patient-physician relationship, as these writers suggest, is one open to learning and characterized by a little distance. Between friends, this therapeutic distance is all too likely to be abridged. A medicine of neighbors, by contrast, encourages it. "Good fences make good neighbors," Robert Frost observed, while calling attention to the forces that work against such order.[10] The image of the physician as neighbor serves as a guide for conduct equally well in a brief encounter or a long chronic illness. Unlike friendship, it does not require, in advance of the rest of late capitalist society, equality of circumstance or (a depressing thought) the reassuring sameness of ethnicity or life experience. A medicine of neighbors does not encourage young physicians, as a medicine of friends does, to return to the suburbs where they grew up in order to practice humanly rewarding medicine. Above all, the physician as neighbor entails a relation to community that itself is caring. This ideal was expressed in our political life more than thirty years ago by Lyndon Johnson in his 1965 inaugural address. Outlining his vision of the Great Society, Johnson invoked medicine's broader mission: "In a land of healing miracles, neighbors must not suffer and die unattended." We have scarcely begun to achieve that goal.

A community of neighbors is no more a closed circle, restricted to similar people, than is the physician-patient relationship. "Who is my neighbor?" Jesus is asked by one of his disciples (Luke: 10:29). In context of the New Testament parable, we know that, whoever it turns out to be, a neighbor is that person one is supposed to "love as yourself." The answer is not a friend or a member of the same tribe or ethnic group, but someone from a different and despised group, the Samaritan. A neighbor is the person who is passing by and stops to help.

A medicine of neighbors has all the virtues of good anthropology. Like physicians, anthropologists do their work between science and subjectivity. Their field is distinctively, if not uniquely, the intellectual discipline that has struggled with the unknowability of the Other and the distortions of colonialization. Their method is, first, to describe what they see as they see it and then to describe what they see as its participants see it. The virtues involved in this reflexive doubleness are respect, open self-presentation, tactful withdrawal when necessary, attentive listening, questions about differences and their meaning, the habit of checking conclusions with the informant, and above all not "going native," eliding differences between observer and observed. Friendship is not among them.[11]

Neighborliness is a virtue that recommends itself for American life in the twenty-first century. It possesses homely qualities that friendship transcends and sometimes can ignore: chief among them are a clear regard and a fundamental respect for the other. In the absence of a "content-full" moral vision, (Englehardt, 1996) these qualities are warmer and more productive than a default libertarianism necessitated by the nation's increasing cultural pluralism. More nearly minimal than friendship and more circumstantial, neighborliness creates occasions for learning. It is does not necessitate self-revelation or an enduring personal bond. Instead, it requires a recognition of both the accidental character of much that befalls us and our common life, our common need, our common fate. Well realized, the patient-physician relationship may *become* something like friendship. Now and then, friendship is the reward for having been a good neighbor. But neither relationship necessarily includes it and neither begins there. Friendship is neither a precondition nor a goal of the patient-physician relationship.

Medicine already is or should be the care of neighbors. This has been a norm that has guided medicine in many times and places before ours. We are challenged now to apply it not just to our enemies but to those we live among: our literal neighbors. We could do worse than to imagine the physician as a neighbor and to evaluate the new health care arrangements to which we are consenting by the degree of neighborliness they permit and encourage.

Northwestern University Medical School
Chicago, Illinois

NOTES

[1]   Veatch, 1972, pp. 5-7. William May, following Paul Ramsey, supplies another model (1983).
[2]   Entralgo, 1969, p. 7. The epigraph, of which I have quoted only the first quarter, is from Seneca's *De Beneficiis*, VI, 16.
[3]   Which of Aristotle's criteria for friendship is met by this notion of therapy is arguable. See Bellah, *et al.*, 1985, p. 115-23. I'm grateful to Tod Chambers for reminding me of Bellah's study and for telling me about Rabinow's struggle with the problem of friendship, cited below.
[4]   The few exceptions are bioethicists who to some degree are allied with the profession; they do not write as patients.
[5]   Illingworth, 1988, pp. 22-36. Her view has been challenged by David N. James, (1989, pp. 142-146) who maintains that the trust essential to good patient-physician relationships resembles friendship closely enough to warrant the exploration of the model.
[6]   I am indebted to Douglas R. Reifler for this observation.
[7]   Ingelfinger, 1980, pp. 1507-11. The problem of consulting friends as physicians may be especially severe for other physicians; see Rosenbaum (1988).
[8]   This may be the civic friendship imagined by Cicero and Aquinas, but as Bellah and his colleagues point out (1985, p. 115), that concept has been all but lost in American life.
[9]   John Berger, 1967, p. 103. Mohr's photographs are a reminder that the subject of W. Eugene Smith's iconic photographs in *Country Doctor* is a neighbor rather than a friend.
[10]   The line, "Something there is that does not love a wall," occurs as often as "Good fences make good neighbors;" Robert Frost, 1915.
[11]   Paul Rabinow (1977, pp. 142-148) wrestles with the problem of friendship between the anthropologist and a member of the group he is studying.

REFERENCES

Aristotle: 1995, *Nicomachean Ethics*, T. Irwin (trans.), Hackett, Indianapolis.
Auden, W.H.: 1991, 'The Art of Healing: In Memoriam David Protetch, MD,' in R. Reynolds and J. Stone (eds.), *On Doctoring: Stories, Poems, Essays,* Simon and Schuster, New York, pp. 168-70.
Bellah, R., Madsen, R., Sullivan, W.M., Swidler, A., Tipton, S.M.: 1985, *Habits of the Heart: Individualism and Commitment in American Life,* University of California Press, Berkeley, pp. 115-123.
Berger, J.: 1967, *A Fortunate Man,* Holt, New York.

Branch, W.T., Suchman, A.: 1990, 'Meaningful experiences in medicine,' *American Journal of Medicine* 88, 56-59.

Brody, H.: 1980, *Placebos and the Philosophy of Medicine: Clinical, Conceptual and Ethical Issues*, University of Chicago Press, Chicago.

Broyard, A.: 1992, *Intoxicated by My Illness*, Clarkson Potter, New York.

Childress, J.F., Siegler, M.: 1984, 'Metaphors and models of doctor-patient relationships: Their implications for autonomy,' *Theoretical Medicine* 5, 17-30.

Callahan, S.: 1988, 'The role of emotion in ethical decision-making,' *Hastings Center Report* 8(3), 9-14.

Connelly, J.: 1998, 'Emotions, ethics, and decisions in primary care,' *Journal of Clinical Ethics* 9(3), 225-234.

Cousins, N. 1979, *Anatomy of an Illness as Perceived by the Patient: Reflections on Healing and Regeneration*, Norton, New York.

Emanuel, E.J., Emanuel, L.L.: 1992, 'Four models of the physician-patient relationship,' *Journal of the American Medical Association* 267, 2221-2226.

Engelhardt, H.T.: 1996, *The Foundations of Bioethics*, 2nd ed., Oxford University Press, New York.

Entralgo, P.L.: 1969, *Doctor and Patient*, F. Partridge (trans.), McGraw-Hill, New York.

Erde, E.L., Jones, A.H.: 1983, 'Diminished capacity, friendship and medical paternalism: Two case studies from fiction,' *Theoretical Medicine* 4, 303-22.

Foucault, M.: 1975, *The Birth of the Clinic: An Archeology of Medical Perception*, A.M.S. Smith (trans.), Vintage, New York.

Frank, A.: 1994, 'Reclaiming an orphan genre: First-person narratives of illness,' *Literature and Medicine* 13, 1-21.

Fried, C.: 1976, 'The lawyer as friend: The moral foundations of the lawyer-client relation,' *Yale Law Review* 85, 1060-89.

Frost, R.: 1915, 'Mending Wall,' *North of Boston*, Holt, New York.

Hawkins, A.H.: 1993, *Reconstructing Illness: Studies in Pathography*, Purdue University Press, West Lafayette, Indiana.

Illingworth, P.M.L.: 1988, 'The friendship model of physician/patient relationship and patient autonomy,' *Bioethics* 2, 22-36.

Ingelfinger, F.: 1980, 'Arrogance,' *New England Journal of Medicine* 303, 1507-1511.

James, D.N.: 1989, 'The friendship model: A reply to Illingworth,' *Bioethics* 3, 42-146.

Johnson, L.B.: 1965, Inaugural Address.

Kleinman, A.: 1989, *Illness Narratives: Suffering, Healing and the Human Condition*, Basic Books, New York.

Lysaught, M.T.: 1992, 'Who is my neighbor?' *Second Opinion* 18 (October), 59-67.

May, W.: 1983, *The Physician's Covenant: Images of the Healer in Medical Ethics*, Westminster, Philadelphia.

Montgomery Hunter, K.: 1996, '"Don't think zebras:" Uncertainty, interpretation, and the place of paradox in clinical education,' *Theoretical Medicine* 5, 1-17.

Montgomery Hunter, K.:, 1991, *Doctors' Stories: The Narrative Structure of Medical Knowledge*, Princeton University Press, Princeton, NJ.

Peabody, F.W. 1927, 'The care of the patient,' *Journal of the American Medical Association* 88, 877-82.

Pellegrino, E.D., Thomasma, D.C.: 1993, *The Virtues in Medical Practice*, Oxford University Press, New York.

Price, R.: 1994, *A Whole New Life: An Illness and a Healing*, Atheneum, New York
Rabinow, P.: 1977, *Reflections on Fieldwork in Morocco*, University of California Press, Berkeley.
Rhodes, R.: 1995, 'Love thy patient: Justice, caring, and the doctor-patient relationship,' *Cambridge Quarterly of Healthcare and Bioethics* 4, 434-447.
Rosenbaum, E.E.: 1988, *A Taste of My Own Medicine: When the Doctor Is the Patient*, New York, Random House.
Rosenbaum, J.R.: 2000, 'Trust, institutions, and the physician-patient relationship: Implications for continuity of care,' this volume.
Rosenberg, C.E.: 1987, *The Care of Strangers: The Rise of America's Hospital System*, Basic Books, New York.
Smith,W.E.: 1948, 'Country doctor,' *Life Magazine*.
Stanford, A.F., King, N.M.P.: 1992, 'Patient stories, doctor stories, and true stories: A cautionary reading.' *Literature and Medicine*, 11 185-11 199.
Stone, J.: 1981, 'He Makes a House Call,' in *All This Rain*, Louisiana State University Press, pp. 4-5.
Taussig, M.T.: 1980, 'Reification and the consciousness of the patient,' *Social Science and Medicine* 143, 12.
Tolstoy, L.: 1960, 'The Death of Ivan Ilych,' in *The Death of Ivan Ilych and Other Stories*, A. Maude (tran.), Signet, New York
Veatch, R.M.: 1983, 'Models for ethical medicine in a revolutionary age,' *Hastings Center Report* 2, 5-7.

JULIE ROTHSTEIN ROSENBAUM

TRUST, INSTITUTIONS, AND THE PHYSICIAN-
PATIENT RELATIONSHIP: IMPLICATIONS FOR
CONTINUITY OF CARE

During this conference, the papers have addressed many challenging aspects of the ever-shifting health care delivery systems and the technologies that contribute to the care given in these arenas. Of course, beyond all of the techniques, all of the machinery, and all of the health care arrangements, is the physician-patient relationship. Current changes in the American health care system are threatening the ideal, trusting physician-patient relationship (Emanuel and Brett, 1993; Emanuel and Dubler, 1995; Orentlicher, 1995; Morreim, 1995). This ideal is difficult to describe, as Dr. Kathryn Montgomery notes, and difficult to achieve, as Dr. Howard Brody emphasizes. However, we seem to know that there is some good notion of physician-patient relationship that we strive to achieve and protect.

When people speak about the ideal physician-patient relationship, two concepts are usually considered important, though there are many that also apply. One is trust and the importance of a trusting relationship, and the other is continuity of care. Trust, long considered a crucial aspect of the health care encounter, is suffering (Siegler, 1993). The causes are multifactorial and complex. They include changing societal attitudes towards authority, the ascendance of patient autonomy, the effects of technology on health care diagnosis, treatment, and decisionmaking, and the growth of new structures for the provision of health care (Pellegrino and Thomasa, 1981; Lyons, 1994; Reiser, 1978). Similarly, continuity of care is also at risk in the new health care environment, as will be discussed below.

This paper will examine how continuity affects trust between physician and patient and how both aspects will fare in the new health care environment. After considering the importance of continuity of care to the physician-patient relationship, two brief cases will be presented as points for discussion. Then the paper will address the importance of trust in general as well as in the context of the individual health care relationship. This is followed by a description of the benefits of trust and the various levels in which trust functions, including the individual and the

Stephen Wear, James J. Bono, Gerald Logue and Adrianne McEvoy (eds.), Ethical Issues in Health Care on the Frontiers of the Twenty-First Century, 221–240.
© 2000 *Kluwer Academic Publishers. Printed in Great Britain.*

institutional components, and how these levels interact with each other. This paper will also address factors that enhance and encourage trust. The cases will then be used to demonstrate the importance of continuity of care to trusting relationships and the implications for future health care relationships – whether between patients and physicians, or patients and health care systems.

I. HOW IMPORTANT IS CONTINUITY?

According to Emanuel and Dubler, continuity is one of the six C's necessary for the ideal conception of the physician-patient relationship (the other five C's are choice, competence, communication, compassion, and {no} conflict of interest.) (Emanuel and Dubler, 1995 p. 323). The American Medical Association's Code of Medical Ethics states that a patient "has the right to continuity of care" (AMA, 1994, XXXIV). Continuous care potentially allows physicians and patients to know each other with a completeness and depth that is absent from other types of relationships. The physician can learn the patient's history, values, idiosyncrasies, medical, social, and psychological needs more completely over a longer period of time. Many believe that relationships that endure over time are more efficient, not only in terms of time spent pouring over the chart familiarizing oneself with a new patient's data, but in terms of decreasing redundant testing (Emanuel and Brett, 1993, p. 880). The quality of the physician-patient interaction and its outcomes may be improved as well. Of course, quality is a term that is very difficult to define when applied to the physician-patient relationship. Is a good relationship with a physician one that makes the patient feel better physically? One that causes short term pain for long term preventative benefit? One in which a provider is only adequate in caring for the patient's physical needs but superior at addressing the patient's emotional problems? Is the best surgeon the one who kills the fewest patients or is this only the surgeon who takes the fewest risky cases?

Despite the difficulties with defining the quality physician-patient relationship, there are components that can be described. Studies can then address the impact of variables on those components of the good physician-patient relationship, including the effect of continuity.

A comprehensive review of numerous studies of continuity found that longitudinal care was shown not only to improve patient but also staff satisfaction with the health care encounters. In addition, compliance with both appointments and medications improved when a patient consistently saw her own provider, instead of seeing multiple providers. Furthermore, disclosure of sensitive information occurred more easily if a consistent provider relationship was present. For example, a mother was more willing to reveal a child's behavior problem to a provider who provided longitudinal care (Dietrich and Marton, 1982).

One further area that many writers consider a crucial component of the physician-patient relationship that can be more easily secured through continuous relationships is trust. Orentlicher writes, "with long-standing relationships, patients are more likely to trust their physicians and feel comfortable seeking care. As a result, patients are more likely to consult their physician early in the course of an illness, before the illness becomes difficult to treat" (Orentlicher, 1995, p 143). The relationship between trust and continuity is also emphasized by Emanuel and Dubler: "Trusting relationships ensure that in times of stress patients can rely on their physicians secure in the knowledge that their history, attachments, values, and feelings are understood. If patients are frequently forced to change physicians, it is hard for them to develop a deep and understanding relationship" (Emanuel and Dubler, 1995, p. 324-325). Therefore, continuity is more likely to result in relationships of trust; and trust, in turn, leads to fuller, deeper, and more effective physician-patient relationships.

The following questions will be addressed in this paper: if continuity contributes to trust, what other factors can also improve the environment for trust? How does trusting an individual differ from trusting an institution? How does continuity fit into discussions of trust in physicians and managed care organizations (MCOs)?

II. TRUSTING THE PHYSICIAN AND TRUSTING THE INSTITUTION

To elaborate upon how trust functions between patients and individual physicians and between patients and institutions, first I will present two case studies which will illustrate certain points about continuous relationships. Both are artificial and idealized, but are intended to demonstrate the issues involved.

CASE A: The Physician

In the first scenario, consider patient Peter Johnson, a man with a history of hypertension, and his continuous relationship with Dr. Susan Lee. In this case, the relationship was initiated when Mr. Johnson needed a new physician because his previous one retired. He sought out Dr. Lee at the recommendation of a friend, presenting to her office as a non-acute initial visit, instead of entering the health care system through the emergency room during a hypertensive crisis. As Mr. Johnson returns to see Dr. Lee over a number of visits for management of his blood pressure, the physician develops a chart on Mr. Johnson, one that includes his past medical history, his medicines, which ones have been tried and failed, his allergies, his social history, his smoking habits, alcohol consumption, and his family and work life. Dr. Lee can develop her own system for how she organizes data in the chart. With each visit, Dr. Lee has to briefly review the chart with a focus on the last appointment in order to refresh her memory about the patient's situation. Because she is the only physician using the chart, she can read her own penmanship without having to struggle with a colleague's handwriting, an unfortunately significant factor in the medical field. In visit after visit, the physician and patient can glean from each other a sense of how they each make medical decisions and what values contribute to such decisions. During regular clinical encounters, much can be learned about how a person feels about health and how his illness affects his life. This information is not frequently articulated in charts but becomes part of the interpersonal realm of experience that two people share and understand about each other.

Mr. Johnson has come to depend greatly upon Dr. Lee, and will only see her, at times delaying needed medical appointments until her schedule can accommodate him, unless there is an emergency.

CASE B: The Institution

Imagine that when one joins the "American Health Care Associates," one receives a digitalized card that summarizes all of her medical history, including medications, hospitalizations, and social history. Each summary is provided in a standardized, easy-to-read, comprehensive format. With each interaction with a new AHCA physician, whether one visit or multiple consecutive visits, the physician takes the card and scans it into the computer and screens through the patient's history. The managed care organization, (MCO) has invested much money into streamlining the information system, as well as investing effectively in other health care professionals to take the patient's vital signs, triage, and draw labs to make the entire visit as efficient as possible. Multiple systemic quality assurance mechanisms are in place to maximize positive clinical outcomes. Through these mechanisms, efficiency is maximized and patient confusion minimized. As satisfactory outcomes are obtained, the patient develops a trust in the institution of AHCA, and weighs continuity with the system as an important consideration in future health care interactions.

III. WHY TRUST MATTERS

Many believe that a trusting physician-patient relationship results in many good consequences, including better compliance, improved communication, decreased health care costs, and potentially better health care outcomes, although the studies which support such specific health outcomes are few and far between.[1] Beyond the beneficial consequences of trust are the less tangible aspects of a trusting relationship that are so desired by physician and patient alike.

Trust performs many functions in our day to day lives. First, it *reduces the complexity* of people's daily existences and provides psychological economy, decreasing their need to worry about the potential consequences of daily exchanges with others. If one were to consider every possible outcome of such exchanges, the future would appear so complex that rational planning would be impossible. To reduce this complexity, one must reduce the number of possible futures. One way to do this is by predicting each outcome and allotting energy only to the more likely alternatives. However, this kind of planning is extremely costly in terms of time and resources. Instead, people trust. ". . . To trust is to live *as if* certain rationally possible futures will not occur. Thus, trust reduces complexity far more quickly, economically, and simply than prediction" (Lewis and Weigert, 1985, p. 969). By trusting, one eliminates from one's mind the possible outcome that the trusted person will take advantage of one's vulnerabilities. Through her confidence in the trusted person, she no longer has to worry about those negative outcomes.

When a physician provides an exhaustive list of possible risks and adverse effects of a particular procedure or medication during the process of obtaining an informed consent, the patient may worry about each of the possible outcomes, regardless of severity or likelihood. However, if the patient has some level of trust in the physician, the patient can focus on only the more likely or more severe side effects, effectively eliminating the unlikely possibilities from her mind because of the trust she has in the physician's statements that such outcomes are extremely unlikely. Similarly, if the physician has reason to trust the patient, the physician can more freely pursue an open discussion with the patient about risks and side effects. With knowledge and trust in a patient, a physician can more easily and appropriately tailor a conversation. The physician may have less reason to hesitate, fearing unnecessarily scaring a patient during their discussion. The physician might be more forthcoming and honest about her personal opinions of the treatment. Perhaps the more effective the communication, the more it minimizes the possible risk of a lawsuit in the case of an adverse outcome. Through the clarification of risks and allowing patients to accept risk with maximal understanding, the physician can more easily establish legal immunity.

Second, trust can enhance trustworthiness in those that one trusts because of the *therapeutic* nature of the trusting act. Therapeutic trust can be thought about in two ways. One positive effect that trust has on the person being trusted is the further encouragement to trust and be trustworthy. Horsburgh, who was strongly influenced by Gandhi, wrote of the promise of attempts to trust as "therapeutic trust." One deliberately puts trust in someone else even though that person is not fully trustworthy. Through this action, the truster encourages a positive sense of an honest and reliable self in the other that may engender more trustworthy actions and attitudes (Horsburgh, 1960) Therapeutic trust presupposes a belief in the possibility of stirring someone's conscience to an extent sufficient to affect her conduct.

Another kind of therapuetic trust is the alleged clinical benefit to the patient. According to James, in medicine, trust is "both an inherent good and an instrumental good, because trust furthers the good of health" (James, 1989, p. 145). A trusting relationship may be crucial to medical practice in so far as it furthers the purpose of physician-patient interactions. Trust may contribute to the placebo effect, broadly defined, which determines how a patient's psyche affects the outcome of her illness. One report claims that the placebo effect is most powerful when a trusted physician enthusiastically offers a patient a new therapy (Roberts, *et al.*, 1993) According to Albert Anderson, "therapeutic trust persists over the ages as the magic in the practice of medicine" (Anderson, 1978, p. 119). If the health care provider can engender trust by ritual, demonstration of strong belief, and by involvement of the patient in the care process, "the probabilities of favorable outcome are enhanced" (Anderson, 1978, p. 120).

Third, trust is important in the *gaining of knowledge* (Hardwig, 1991). Having become familiar with another person, one may be more likely to trust that individual than a stranger, but to gain information at all about someone or something, one must first trust the sources of the evidence and information. Yet in situations where there is a dearth of information, people are most vulnerable. These are precisely the situations when people most acutely need to place themselves at risk in order to benefit.

"Trust is necessary if one wishes to have knowledge of anything beyond one's own immediate experience ... . If I do not so freely borrow, I shall be hopelessly imprisoned in an impoverished set of beliefs about only those things which I have experienced and can remember ... . This is a debilitating form of skepticism." (Webb, 1993, p. 261). Because one cannot take the time and energy to verify all that she hears, she must trust others. All she can see for herself without verification is the link between testimony and the subject of the testimony.

Physicians and patients depend upon information from each other to determine treatment goals and choices. The physician learns what she can from the patient about her best interests and about the patient's perception of her best interests as well as the symptoms and condition. At the same time, the patient learns about the relevant medical information from the physician, as well as the physician's values and recommendations. Each party presents its own body of knowledge with an important point of view. According to Webb, the subjectivity of this exchange makes the need to trust more acute.

By acknowledging the boundaries of medical knowledge forthrightly with patients, physicians and patients can together negotiate the uncertainty. Coming to terms together with the dearth of knowledge creates a foundation for trust. Where trust flourishes, knowledge can grow in new ways, for example, through improved communication between physician and patient. Doctors and patients can share more regarding their values, motives, and reasoning as they depend upon each other to make health care decisions in an uncertain environment.[2]

IV. LEVELS OF TRUST

To evaluate how trust functions in the above examples, let us step back for a moment to discuss how trust functions in our everyday lives. This discussion will focus on relationships between individuals and between individuals and institutions.

A. Trust Between Individuals: Promise, Role, and Actor

Some types of trust, including self-trust (Govier, 1993) and infantile trust (Erikson, 1950) contribute to the basic environment in which each individual exists, interacts, and communicates with others. Both of these forms are not fully conscious. It is important also to consider the types of trust that depend more completely on conscious choice or intention. Often times a choice to make oneself vulnerable to another depends upon layer upon layer of unconscious motive and experience that shapes one's own ability to trust.

Self-trust and infantile trust are part of the foundation upon which trust develops between individuals. An infant learns that her cries will be received by the mother with the offer of a breast and being held. A teenager learns through her own choices and their outcomes how to trust in her own judgment. Informal kinds of promises exist between strangers during day to day interactions. I expect that the stranger will in fact stop her car at the stop sign and not plow through and damage my car or person. Baier (1986) distinguishes reliance from trust by saying that reliance is simple dependence upon the predicted activity or attitude of another. One relies on others in all areas of everyday life, for other drivers not to hit one's car, for example. However, criminals, comedians, and blackmailers also rely upon certain human behaviors and attitudes to perform certain tasks. A blackmailer may predict that a person will pay a lot of money to prevent the publication of embarrassing photos. Yet we might feel that this is something different from trust. Accordingly, Baier writes that trust is the reliance on the *good will* of another. One relies on friends, physicians, family, and Samaritans. This is not the same as relying upon the store owner not to put poison in the food. Trust is reliance upon the good will of others as opposed to relying upon habits, attitudes, and reactions for success (Baier, 1986, p. 234-235). One might see trusting as expecting an active response, whereas reliance might involve depending upon a more passive failure to harm.

A promise is the basic form of artificial formal device which stabilize conditions in which mutually suspicious, risk-averse strangers can cooperate. It is a manifestation of a kind of intentional trust. However, people cannot depend completely upon promises or contracts. The explicitness of contracts works only insofar as one can predict the contingencies of a given situation. Given the complexity of everyday life, people cannot create lists of all of the possible outcomes and alternatives of their interactions and a person is more likely to focus upon the contingencies that she can imagine might arise: "Until I become aware of a terrorist plumber I am unlikely, even if I should insist upon a contract before giving plumbers access to my drains, to extract a solemn agreement saying that they not blow me up" (Baier, 1986, p. 250). For the other relationships in life, people depend upon trust instead of and beyond promises and contracts because of the "indefiniteness of what we are counting on them to do" (Baier, 1986, p. 251).

Besides promises and contracts, other forms of formal devices of cooperation exist that form part of the fabric of interpersonal trust. Some relationships have formalized duties and obligations arising out of their very nature. Marriage is one kind of relationship in which specific social expectations provide certain types of requirements and obligations between two individuals. Another example of this is the fiduciary relationship. Consider the physician-patient relationship as an example.

The basis of the physician-patient relationship is that the former is a professional who is learned, skilled, and experienced in those subjects about which the latter ordinarily knows relatively little. Because of this knowledge imbalance, the patient must place "great reliance, faith, and confidence in the professional word, advice, and acts of the physician or other practitioner... Being a fiduciary relationship, mutual trust and confidence are essential" (61 *Am Jur* 2d, SS166, p. 298). The fiduciary relationship and the obligations that it involves allow two parties to cooperate with some sense of common understanding of their relationship and roles.

Although roles themselves can provide a basis upon which trust can grow, the trust that arises from the formal construct must be tempered by how a given individual fulfills her role. When someone is deciding whether to make herself vulnerable to another, she considers the role that the individual is playing. The role can allow a person to trust in an individual because of the status inherent in the role. In deciding to make an appointment to see a physician for a bad cough, one realizes that any physician has certain qualifications and duties that arise simply from her role as physician. One can have confidence that she has completed a certain level of training and displayed competence in a way that led the relevant governing organizations to grant her a license to practice. One knows that legal and ethical regulations guide the physician's behavior, including her fiduciary obligations.

However, there is a level of trust that functions beyond understanding the individual within her formal role, and this level involves understanding how the individual acts within the boundaries of the role. In addition to the status trust that one might earn because of her defined role, there is a separate level of merit trust that the individual must earn in her role. Bernard Williams' distinctions of macro- versus micro-motivation illuminate these differences (Williams, 1988, pp. 3-13)

Macro-motivation is socially or role defined. One trusts the pilot's motivation to get the plane safely from place to place. Micro-motivation considers the context and peculiarities of different individuals and situations. One may have less reason to trust the particular pilot who staggers toward the gate and has alcohol on her breath (Williams, 1988, p. 6). One might have similar hesitations about a licensed physician who has alcohol on her breath or one who has a bad reputation according to the patient's friends.

When making a personal decision regarding trust in a particular context, people make judgments based upon what a person has said or done. Useful evidence arises from personal experience or evidence conveyed by already trusted individuals.

When trusting others, one can trust them in a narrow sense, in which one is expected to fulfill minimally the duties of her role. Many times, however, people desire a person to go beyond the minimal expectations. Many trust not only that their primary care physicians will prescribe the right medicine to treat a cough, but also that their physicians will care for them as individuals. This means valuing the patient as an individual and approaching her with an attitude of good will.

Others focus on the trustworthiness of the individual as a whole, beyond that suggested by the social role the individual occupies. Especially significant for trustworthiness are sincerity, promise-keeping, reliability, competence, and concern for others. These factors can provide an overall sense of character, which the patient can then project onto the circumstances of concern to her (Govier, 1993b). Some might be concerned about a clinically excellent physician who cheats on her spouse. Some might say that this detracts from her overall character. Of course, the more areas in which a person trusts someone, the more the trust can be damaged. If a doubt arises in one area of conduct, doubt can spread to other areas of conduct. On the other hand, positive evidence in one area may have a reinforcing effect on perceived trustworthiness.

In summary, one can trust another in a limited way by trusting a person to fulfill a role. In this case, the role determines the obligations. By trusting in this narrow way, people allow the trusted latitude and discretion only in certain areas, for example only in a limited medical area. Often a person may focus on the particulars of a given individual and situation in deciding whether to trust an individual who is in a certain role. Others may choose to yield discretion more broadly, and trust in the individual's character and good will toward the truster. This may be how many think of their long-term relationships with a trusted physician. In this circumstance, the truster expects something beyond reliable and competent medical care.

B. Systemic and Institutional Trust

Beyond fundamental and unconscious levels of trust, beyond individual interpersonal relationships, trust functions at the level of systems and institutions. At any of the aforementioned levels, Baier notes that "when things go well, there can be trust and mutual trust; when the system is faulty, or its human operators incompetent or lacking good will (and the system must *be* faulty if such persons find and keep places within it), then distrust will be appropriate." Baier goes on to say that unless the system is so vicious that the victims of untrustworthiness do not realize that they are victims, the system "will eventually be disrupted by creeping distrust" (Baier, 1992. p. 17).

Systems are often designed to minimize occurrences that would threaten trust. Consider the United States government. The Constitution provides checks and balances that allow accountability, stability, and flexibility. Judicial, executive, and legislative branches work together and balance each other to provide a functioning and effective government. The effectiveness and accountability that the system is intended to provide should decrease reasons for distrust in the institution.

One can see many similarities between the government and the health care system. The public endows the medical profession with a trust that is balanced by legal monitoring, self-regulation by such organizations as medical societies, and market forces.

C. The Web of Levels

When defining the level of trust in any relationship, the complicated social, cultural, and psychological network underlying that relationship deeply affects the individuals and systems involved. Trusting, whether conscious or not, depends upon the interaction of a given individual or institution and the multitudinous levels of trust that may have an impact on an attitude of trust. Trusting then, conversely, has an impact on the involved individuals and institutions in a dynamic, ever-evolving expansion and contraction of trusting relationships.

As each of the levels of trust interacts with and affects each other, some levels may be more important as indicators or effectors of trust than others. According to Niklas Luhmann, system trust ultimately depends upon personal trust (Lewis and Weigert, 1985, p. 975). It may be very hard for a person who through personal experience finds very little reason to trust individuals, whether friends or family or salespersons, to be able to trust the institutions of society or society itself. Conversely, Durkheim noted that institutional trust underwrites interpersonal trust. Accordingly, as trust in common institutions erodes, one might expect that to some extent people would lose trust in other individuals as well (Lewis and Weigert, 1985 p. 974).

When considering relationships within medical institutions, a patient can invest her trust in either her personal physician or the MCO or both. The interaction between each of these levels is complex. The individual physician through her role and her behavior affects the patient's evaluation of her. The patient judges that physician both in terms of her position as a professional, through which she might gain status trust, and in terms of how she specifically acts within her professional role, and so may gain merit trust. The physician also reflects, positively or negatively, on the institution.

Conversely, the MCO, through its policies, whether through incentive plans or the much maligned anti-disparagement clauses, can affect the perceived trustworthiness of the provider. The efficiency and bureaucracy of the institution will similarly have an effect on the patient's willingness to trust those individuals who represent it.

V. ENHANCING THE ENVIRONMENT OF AND FOR TRUST

Having discussed the benefits of trust and a few major types of trusting relationships (individuals and institutions), let us focus now on what encourages trust in these various settings. How are these factors related to continuity of care?

Trust may grow as a result of rather than as a precondition of cooperation. Some look to social arrangements which may provide incentives for people to take risks, and which can then provide opportunities for trust to prosper (Gambetta, 1988). David Good (Good, 1988 p. 31-48) provides a summary of the empirical literature from psychologists regarding trusting behaviors in humans. Good's summary focuses on psychological games that reproduce situations that provide incentives to cooperate. In one game, each player had a toy truck that was to transport "goods" back and forth along a play common road between the two players. Each player had one gate that she could use to control access to the road. Each player desired free access to the road to gain "wealth." To win the game, the player had to gain more wealth than the other. One way to win was to block the road strategically to obstruct the other player's path. If each removed the gate altogether, the players had less opportunity to block the other player. However, the players discovered that by removing the gates, they could accumulate higher "wealth" through better access back and forth on the roads. As a result, there was a much higher degree of cooperation and success, as measured by the total "wealth" generated by each player.

In addition, changing the initial conditions of the games could also improve cooperation. If the investigators slowly increased the level of reward for each player from a small sum instead of starting from a very high level, then there was increased collaboration. Instead of initially transporting a large amount of beans back and forth in the toy trucks, the psychological game began with only a few beans. In later rounds, the players transported increasing amounts of beans. This led to a higher level of cooperation than when the players transported more valuable amounts from the onset. Furthermore, in situations where the players had to think about long-term outcomes, collaboration grew. When players realized that they would be interacting with each other for more than one game or event, they had a greater chance of cooperating.

Another study found that if the players could communicate with each other, then there was increased cooperation. There was also a greater likelihood of a mutually beneficial outcome. If the players could not talk to each other, then they were consumed with guessing the motivations and next moves of their opponent. When the players could talk and negotiate openly, the conditions for cooperation improved. Of course, this claim regarding the connection between communication and cooperation must be qualified because if it were a situation in which the level of ambiguity and uncertainty was high, then one player could use communication to exploit the other. If there was a minimum of ambiguity, then such an opportunity for advantage disappeared (Good, 1988).

In summary, there are many factors that might increase cooperative behavior and, therefore, the level of trust. What encourages people to put themselves in positions that build trust? One factor that will improve the likelihood of trust is initiating the interaction when the *stakes are smaller* and less valuable. In the health care setting, a more relaxed setting may be more conducive to a trusting relationship. This may be more difficult when there is an acute medical problem that detracts from the rest of the human interaction and communication between the physician and patient. People who have little access to primary care often enter the medical system when they are in worse condition, when the stakes are higher. If trust can develop in a more comfortable environment when the risks are not so great, then increasing access to primary health care may have deeper advantages than simply improving attempts at preventive medicine.

Good's summary suggests that *shared long-term interests* improve the atmosphere for trust. When a patient and physician meet each other in an acute setting or in a walk-in clinic in which there is no expectation of a continued relationship, the participants may be less willing to get to know one another on deeper levels. If the patient perceives the encounter poorly, she does not have to see that provider again, and so there is less incentive to create the foundation for a true relationship. The physician also takes the ephemeral nature of the encounter into account in the way she interacts with the patient. However, in the context of a continuous relationship, both individuals realize that a certain level of etiquette, consideration, and care is necessary because they will meet again and will want to minimize the discomfort and antipathy between them. By ensuring the establishment of longer lasting health care relationships, trust may be fostered.

Once two individuals embark on an attempt to cooperate, improving *communication* will enrich the understanding of motives and possibilities. By cooperating, the partners can each maximize their opportunities. All forms of communication, including facilitative gestures and tone can improve the likelihood of productive exchange and cooperation. Communication can also decrease the potential for danger and threat by *reducing ambiguity* and demonstrating that the opportunity for exploitation will not be taken. Approaching an ethical ideal of informed consent allows both patient and physician to understand the values and knowledge that each can bring to the relationship. By being forthcoming about diagnostic, therapeutic, and prognostic information and options, the physician can reduce the level of ambiguity and ease the patient fear of exploitation. With more knowledge, the patient will be better able to determine the type of care she desires. By being forthcoming about symptoms, values, and other factors that have an impact on her life and health, a patient helps the doctor to formulate the most appropriate options for care and recommend the most appropriate plan. This, in turn, reduces the ambiguity for the physician and allows greater security within the relationship.

VI. LEVELS OF TRUST AND THE NEW HEALTH CARE ENVIRONMENT

Now, consider how what we've learned from Good's summary affects how we think about trust in the health care setting. In case A, the relationship is initiated outside of the acute situation, allowing energy to be directed toward the interpersonal aspects and the grounds for long term trust, instead of medical necessity alone. Because the two individuals understand that Dr. Lee will ideally be there as long as Mr. Johnson needs her, the two can develop a sense of shared long term interests.

According to the studies mentioned at the beginning of the talk, this continuous relationship will yield better communication, reduce confusion, and increase compliance and patient disclosures. Additionally, the relationship may result in more efficient use of the medical system and controlled costs. By initiating the relationship when the stakes are low, by maintaining a focus on mutual long-term interests, and maximizing communication, the doctor and patient create a social and professional milieu conducive to the development of trust. Consistent communication can clarify diagnoses, prognoses, and the values that contribute to medical decision making. The continuity should therefore contribute to the development of trust.

The relationship was described in isolation, without regard to payment or organizational structures. How does such a relationship fare in the new world of managed care organizations? The relationship between patient and physician of necessity now involves a relationship between the patient and the health care system or MCO. Today, instead of simply choosing a physician or a group practice, patients are often forced by employers or other financial considerations to choose an MCO. The realities of the current health care delivery system force patients to consider not only their physicians but the system through which the physicians provide care. The health care consumer takes into consideration cost, the services provided, the quality and quantity of physicians involved in the MCO, and how well her health care needs will be met.

Even though the patient, whether because of employer or financial constraints, is not often completely free to select which MCO to belong to, a relationship can and should be established between the patient and the institution. The relationship consists of a person in a vulnerable position because of her health care needs who depends upon the will of the institution to provide for those needs in an effective and affordable way.

Of course, the relationship with the MCO has a huge impact on the relationship between the physician and patient. Much has been written about the conflicts of interest that are inherent when physicians become employees of MCO and how these conflicts affect patient care (Mechanic and Schlesinger, 1996). How does managed care affect continuity between patient and physician? David Orentlicher has outlined four threats to continuity between physician and patient. First of all, when a patient initially chooses an HMO, she may find that some or none of her current physicians are on the HMO's panel. As a result, the patient may have to sever ties with old physicians and switch to new physicians for care. Second, if a patient becomes dissatisfied with a particular HMO and switches to another one, she may also have to switch to a different panel of physicians, potentially having to terminate existing relationships. Third, a patient may be satisfied with her current physicians and HMO, but if one physician transfers out of a given HMO, then the patient must choose whether to switch HMOs also or stay with the remaining caretakers at the expense of the other relationship. Fourth, a patient may receive health care as a benefit of employment and find that her employer no longer offers a particular HMO. In order to keep the employer's contribution to her health care coverage, the patient might have to switch (Orentlicher, 1995, pp. 145-146).

Because of the increasing importance of the relationship between the patient and the MCO, considerations of continuity take on a new dimension. As a consumer finds a health care plan that provides the best balance of services, quality, and cost-effectiveness, a consumer might come to value continuity with the MCO over continuity with individual practitioners. One study reaches the following conclusions: "A high degree of treatment continuity appears to be an important determinant of trust *only* to the extent that the one-on-one encounters have been perceived as successful by the client" (Caterinicchio, 1979, p. 95). Patients who saw different providers at each visit to a particular clinic and experienced successful outcomes expressed feelings of trust for the clinic, viewing it as a "collection of trustworthy physicians."

Can a patient's continuity with an MCO provide the vehicle for trust that is so important to medical encounters? Consider case B, American Health Care Associates. MCOs often include an initial visit and annual visits with a primary care physician as part of the package. By mandating that health care relationships are established when the stakes are smaller and less urgent, the MCO is in effect creating an environment more likely to encourage trust. By focusing on shared long-term interests, namely cost-effective, high quality medical care with an eye toward preventive medicine, the interests of the patient and the institution are more likely to coincide. For example, the patient may actually benefit from a breach in individual continuity by being able to see another MCO physician in more urgent circumstances when her own primary care physician is unavailable.

Finally, trust will be enhanced between patient and institution if effective communication can be maximized. If ensuring that communication regarding medical histories, test results, and appointment times is a priority within an MCO, then trust will also be more likely to grow.

VII. CONTINUITY AND TRUST

If trust is a prized feature of the physician-patient relationship and one of the rewards of a continuous relationship, can it really be so easily transposed onto a health care institution, perhaps at the expense of the continuous individual relationship? The descriptions above include highly idealized continuous relationships between patients and physicians and patients and institutions. The delineation is somewhat artificial in that quite often MCOs (as well as patients) attempt to ensure a continuous individual relationship within the continuous relationship with the institution. Whether such idealized visions can be realized in practice is a question that cannot be answered at this time. It also remains to be seen whether organizations can provide the rich benefits which previously had been solely the domain of the individual continuous relationship.

In college I had multiple conversations with my thesis advisor, a physician, about whether we would be more satisfied stepping into a perfect machine that coldly though proficiently provided whatever medical care we needed or whether we would prefer to still have the personal, interactive encounter with the human, though at times flawed, physician. We both leaned toward the human option, perhaps depending upon the acuity or chronicity of the medical problem. However, a new powerful machine is rising on medicine's horizon. Institutional health care systems are forcing us again to examine what is useful, beneficial, and special about the individualized physician-patient relationship. As the system takes on more and more of what individual physicians could not do alone or in small groups, the individual physician-patient relationship, in order to justify its existence, will need to reestablish its importance and uniqueness as a place where continuous relationships can flourish in trusting environments. These should be places where physicians can earn both status and merit trust. However, certain aspects that contribute to positive health outcomes may be more easily attained in well-functioning health care systems, including continued access to medical services. Just as MCOs might compromise certain aspects of continuity and therefore trusting relationships, patients may also lose out by depending too much upon an individual practitioner when the institution can provide what one physician cannot provide alone. As MCOs continue to grow and expand in importance and influence as molders of physician-patient relationships, individual physicians and MCOs must attend to continuity and trust as crucial factors in the health care encounter, not only for the sake of cost-effectiveness, but for the quality of medicine and the lives of patients.

I want to thank Robert J. Levine, James L. Nelson, Hilde L. Nelson, Joseph J. Fins, and Richard Hatchett for their time and contributions to this paper.

Department of Medicine, Yale University School of Medicine
New Haven, Connecticut

NOTES

[1] Caterinicchio tested and found support for the following hypothesis: 1) the greater the frequency of successful treatment, the greater the degree of trust in the physician; 2) the greater the frequency of the successful treatment, the greater the degree of perceived positive health gains from treatment; 3) the greater the degree of trust in the physician, the greater the degree of perceived positive health gains from treatment, which was correlated with lower levels of treatment anxiety and greater tolerance for treatment pain intensity (Caterinicchio, 1976).

[2] Consider Howard Brody's transparency model of informed consent in which a more complete consent is obtained when a physician discloses his rationale and reasoning behind a particular medical conclusion or suggestion, whether diagnostic, prognostic, or treatment related (Brody, 1989).

REFERENCES

61 *Am Jur* 2d, SS166, p. 298.

Anderson, A.D.: 1978, 'Therapeutic trust: The magic in medicine,' *Man and Medicine* 3, 119-26.

Baier, A.C.: 1986, 'Antitrust,' *Ethics* 96, 231-260.

Baier, A.C.: 1992, 'Alternative offerings to Asclepius?' *Medical Humanities Review* 6, 1744.

Brody, H.: 1989, 'Transparency: Informed consent in primary care,' *Hastings Center Report* 19, 5-9.

Caterinicchio, R.P.: 1976, 'Interpersonal trust in the physician, and the tolerance of treatment-7 induced pain: A multivariate analysis in a natural clinical setting,' *Dissertation Abstracts International* 36(8-A), 5578.

Caterinicchio, R.P.: 1979, 'Testing plausible path models of interpersonal trust in patient-physician treatment relationships,' *Social Science and Medicine* 13A, 95.

Council on Ethical and Judicial Affairs, American Medical Association: 1994, *Code of Medical Ethics: Current Opinions with Annotations*, a1t XXXIV.

Dietrich, A.J., and Marton, K.I.: 1982, 'Does continuous care from a physician make a difference?' *The Journal of Family Practice* 15, 929-37.

Durkheim, E.: 1964, The Division of Labor in Society cited in J.D. Lewis, 974.

240   JULIE ROTHSTEIN ROSENBAUM

Emanuel, E.J., Brett, A.S.: 1993, 'Managed competition and the patient-physician relationship,' *New England Journal of Medicine* 329, 879-82.
Emanuel, E.J., Dubler, N.N.: 1995, 'Preserving the physician-patient relationship in the era of managed care,' *Journal of the American Medical Association* 273, 323-29.
Erikson, E.H.: 1950, 'Growth and crises of the healthy personality,' in Milton J.E. Senn (ed.), *Symposium on the Healthy Personality*, Yale University Press, New Haven, pp. 91-146.
Gambetta, D.: 1988, 'Can we trust trust?' in D. Gambetta (ed.), *Trust: Making and Breaking Cooperative Relations*, Basil Blackwell, London, pp. 225-235.
Good, D.: 1988, 'Individuals, interpersonal relations, and trust,' in D. Gambetta (1988), pp. 31-48.
Govier, T.: 1993, 'An epistemology of trust,' *International Journal of Moral Studies* 8, 155-174.
Govier, T.: 1993, 'Self-trust, autonomy, and self-esteem,' *Hypatia* 8, 99-120.
Hardwig, J.: 1991, 'The role of trust in knowledge,' *Journal of Philosophy* 88, 693-708.
Horsburgh, H.J.N.: 1960, 'The ethics of trust,' *The Philosophical Quarterly* 10, 343-354.
James, D.N.: 1989, 'The friendship model: A reply to Illingworth,' *Bioethics* 3, 142-146.
Lewis, J.D., Weigert, A.: 1985, 'Trust as a social reality,' *Social Forces* 63, 967-985.
Lyons, J.P.: 1994, 'The American medical doctor in the current milieu: A matter of trust,' *Perspectives in Biology and Medicine* 37, 442-459.
Mechanic, D., Schlesinger, M.: 1996, 'The impact of managed care on patients' trust in medical care and their physicians,' *Journal of the American Medical Association* 275, 1693-1697.
Morreim, E.H.: 1995, 'The ethics of incentives in managed care,' *Trends in Health Care, Law & Ethics* 10, 56-62.
Orentlicher, D: 1995, 'Health care reform and the physician-patient relationship,' *Health Matrix* 5, 141-180.
Orentlicher, D., Woolhandler, S., Himmelstein, D.U.: 1995, 'Extreme risk – the new corporate proposition for physicians,' *New England Journal of Medicine* 333, 1706-1708.
Pellegrino, E.D, Thomasma, D.C: 1981, *A Philosophical Basis of Medical Practice: Toward a Philosophy and Ethic of the Healing Professions*, Oxford University Press, Oxford.
Reiser, S.D: 1978, *Medicine and The Reign of Technology*, Cambridge University Press, Cambridge, Massachusetts
Roberts, A.H., Kewman, D.G., Mercier, L., Hovell, M.: 1993, 'The power of nonspecific effects in healing: Implications for psychosocial and biological treatments,' *Clinical Psychology Review* 13, 375-391.
Siegler, M: 1993, 'Falling off the pedestal: What is happening to the traditional doctor-patient relationship?' *Mayo Clinic Proceedings* 68, 461-467.
Webb, M.O.: 1993, 'Why I know about as much as you: A reply to Hardwig,' *Journal of Philosophy* 90, 260-270.
Williams, B.: 1988, 'Trust considered: formal structures and social reality,' in D. Gambetta (1988).

HOWARD BRODY

CAN RELATIONSHIPS HEAL –
AT A REASONABLE COST?

A few years ago (Brody, 1994). I suggested that there are a number of important values that we might legitimately seek to maximize in caring for patients:
Scientific health care
Ethical health care
Humane health care
Effective health care
Cost-conscious health care
This list might at first glance show why we have serious problems today, amounting in many people's estimation to a health care crisis. "Scientific" medicine has come to be seen as synonymous with high-technology medicine, which is in turn viewed increasingly as both excessively costly and inhumane. To many physicians, "humane" medicine implies the old notion of "bedside manner," which may be compassionate but seems in today's world hardly a way to deliver *effective,* let alone *scientific* medical care. We have tended to define "ethical" medicine in terms primarily of the principle of patient autonomy. That implies that if patients truly want a lot of medical technology (as Americans seem prone to do), then any system which limits their access to this technology in the name of cost containment is going to be unethical (McCullough, 1994).

These observations contain a good deal of truth. I do not wish to defend the notion that we can have everything we want in health care, at an affordable cost, without facing tradeoffs which in many cases will be difficult and in a few cases even tragic.

However, I do wish to suggest that the debates over health care policy have tended to slight some dimensions of practice which hold out hope for promoting the convergence of the important values on our list. I wish to look especially at primary care, at the importance of continuity of care as a critical dimension of primary care, and at narrative approaches to the patient-physician relationship. I wish to argue that a health system which stresses primary care generally and continuity of care specifically, and

Stephen Wear, James J. Bono, Gerald Logue and Adrianne McEvoy (eds.), Ethical Issues in Health Care on the Frontiers of the Twenty-First Century, 241–257.
© 2000 *Kluwer Academic Publishers. Printed in Great Britain.*

which emphasizes a narrative, relational approach, can be scientific, ethical, humane, effective, and cost-conscious all at the same time.

I. PRIMARY CARE AND CONTINUITY

Primary health care may be generally defined as first-contact care which is both comprehensive and continuous. My primary care physician is prepared to be the first provider who sees me for the vast majority of ills that I may suffer. He is prepared to offer me a comprehensive array of skills and services to treat those ills, so that approximately 90 percent of the time his care will be fully adequate to resolve the problem, without referral elsewhere. And he offers me continuous care which is not limited by the specific sort of illness I suffer from, the organ system which causes the illness, or even whether I have any illness at all (Starfield, 1992).

Within medicine, family physicians, general internists, and general pediatricians are trained in post-graduate programs that specifically focus upon primary care skills and attitudes. Physician assistants and nurse practitioners receive similar training and perform very effectively in primary care roles. This does not mean, however, that other medical specialists need be excluded from whatever benefits a primary care approach might bring to practice. As we will see below, primary care has much more to do with particular skills and attitudes and much less to do with what board exam one has taken and what letters appear after one's name. (In this sense, at least some cardiologists and endocrinologists are better primary care physicians than are some family physicians.) Nonetheless, given how physicians are trained today, it will be harder for the aspiring cardiologist to learn primary care skills and attitudes that it would be for, say, the general internist.

As a percentage of all the health care research conducted in the U.S., research on primary care has been relatively sparse. In trying to create a scientific basis for any statements or conclusions about primary care and physician-patient relationships, one must use caution, selecting studies with some care and frequently encountering gaps in the available research. While much more research would be desirable, and would indeed become a high national priority if the case I am arguing for here should be accepted, I believe that the statements I will make below about primary care can be defended as reasonable summations of the best available research.

1. When we compare the costs of health care in the U.S. to that of other developed nations, at least some of the excessive costs of U.S. medicine seem to correlate with the lesser proportion of U.S. physicians in primary care specialties. Moreover, when major discrepancies are found in the costs of care among regions within the U.S., there is a direct relationship between a higher cost of care in a region, and a lesser percentage of physicians in that region in primary care practice (Welch *et al.*, 1993).

2. When primary care physicians care for the same medical conditions as other specialists, the primary physicians generate lower costs but produce roughly equivalent outcomes (Greenfield *et al.*, 1995).

3. Managed care usually works to lower costs by insisting that patients get more of their care from primary physicians, and by placing some administrative barriers in the way of referral to specialists. According to conventional wisdom, this makes managed care ethically problematic, because it rewards physicians for providing less care, and would thus seem a direct threat to the patient's health. It is therefore of some interest that the available literature comparing outcomes in managed care and in other payment systems *fails almost totally* to document any worsening of outcomes in managed care, and in some instances demonstrates superior outcomes in managed care (Franks *et al.*, 1992; Miles and Koepp, 1995).

This statement requires an important exception and an important reservation. The exception is that in one early study, poor patients randomized to managed care instead of traditional fee-for-service care fared worse, while middle class patient fared better (McCullough, 1994). In a more recent study, both the poor and the elderly appeared to suffer from being assigned to managed care systems (Ware *et al.*, 1996). Exactly why that was so, and what implications this may have for future health policy, needs to be determined. The reservation is that as the marketplace ratchets down ever more tightly on costs, there would have to come a point where managed care systems start to produce worse patient outcomes – not because managed care is inherently bad, but because any care system is bad if it fails to provide enough funds for necessary and beneficial care. Even though no such studies have yet appeared in the literature, to my knowledge, for the non-poor and non-elderly, anecdotal evidence suggests that such studies may start to appear soon. The studies already published will then be important because they will prove that the problem is underfunding and perverse incentives, not a system of primary care gatekeepers *per se.*

4. Of all dimensions of primary care, continuity of care has been even less well studied than some others. Some data show that continuity of care is good for patient satisfaction, and that in some situations it saves money and produces superior clinical outcomes (Becker, *et al.*, 1974; Bindman, *et al.*, 1996; Chao, 1988; Dietrich and Marton, 1982; Wasson *et al.*, 1984; Weiss and Blustein, 1996). Given the data available, the authors of the influential Medical Outcomes Study did not hesitate to name continuity as that dimension of primary care which was most closely related to its lower cost and equal efficacy (Safran, *et al.*, 1994).

5. If one looks at the sort of encounter which is typical in primary care practice, one can identify a number of elements which have individually been linked to enhanced patient outcomes (Novack, 1987). These include the patient feeling fully "listened to" when first presenting with a complaint (Bass, Buck, *et al.*, 1986; Bass, McWhinney, *et al.*, 1986; Starfield, *et al.* 1981); the patient sensing care and concern among those present in the healing encounter (Kennell, *et al.*, 1991; Sosa, *et al.*, 1980); and the patient feeling more empowered to control the symptoms as a result of the encounter (Greenfield, *et al.*, 1985; Kaplan, *et al.*, 1989; Malterud, 1994; Roter, *et al.* 1995). By combining these elements, one can sketch out in some detail an "optimal primary care visit" which would seem, statistically, to have the highest probability of producing a healing outcome for the patient – *independent of* the technical quality of the other aspects of the encounter, including medical history, physical and laboratory examination, differential diagnosis, and pharmacotherapeutic selection and management.

Another way of making this point is to say that the optimal primary care visit is concerned not merely with curing but also with healing. American medicine has come under increasing criticism recently for its exclusive fixation on curing to the exclusion of healing (Fox, 1997). Curing focuses on the disease while healing addresses the whole person; and healing becomes especially important when more and more patients suffer from chronic diseases where a cure is simply not possible. Healing aims at the reduction in suffering and the restoration of wholeness, including bodily integrity and function and reconnection with the life of the community (Cassell, 1991; Frank, 1995). Imagine, for instance, the heart attack victim who has had successful reperfusion of the myocardium by angioplasty, but who now remains fearful of overexertion, has been unable to return to work, and cannot have sex with his wife. This patient, in a sense, has been cured but not healed. The invasive cardiologist might view this case as a great success; the primary care practitioner would see much work yet to be done.

So far, then, we seem to have a partial prescription for effective, cost-conscious medical care – emphasize wherever possible the use of primary care practitioners over other specialists; and investigate further how to optimize those elements of the primary care encounter which could lead to increased healing efficacy. The latter strategy has additional payoffs, since both primary care physicians and subspecialists can incorporate the fruits of this further research into their practices.

It is important in this discussion not to become overly romantic about primary care. There are, of course, incompetent primary care practitioners – some who are concerned a great deal with the patient's emotions and social life but who cannot accurately diagnose and treat disease, some who can do a great job of diagnosis and treatment but who neglect the patient as a person, and some with both sets of flaws. Calling for more utilization of primary care, and more research into primary care, does not negate the pressing need to do a better job of measuring competence and quality throughout the system (Brook, 1997). Indeed, the research called for here ought to provide much more measurable benchmarks as to what counts as competent practice in primary care.

II. THE PLACEBO RESPONSE AND MEANING

The scientific line of inquiry which might identify how to optimize the healing potential of the primary care encounter closely parallels (if it does not completely overlap) the research area which has up to now been called the placebo response. This area, again, has been relatively undeveloped. The placebo effect is usually treated as a nuisance variable, worthy of attention only insofar as it can interfere with our understanding of the effects of "real" treatments like drugs and surgery. Fewer studies have looked at the placebo response for is own sake, as an important contributor to healing. Those studies that do so, however, often turn up impressive results, showing that for at least some patients on some occasions, placebos can improve both the symptoms and physiological changes associated with a wide variety of illnesses (Harrington, 1997; White, et al., 1985).

We can identify three general sorts of reasons why a patient might get better (or worse, for that matter) after seeing a physician. One is the "specific" therapy administered during or after the visit. "Specific" therapy is easy to understand when it comes in the forms of medication or surgery, for instance, but conceptually it is rather a complex and somewhat contradictory notion (Grunbaum, 1985). For instance, is dietary advice or patient education "specific" treatment for diabetes? A second reason a patient might improve is the natural history of the disease, independent of the intervention of a physician. The term "placebo response," I would suggest, is properly reserved for "neither of the above" healing effects – which are, by hypothesis, the result of the emotional or symbolic aspects of the healing encounter. (This of course assumes that all such aspects of the encounter are "nonspecific" – which may be true only to the extent that we do not study them and try to increase their frequency and potency (Brody, 1985)).

We may learn more about placebo or nonspecific effects if we discover more about their biochemistry, and find out precisely which intermediate pathways account for the translation of an emotional or symbolic influence into end-organ results. Thus there has been interest in tracing the relationship between placebo responses and endorphin pathways, catecholamine release, and neuroimmune function (Harrington, 1997; White, et al., 1985). But the primary physician is probably not going to be able to stimulate any of these pathways via direct intervention; from her viewpoint, the question of biochemical mediators is likely to remain a black box. As Bulger has elegantly stated, the placebo response may best be viewed as a method by which the patient unleashes her already existing internal pharmacy and puts it to work on the illness or symptom in question (Bulger, 1990); and precisely what pharmacologic agents that internal pharmacy keeps in stock need not concern the practitioner as much as the research scientist. The research question for primary care is, therefore, what aspects of a healing encounter, or a healing relationship, are of sufficient symbolic import and impact to stimulate the correct biochemical pathways and thus lead eventually to enhanced healing.

While much more remains to be learned, a possible way to organize what we know so far is that physicians seem to be able to produce positive placebo responses when they are able to change the *meaning* of the illness experience for the patient in a more positive direction. For our purposes, there are two major things to say about meaning. The first is that we could construe the "healing" elements of the encounter, which we listed in the previous section, as elements of positive meaning.

Specifically, the patient may experience a more positive meaning for the illness experience when three things occur:

1. The patient is provided with an explanation for the illness, its causes, and its consequences which is meaningful and consistent.
2. The patient senses care and concern from the immediate social circle.
3. The patient gains an enhanced sense of mastery or control over the illness or its symptoms (Brody, 1986; Brody, 1994).

By this approach we tend to look at meaning analytically, breaking it down into component behaviors within the healing encounter. But, for purposes of therapy, we would prefer to look at meaning holistically, taking advantage of the fact that the separate elements may be present simultaneously and may be mutually reinforcing.

III. MEANING AND NARRATIVE

If we presume that a positive placebo effect is most likely to occur when the meaning of the illness experience for the patient is altered in a positive direction, we are next led to ask: exactly how do humans generally attach meaning to experience? The most promising answer is that we tell stories about the experience, and it is through narrative that we attach meaning to an otherwise complex and seamless existence (Bruner, 1986). For a negative, frightening experience to be altered into a more positive experience, we must tell a different story about it. This links our concern with the therapeutic aspects of the physician-patient encounter, via the science of the placebo response, directly to an interest in how physicians and patients construct illness narratives and how these narratives may best be understood and altered (Brody, 1987; Frank, 1995; Hunter, 1991; Kleinman, 1988).

The notion of a story of an illness experience looks at first glance to be simple – so simple, in fact, that nothing of real scientific or therapeutic importance could possibly follow. Upon investigation, the notion turns out instead to be dauntingly complex. Physicians construct medical narratives as a way of communicating with each other and of carrying out medical work (Hunter, 1991). Sometimes, these medical stories may become the patient's own story and may prove helpful and therapeutic for the patient (Brody, 1987; Kleinman, 1988). Other times, these stories actually impede the patient's healing, by taking away control and by stifling the patient's own voice (Frank, 1995). The telling of stories implies a listener; and in turn this implies a community and a culture, since all storytelling occurs as a matter of social and linguistic convention (Fisher, 1984). Storytelling implies expanding circles of reflexivity – when I tell you my story, I also am listening to my story as I am telling it, and this experience of retelling a story I have told before (or perhaps, of telling a story for the first time) will perhaps cause me to see events in a way I have never before appreciated them, and even to attach a different meaning to them. Similarly, when you are listening to my story, you do not actually "hear" my story as much as you hear my words and construct a story in your own mind to explain to you what those words mean. In the process, each inevitable gap in my narrative will be filled in by you with the meaning that *you* attach to events of that sort, which may be quite different than the meaning *I* attach to that type of event.

Despite these complexities, your listening to my story is a communal act of confirmation and validation, and is thus a moral act (Booth, 1988; Fisher, 1984; Frank, 1995). The accounts of the ill persons themselves suggest at times that the single most important event in the process of healing, or of relief of suffering (especially in the face of chronic or terminal illness, where no technical "fix" is possible), is to feel at some point that one has really told one's story and that someone else has really listened. It is a sad commentary on the health professions that this listener is seldom a trained "healer" and may be much more likely to be a fellow sufferer from the same or a similar illness. Indeed, Frank has suggested that the chronically ill ought more or less to abandon hopes that they will be healed by physicians, and instead band together in the communal enterprise of telling and listening to each other's stories. I may help to heal you when I listen to your story of suffering with openness and vulnerability. In turn, I may find myself ill at any time in the future, and then I may be surprised to find that the story you have told is helping to heal me (Frank, 1995).

Borkan and colleagues, in a paper which might serve as a template for sensitive, qualitative research into these questions on a disease-specific basis, interviewed elderly hip-fracture patients and related the meaning that they attached to the event with their eventual clinical outcomes (Borkan, *et al.*, 1991). They found that those who told a story about a basically healthy person who fell and suffered a purely mechanical problem, demanding in turn a mechanical solution, recovered faster and reached a higher level of functional status, compared to those who told a story of a basically unhealthy and already severely damaged person who fell as a result of those pre-existing problems. This study, of course, leaves a number of questions unanswered. We cannot be certain that those who told a "sicker" story were not in fact sicker to start with and hence had less recovery for that reason. And we do not know whether any intervention could have changed the stories of any of these patients, and if so, what if any would have been the change in outcome. But further research could address all of these issues and begin to deepen our understanding of the clinical importance of meaning and narrative.

If physicians today seem too little attuned to their patients' stories, and too busy or preoccupied to do the most important work of healing by really listening to those stories, this state of affairs is sad but by no means inevitable. We could ask, if any sort of relationship with a physician is more likely than others to lead to this sort of openness to listening, what would it be? And I would conclude that the ideal primary care relationship seems to be the best approximation we could envision today. If we can promote real continuity of care, there is a sense that the relationship itself is an ongoing story, an unfolding conversation; each visit is a new chapter in the story, a picking up of the conversation temporarily suspended. If the relationship is truly a primary care relationship, its continuation is predicated only upon the personhood and ongoing need of the patient – it is not limited by the types of symptoms, or mode of treatment. Finally, of all physicians today besides psychiatrists, primary care physicians are best trained in the psychological, behavioral, and interviewing skills necessary to listen to stories nonjudgmentally and with therapeutic understanding (Smith and Hoppe, 1991).

IV. FROM CONTINUITY TO SUSTAINED PARTNERSHIP

I have already alluded to the dearth of solid research on the effects of continuity of care in the physician-patient relationship. There are two major reasons for this gap in our knowledge. First, it has traditionally been only primary care physicians who cared about this subject, if anyone; and there is no primary care research lobby at the Federal level capable of channeling large amounts of money and academic resources into research of that sort. Second, such research is notoriously hard to conduct, because the "crucial experiment" in which continuity is the only variable that is altered, while all other relevant aspects of care are held constant, is almost impossible to design; and if it were to be designed, it would probably constitute a laboratory model so far removed from the world of actual primary care practice as to make any generalization of its findings impossible.

While continuity is relatively little studied, there exists even less research upon an arguably more important concept, "sustained partnership," which Leopold and colleagues have reviewed comprehensively (Leopold, et al., 1996). They note that the Institute of Medicine, expanding upon an earlier definition of primary care developed in 1978, added in 1994 the notion that primary care involves a sustained partnership between primary physicians and their patients (Institute of Medicine, 1994). Continuity of care is necessary but not sufficient for this sustained partnership. Leopold et al., drawing upon studies of health outcomes, patient satisfaction, physician-patient communication, and medical ethics, suggest that sustained partnership has these components:

1. Whole-person focus
2. Physician's knowledge of the patient
3. Caring and empathy
4. Patient trust of physician
5. Care appropriately adapted to conform with the patient's goals, expectations, beliefs, and values
6. Patient participation and shared decision-making

The authors note in passing that each of these elements, taken individually, has been shown in the available research studies to enhance patient satisfaction, health outcomes, or both.

From our earlier discussion we can see that these elements of shared partnerships in primary care cohere closely both with the meaning model of the placebo response, and also with the essential preconditions for a narrative approach to the patient's illness experience. To the extent that patients trust physicians, physicians display caring and empathy, the care coheres with the patient's beliefs and expectations, and the patient feels added mastery and control, we would expect that the largest possible positive placebo response would be elicited. The physician who seeks to know the patient as a whole person and not merely as a medical history and review of systems, and who is willing to share control of the interview to make the patient a partner in deciding what topics will be discussed, is the physician most open to hearing the patient's story.

Taken as a whole, the shared partnership model increases the likelihood that the patient will tell a full and rich story to the physician about what the illness means in the context of the patient's life. This in turn would make the meaning that the patient attaches to the illness experience fully accessible to the physician. The partnership could then move toward the re-evaluation and re-negotiation of that meaning, as a fully shared enterprise in which the patient feels real ownership. If room is found to replace a negative meaning with a more positive one, a number of things might happen – the patient may feel better as a result of an enhanced placebo response, and the patient may be motivated to undertake a variety of healing actions such as life-style changes and strict medication regimens.

V. SHARED PARTNERSHIP VS. FISCAL CONSTRAINTS

The skeptic will now insist that it is time to wake up and smell the coffee. All this talk of stories and shared partnerships has been a pleasant exercise in medical nostalgia, akin to memories of the old small-town GP making house calls and accepting a basket of eggs in payment. Clearly, the system we have evolved to finance health care in the U.S. as we enter the next century will not allow any such indulgences. Driven by managed care contracts and quotas, primary care physicians (if they have not already been replaced by lower-salaried physician extenders) have no choice but to churn out visits along the assembly line, spending too little time to hear any sort of complete story, and hardly ever forming close personal relationships.

Like the idea that managed care inevitably underserves patients and thus leads to worse health outcomes – which, as we saw, is not true – this view of managed care as assembly-line has been repeated so often as to become almost an article of faith. Many demoralized physicians seem to feel today that this is simply the reality of the future of practice and one must either accommodate or perish. And at least some anecdotal data suggest that much managed care is indeed moving in this direction.

There is no guarantee, however – especially in an industry which is rapidly changing and is in considerable turmoil – that the sorts of management decisions being made today will necessarily lead to the greatest long-term financial stability in the health care marketplace. It may not be simply that managed care, in some of its forms, is acting in ways contrary to traditional medical and patient-centered values; it may also be true that managed care is shooting itself in the foot so far as its own long-range interests are concerned (Brody and Goold; Clancy and Brody, 1995).

Presumably, a managed care plan will best corner its local market if it can provide high-quality care at lower cost and still promote a high level of patient satisfaction. And, if we review again the research conclusions summarized in previous sections, any managers who think they can do this, and at the same time dispense with continuity of care in primary care, are simply incompetent.

A few extra minutes of the time of the primary physician is far from the most expensive commodity that the health system can dispense. If those extra minutes lead to careful listening to the patient's story and the creation of an effective shared partnership, the outcomes and increased patient satisfaction can be well worth the extra investment. Moreover, if those few extra minutes are not spent, there is a much greater likelihood that the plan will end up spending much more money later on unnecessary tests, x-rays, and consultations with other specialists.

Physicians distressed and disoriented by rapid changes in the practice environment may prefer to speak of managed care as having only the potential to do harm; but in fact, for most elements of a good primary care relationship, managed care as a system or a concept (as opposed to any particular managed care plan) has the potential to improve care as well (Emanuel and Dubler, 1995). Managed care plans have both the resources and the need to track data from physician visits in a comprehensive and systematic way that was impossible in fee-for-service practice. If a plan is short-sighted and bottom-line-oriented, it may choose to track only cost data and to reward physicians only for keeping immediate costs low. But a plan with slightly more vision may instead try to gather data on the quality and the continuity within primary care relationships, including patient satisfaction with specific elements of the encounter, and selectively reward physicians who score highest on those measures. That would, in turn, stimulate a much more active research program around the shared-partnership concept, so that managed care could best determine what to measure and how (Leopold, et al., 1996).

Admittedly, the present marketplace may show little evidence for this sort of vision on the part of industry. When an industry whose mission is, supposedly, caring for patients, elects to label the amount of money it spends on actually caring for patients its "medical loss ratio," one may well despair of both its ethical values and its business acumen. But there is also reason to hope that much of the turmoil in the present marketplace is a transitional rather than a mature phase of what managed care might look like – even if ultimately, more government regulation will be required to bring patient-centered values into proportion with the profit motive (Brody and Goold).

VI. CONCLUSION

I have tried to show that physicians can heal patients most cost-effectively when they adopt a primary care approach with a sustained partnership model, and take the patient's story very seriously as the way to reconstruct the meaning of the illness experience in a positive manner. This sort of care supplements and enhances the other, purely technical biomedical skills which make up modern medical practice. Moreover, such an approach need hardly be restricted to primary physicians; specialists in all fields can readily learn and adopt the same skills and attitudes, so long as they decide that it is worth while to do so.

"Healing patients cheaply" may at first sound like an inversion of the usual ethical presumption in medicine, which holds that the patient's well-being comes first, and concerns about finances come way down the list. Therefore it is important to review the elements of the sustained partnership model:

1. Whole-person focus
2. Physician's knowledge of the patient
3. Caring and empathy
4. Patient trust of physician
5. Care appropriately adapted to conform with the patient's goals, expectations, beliefs, and values
6. Patient participation and shared decision-making

This list is hardly in opposition to medical-ethical values as we currently understand them; and instead it heightens the likelihood that the patient will view the encounter with the physician as maximally humane and properly patient-centered. Patient autonomy, the focus of most discussions in medical ethics over the past 30 years, seems more secure in this sort of relationship than it arguably is in most encounters today within the U.S. health care system (SUPPORT, 1995). And there is much less chance that this sort of relationship will lead to a mechanical,

HOWARD BRODY

inhumane overemphasis on autonomy, such as some bioethicists fear (Callahan, 1984).

If we assume that some version of managed care will dominate the medical marketplace for the foreseeable future, then whether a sustained partnership model and a focus upon the patient's story become dominant will depend on two critical questions. One is whether the managed care industry will be led by managers who truly understand their business. The other is whether American physicians will rediscover what it means to be true leaders in health reform (Berwick, 1994), instead of falling into the demoralized whining or the greedy self-protection that seem rather too prevalent today.

Winston Churchill is supposed to have said once that the American people can be counted on to do the right thing – after they have exhausted all other possibilities. I hope that in the management of our nation's health care, we are getting ready to do the right thing soon, and that we will soon be done with the stage of exhausting the other possibilities in the meanwhile.

Michigan State University
East Lansing, Michigan

REFERENCES

Bass, M.J., Buck, C., Turner, L., *et al.*: 1986, 'The physician's actions and the outcome of illness in family practice,' *Journal of Family Practice* 23, 43-47.

Bass, M.J., McWhinney, I.R., Dempsey, J.B., *et al.*: 1986, 'Predictors of outcome in headache patients presenting to family physicians – a one year prospective study,' *Headache Journal,* 26, 285-294.

Becker, M.H., Drachman, R.H., Kirscht, J.P.: 1974, 'A field experiment to evaluate various outcomes of continuity of physician care,' *American Journal of Public Health* 64, 1062-1070.

Berwick, D.M.: 1994, 'Eleven worthy aims for clinical leadership of health system reform,' *Journal of the American Medical Association* 272, 797-802.

Bindman, A.B., Grumbach, K., Osmond, D., Vranizan, K., Stewart, A.L.: 1996, 'Primary care and the receipt of preventive service,' *Journal of General Internal Medicine* 11, 269-276.

Booth, W.C.: 1988, *The Company We Keep: An Ethics of Fiction,* University of California Press, Berkeley.

Borkan, J.M., Quirk, M., Sullivan, M.: 1991, 'Finding meaning after the fall: Injury narratives from elderly hip fracture patients,' *Social Science Medicine* 33, 947-957.

Brody, H.: 1985, 'Placebo effect: An examination of Grunbaum's definition,' in L. White, B. Tursky, and G.E. Schwartz (eds.), *Placebo: Theory, Research, and Mechanisms,* Guilford, New York.

Brody, H.: 1986, 'The placebo response. Part I. Exploring the myths. Part 2. Use in clinical practice,' *Drug Therapy* 16(7), 106-131.

Brody, H.: 1987, *Stories of Sickness*, Yale University Press, New Haven.

Brody, H: 1994, 'My story is broken, can you help me fix it? Medical ethics and the joint construction of narrative,' *Literary Medicine* 13, 79-92.

Brody, H., Goold, S.D. (in preparation): 'Jekyll, Hyde, and Gresham: The future of managed care in an unregulated market.'

Brook, R.H.: 1997, 'Managed care is not the problem, quality is,' *Journal of the American Medical Association* 278, 1612-1614.

Bruner, J.: 1986, *Actual Minds, Possible Worlds*, Harvard University Press, Cambridge, MA.

Bulger, R.J.: 1990, 'The demise of the placebo effect in the practice of scientific medicine – a natural progression or an undesirable aberration?' *Transactions of the American Clinical and Climatological Association* 102, 285-293.

Callahan, D.: 1984, 'Autonomy: A moral good, not a moral obsession,' *Hastings Center Report* 14(5), 40-42.

Cassell, E.J.: 1991, *The Nature of Suffering and the Goals of Medicine*, Oxford University Press, New York.

Chao, J.: 1988, 'Continuity of care: Incorporating patient perceptions,' *Family Medicine* 20, 333-337.

Clancy, C.M., Brody, H.: 1995, 'Managed care: Jekyll or Hyde?' *Journal of the American Medical Association* 273, 338-339.

Dietrich, A.J., Marton, K.I.: 1982, 'Does continuous care from a physician make a difference?' *Journal of Family Practice* 15, 929-937.

Emanuel, E.J., Dubler, N.N.: 1995, 'Preserving the physician-patient relationship in the era of managed care,' *Journal of the American Medical Association* 273, 323-329.

Fisher, W.R.: 1984, 'Narration as a human communication paradigm: The case of public moral argument,' *Communication Monographs* 51, 1-22.

Fox, E.: 1997, 'Predominance of the curative model of medical care: A residual problem,' *Journal of the American Medical Association* 278, 761-763.

Frank, A.W.: 1995, *The Wounded Storyteller: Body, Illness, and Ethics*, University of Chicago Press, Chicago.

Franks, P., Clancy, C.M., Nutting, P.A.: 1992, 'Gatekeeping revisited – protecting patients from overtreatment,' *New England Journal of Medicine* 327, 424-429.

Greenfield, S., Kaplan, S., Ware, J.E.: 1985, 'Expanding patient involvement in care: Effects on patient outcomes,' *Annals of Internal Medicine* 102, 520-528.

Greenfield, S., Rogers, W., Mangotich, M., Carney, M.F., Tarlov, A.R.: 1995, 'Outcomes of patients with hypertension and non-insulin-dependent diabetes mellitus treated by different systems and specialists: results from the Medical Outcomes Study,' *Journal of the American Medical Association* 274, 1436-1444.

Grunbaum, A.: 1985, 'Explication and implications of the placebo concept,' in L. White, B. Tursky, and G.E. Schwartz (eds.), *Placebo: Theory, Research, and Mechanisms*, Guilford, New York, pp. 9-36.

Harrington, A. (ed.): 1997, *The Placebo Effect: An Interdisciplinary Exploration*, Harvard University Press, Cambridge, Massachusetts.

Hunter, K.M.: 1991, *Doctor's Stories: the Narrative Structure of Medical Knowledge*, Princeton University Press, Princeton, New Jersey.

Institute of Medicine: 1994, *Defining Primary Care: An Interim Report*, National Academy Press, Washington, D.C.

Kaplan, S.H., Greenfield, S., Ware, J.E.: (1989): 'Assessing the effects of physician-patient interactions on the outcomes of chronic disease,' *Medical Care* 27, S110-S127.

Kennell, J.H., Klaus, M.H., McGrath, S., *et al.*: 1991, 'Continuous emotional support during labor in a U.S. hospital,' *Journal of the American Medical Association* 265, 2197-2201.

Kleinman, A.: 1988, *The Illness Narratives: Suffering, Healing and the Human Condition*, Basic Books, New York.

Leopold, N., Cooper, J., Clancy, C.: 1996, 'Sustained partnership in primary care,' *Journal of Family Practice* 42, 129-137.

Malterud, K.: 1994, 'Key questions – a strategy for modifying clinical communication. Transforming tacit skills into a clinical method,' *Scandinavian Journal of Primary Health Care* 12, 121-127.

McCullough, L.B.; 1994, 'Should we create a health care system in the United States?' *Journal of Medicine and Philosophy* 19, 483-490.

Miles, S.H., Koepp, R.: 1995. 'Comments on the AMA report "Ethical Issues in Managed Care",' *Journal of Clinical Ethics* 6, 306-311.

Novack, D.H.: 1987, 'Therapeutic aspects of the clinical encounter,' *Journal of General Internal Medicine* 2, 346-355.

Roter, D.L., Hall, J.A., Kern, D.E., Barker, L.R., Cole, K.A., Roca, R.P.: 1995, 'Improving physicians' interviewing skills and reducing patients' emotional distress. A randomized clinical trial,' *Archives of Internal Medicine* 155,. 1877-1884.

Safran, D.G., Tarlov, A.R., Rogers, W.H.: 1994 'Primary-care performance in fee-for-service and prepaid health care systems,' *Journal of the American Medical Association* 271, 1579-1586.

Smith, R.C., Hoppe, R.B.: 1991, 'The patient's story: Integrating the patient- and physician-centered approaches to interviewing,' *Annals of Internal Medicine* 115, 470-477.

Sosa, R., Kennell, J.H., Robertson, S., *et al.*,: 1980, 'The effect of a supportive companion on perinatal problems, length of labor and mother-infant interaction,' *New England Journal of Medicine* 303, 597-600.

Starfield, B: 1992, *Primary Care: Concept, Education, and Policy*, Oxford University Press, New York.

Starfield, B., Wray, C., Hess, K., Gross, R., Birk, P.S., D'Lugoff, B.C.: 1981, 'The influence of patient-practitioner agreement on outcome of care,' *American Journal of Public Health* 71, 127-132.

The SUPPORT Principal Investigators: 1995, 'A controlled trial to improve care for seriously ill hospitalized patients: The study to understand prognoses and preferences for outcomes and risks of treatment (SUPPORT),' *Journal of the American Medical Association* 274, 1591-1598.

Ware, J.E., Bayliss, M.S., Rogers, W.H., Kosinski, M., Tarlov, A.R.: 1996, 'Difference in 4-year health outcomes of elderly and poor, chronically ill patients treated in HMO and fee-for-service systems: Results from the Medical Outcomes Study,' *Journal of the American Medical Association* 276, 1039-1047.

Wasson, J.H., Sauvigne, A.E., Mogielnicki, R.P., Frey, W.G., Sox, C.H., Gaudette, C., Rockwell, A.: 1984, 'Continuity of outpatient medical care in elderly men: A randomized trial,' *Journal of the American Medical Association* 252, 2413-2417.

Weiss, L.J., Blustein, J.,: 1996, 'Faithful patients: The effect of long-term physician-patient relationships on the costs and use of health care by older Americans,' *American. Journal of Public Health* 86, 1742-1747.

Welch, W.P., Miller, M.E., Welch, H.G., Fisher, E.S., Wennberg, J.E.: 1993, 'Geographic variations in expenditures for physicians' services in the United States,' *New England Journal of Medicine* 328, 621-627.

White, L., Tursky, B., Schwartz, G.E. (eds.): 1985, *Placebo: Theory, Research, and Mechanisms*, Guilford Press, New York.

SCOTT DEVITO

VALUES AND THE PATIENT-PHYSICIAN
RELATIONSHIP

I. INTRODUCTION: QUESTIONS, ANSWERS, REJECTIONS

What is the ideal patient-physician relationship (PPR)? Three of the
papers presented at this conference give three different answers to this
question. Julie Rothstein Rosenbaum argues that the PPR must include
continuity of care and trust. Howard Brody argues for a PPR that
maximizes healing (while lowering costs) by helping the physician to
induce the placebo effect in her patients. Finally, Katherine Montgomery
contends that the ideal PPR is one that maximizes a physician's ability to
treat patients by maintaining a therapeutic distance between patient and
physician. Much of what these three authors suggest conforms to our
intuitions about the ideal PPR. It certainly seems like a good idea to have
trust, continuity of care, increased rate of healing, lower costs, and
keeping physicians focused on the patient as aspects of the ideal PPR.

But is it necessarily the case that the PPR must have these
characteristics? Can we reject trust, continuity of care, low costs, etc. and
still have a good PPR? And, what is the price of modifying the PPR to
incorporate trust, continuity of care, etc.? If we adopt the views expressed
by Engelhardt in his keynote address, then our answer to these questions
must be, respectively: no, yes, and possibly high. In the remainder of this
paper, I take a look at the arguments provided by Rosenbaum, Brody, and
Montgomery in their papers on the PPR with the goal of seeing how well
their work fits with the claims Engelhardt makes in his keynote address.
To do so, I first give a brief description of Engelhardt's views. I then turn
to the arguments provided by Rosenbaum, Brody, and Montgomery in
their papers at this conference. I explain their views and describe how
they fit with Engelhardt. I conclude that they do not fit well with
Engelhardt's views about morality in a liberal secular society because
they ignore Engelhardt's main point - that in a liberal secular society
there can be no privileged moral viewpoint and associated (ordered) set of
values.

*Stephen Wear, James J. Bono, Gerald Logue and Adrianne McEvoy (eds.), Ethical Issues in
Health Care on the Frontiers of the Twenty-First Century, 259–274.*

II. MORAL STRANGERS

At the core of Engelhardt's understanding of any ethical problem is his view that members of a liberal secular society are moral strangers. Two persons are moral strangers when they do not hold the same basic moral principles or they have differing rules of evidence and inference and, as a result, they can not resolve moral dilemmas through sound rational argument. In addition, moral strangers, unlike moral friends, can not resolve their moral disputes by appeal to a canonical source, whether that source is an institution (like the Church) or an individual (like the Pope) (Englehardt, 1996, p. 6).

Moral friends, on the other hand, either agree upon the same basic moral premises, have the same rules of evidence and inference, or have a canonical source upon whom they may call to resolve disputes. In other words, members of the same moral community can not have irresolvable moral disputes. Members of the same religious community turn out to be good examples of moral friends. They have, for the most part, the same moral premises and when they can not resolve ethical disputes through reason (using those premises) they can turn to an agreed upon authority (e.g., the Pope) to resolve those disputes.

But in a modern, liberal, secular state, we can not be assured that people will meet Engelhardt's three criteria of moral friendship. It is a key feature of these societies that their members may disagree about what are the basic moral principles, the rules of (moral) evidence and inference, and upon sources to be used to resolve moral disputes. So, when two people meet they may have very different moral beliefs and value systems. Depending on how different their moral systems are, these moral strangers may not be able to resolve their moral disagreement through either reason or appeal to an authority. As a result, Engelhardt argues, in a liberal secular state, the only recourse two citizens have, when they have a moral dispute (and they are not moral friends), is to establish a non-coercive agreement stipulating what each will allow the other to do to the other.

This moral estrangement raises particularly interesting problems for patients and physicians. In liberal secular societies, patients may encounter physicians who do not share the same basic moral values as their patients. Consequently, when patient and physician first meet, neither can be certain what value system the other holds. This can be a problem because while both patient and physician agree that the patient has arrived for some form of treatment, they may disagree more than they agree. For example, the patient's definition of wellness may differ from that of the physician, the willingness to risk certain side-effects may differ between patient and physician, aspects of the disease may be of differing importance to each, and so on.

Given the possibility that patients and their physicians may be moral strangers we can ask, "What can be done to prevent physicians from acting improperly towards their patients and to prevent patients from acting improperly towards their physicians?" The three papers presented in this conference, on the patient-physician relationship, can be seen as attempting to provide part of the answer this question. Unfortunately, because they rely upon implicit value systems, they fail to meet the strong requirements set by Engelhardt.

III. CONTINUITY OF CARE AND TRUST

In 'Trust, institutions, and the physician-patient relationship: Implications for continuity of care,' Julie Rothstein-Rosenbaum argues that continuity of care is a necessary component of the PPR because it is a powerful method for building trust and because a PPR based in trust has positive outcomes for both patients and physicians. According to Rosenbaum, trust plays a number of useful roles in the PPR. For example, trust *reduces* the *complexity* of people's daily existences and provides psychological economy (Roberts, 1996). Trust can enhance trustworthiness in those that one trusts because of the *therapeutic* nature of the trusting act (Rosenbaum, this volume), and trust is important in the *gaining of knowledge* (Wolf, 1996).

We can see how this makes *prima facie* sense when we think of an encounter between a physician and a new patient. The first time that the patient meets the physician, the patient does not know what to expect from the physician and the physician does not know what to expect from the patient. In such a case there is neither trust nor distrust. In this situation both the patient and the physician must be on their guard. Neither knows the values of the other (so it is difficult to determine what will be important or safe to divulge) and neither knows whether the other will take advantage of their vulnerabilities. The first time a patient sees a physician, the patient must concern himself with thinking about the outcome of the patient-physician interaction and must determine the dangers and benefits of listening to the physician and divulging (potentially damaging or embarrassing) information to the physician. Similarly, the physician must listen carefully to the patient in order to recognize if the patient is lying, leaving important details out, etc. And, when the physician discusses an intervention for the patient's problem, the physician must clearly outline all of the ramifications of the treatment in order to protect herself from the possibility that what is not important to the physician will be important to the patient.

Since neither the patient nor the physician inherently trusts the other, they are forced to think about what is said and done by each other to a greater degree than if the patient and physician trust each other. (In effect, they have not been able, as of yet, to hammer out an Engelhardtian contract.) As Rosenbaum argues, trust is a vital component of the PPR, because it helps reduce the psychological complexity of the relationship. When two people know each other and trust each other this psychological second-guessing can be eliminated (or at least minimized).[1] The patient who trusts his physician will divulge relevant information, even if it is embarrassing or potentially damaging, because the patient does not have to concern himself with the thought that the physician will take advantage of the patient's divulging of this information. In addition, when the physician recommends a particular therapy, the patient who trusts the physician will not have to worry that the therapy contains hidden dangers for the patient, because the patient believes that the physician would tell him if there were any dangers that the patient should know about. Similarly, when a physician trusts her patient, she can freely discuss the risks and side effects of treatment, more easily tailor the discussion of the treatment to the patient and be more forthcoming about her personal opinion of the intervention. So trust allows the patient and the physician to interact in an easier manner than when trust is absent. In addition, trust allows the smooth transfer of information between patient and physician.[2] Finally, trust contributes to the placebo effect. In the placebo effect merely being medically treated (even if the treatment is actually inert) can both lead to healing and increase the rate of the patient's healing by stimulating the patient's internal immunological armory to address the illness.

Once Rosenbaum establishes the importance of trust as a component of the PPR, she turns to an examination of trust-building. In particular, she asks what aspect of the PPR plays a vital role in developing trust. Her answer is that continuity of treatment helps in the development of trust between the patient and the physician. According to Rosenbaum, continuity of care should begin either early in the patient's illness or before the patient becomes ill, because at those points in time the stakes are small and, as studies have shown, trust is easier to build when what is at risk has lower value. The problem with entering the system when much is at stake can be seen when we think of those people who have little access to primary care. They often enter the health care system when they are in more desperate need. Such people are less likely to have a physician examine them when their illness has just begun and treatment is quick, cheap, minimally disruptive, safe and likely to effective. Instead, they enter the system when the illness has become severe and they are seriously ill. Treatment is no longer quick, cheap, minimally disruptive and safe nor is it as likely that there will be a good outcome. Such a situation does not engender trust for two reasons: (1) high stakes are inherently less likely to increase trust in a relationship than low stakes and (2) neither the physician nor the patient have any stake in building a relationship because there is no expectation of a continuing relationship.

IV. CONTINUITY OF CARE AND MORAL CONTRACTS

Returning to Engelhardt's notion of moral strangers and his contract-based morality, we see that continuity of care strengthens the contract by allowing the participants to know more about the other's moral community. The patient and physician who have had a continuous relationship know more about the other than the patient and physician who meet once or just a few times during a crisis period. This facilitates their interaction and decreases the likelihood that either will act in a way towards the other that would be inconsistent with the other's moral viewpoint. It also permits the patient to know *when* the physician might act in a way that would be inconsistent with his beliefs and to defend himself against that behavior. Similarly, continuity of care allows the physician to know, e.g., when a patient might not directly divulge information that is important to properly treating the patient (where properly is defined according to the physician's viewpoint). This allows the physician to modify her behavior so that she can treat the patient in a way that the physician believes is appropriate[3] *even if* the patient does not provide the information. In sum, we can say that with continuity of care patients and physicians are able to establish clearer, stronger and more acceptable (to patient and physician) Engelhardtian moral contracts.

V. THE VALUE-LADEN-NESS OF TRUST AND CONTINUITY OF CARE

While continuity of care can be seen as allowing patients and physicians to build stronger Engelhardtean contracts, there are problems with Rosenbaum's arguments for trust and continuity of care. The central problem is that Rosenbaum's arguments, for increasing the amount of trust in PPRs, exposes a negative relationship between trust and informed consent and between continuity of care and informed consent. In essence, as continuity of care and trust increase patient information decreases, which results in less *informed* consent. Because Rosenbaum is arguing for a PPR in which there is more trust between physician and patient and there is a negative relation between trust and informed consent, she is also arguing for *less* informed consent. Consequently, as I show, Rosenbaum's arguments for trust depend upon her establishing a preferred value-ordering (trust over informed consent) and, as any preferred value-ordering is not compatible with Engelhardt's view, attempting to modify the PPR so that there is more patient-physician trust can not be justified in an Engelhardtian post-enlightenment world in the manner described by Rosenbaum.

The negative relationship between trust and informed consent comes out in Rosenbaum's psychological-simplicity argument for trust playing a role in the good PPR. According to Rosenbaum, by trusting, one eliminates from one's mind the possible outcome that the trusted person will take advantage of one's vulnerabilities. Through her confidence in the trusted person, she (the patient) no longer has to worry about these negative outcomes. In addition, with knowledge and trust in a patient, a physician can more easily and appropriately *tailor* a conversation.

These statements implicitly express the negative correlation between trust and informed consent. If we trust, then we have less reason to ask questions, worry that the other will fail to give relevant information, etc. In other words, a patient who trusts his physician need not be concerned that the physician has left out important information or that the physician has underemphasized certain aspects of the treatment, safe in (the patient's) knowledge that the physician would not act to harm the patient.

In addition, the physician who trusts and knows her patient can exclude information about treatment outcomes that the physician deems irrelevant or unlikely, can focus on certain aspects of the treatment rather than other aspects, and so on, safe in the physician's knowledge of the patient and trust in the patient to act appropriately.

But this safety belies another danger. A patient who does not carefully and fully examine the possible consequences of treatment decreases his ability to give full and informed consent. At the same time, the physician who knows her patient can (to use Rosenbaum's word) *tailor*[4] the discussion of treatment consequences in a way that decreases both patient information about certain treatment outcomes and patient interest in those outcomes. Doing so decreases the patient's knowledge of the treatment and treatment outcomes. Even if the physician actually understands the patient's values and applies them correctly in her discussion of the treatment, *the patient* has been *denied the opportunity* to make the choice about his treatment in a fully informed manner. While some (even most) patients may prefer such a situation, there are others who might prefer genuinely *full* and informed consent over less-than-fully informed consent.

Rosenbaum's psychological simplicity argument implicitly relies upon a given set of values – that trust is more important than informed consent (or, at minimum, that trust is more important than it is commonly held to be and that informed consent is less important than it is commonly held to be). The patient who trusts his physician can assume that the physician has provided relevant information about the treatment to the patient and, as a result, turns over some (not all) decision making power to that physician. Furthermore, it is vital to Rosenbaum's psychological simplicity argument that patients actually *do* rely more upon their trusted physician to make decisions about their treatment. After all, if trust only had the capacity to reduce psychological complexity but rarely or never did, then the argument for trust from psychological simplicity would be irrelevant. So, Rosenbaum must believe that patients who trust their physicians actually do rely more upon the physician to make certain decisions about treatment (such as the decision not to discuss or emphasize certain possible outcomes of treatment). This can only be a good outcome, if trust is (in some sense) more valuable than informed consent.

But, if trust fits into the PPR in the way and for the reasons Rosenbaum describes, then informed consent must give way. This implies a preference ordering of values (trust over informed consent) not argued for by Rosenbaum. In addition, Engelhardt has explicitly argued that there can be no *preferred* value-ordering in a liberal secular society. This leaves Rosenbaum with two (mutually incompatible) responses: (1) she could provide an argument for her preference ordering of values and reject Engelhardt's argument that members of liberal secular societies have the capacity to be moral strangers or (2) she could bite the bullet and reject her psychological simplicity argument. At present, since no argument for (1) has been given, *we* must assume that (2) is true and, as a result, recognize that trust need not play the role Rosenbaum envisions it playing in the ideal PPR.

In addition, as Rosenbaum has shown, continuity of care is a powerful trust-building tool. Consequently, continuity of care is also a powerful informed-consent-decreasing tool. Thus, Rosenbaum's argument for continuity of care also contains a hidden preference ranking: continuity of care over informed consent. Such a preference ordering is also not compatible with Engelhardt's post-enlightenment world.

VI. CREATING THE PLACEBO EFFECT AS A MEASURE OF THE IDEAL PHYSICIAN-PATIENT RELATIONSHIP

Rosenbaum sought to argue that trust and continuity of care are important aspects of the ideal PPR. As I have shown, her argument for the role (or degree of importance) of trust and continuity of care in the PPR is undermined by Engelhardt's arguments that members of liberal secular societies are moral strangers. In 'Can relationships heal – at a reasonable cost?', Howard Brody also argues for continuity in the PPR, but he argues for continuity of care based upon its impact on patient healing instead of its ability to increase trust. According to Brody, one way to optimize patient healing (while keeping costs down) is to get the patient to experience the placebo effect while the patient undergoes the standard therapy. The placebo effect results in the body increasing its ability to heal itself and, as a result, inducing the placebo effect would result in faster and cheaper healing. Since a patient experiencing the placebo effect, in addition to the standard treatment, will fare better than a similar patient who does not experience the placebo effect, Brody infers that the *ideal* PPR must be structured so that the placebo effect is taken advantage of in the healing process.

The physician produces a positive placebo response when the physician changes the meaning of the illness experience for the patient in a more positive direction, where meaning is thought to attach to experience through stories (Brody, p. 224). Consequently, in the ideal PPR, the physician must help patients tell themselves a more positive story about their problem. For example, if an elderly patient, who has a broken hip, tells a story about herself as a woman who is on a downhill course and has just taken another step, the physician must help (where possible) that woman change her story to view the hip injury in a more positive light. Perhaps, the story becomes one of a healthy person who just had an accident and will return to being healthy again. Doing so will help the woman to recover more quickly, more completely, and more inexpensively.

VII. VALUES AND STORY CHANGING

The most problematic aspect of Brody's argument, in an Engelhardtian post-Enlightenment world, is Brody's view that making a patient's story more positive (in order to induce the placebo effect) is a good. Brody believes that doing so is a good because doing so can induce the placebo effect and result in faster, cheaper, and superior healing. The problem is that in order to buy Brody's argument one must also buy the view that inducing the placebo effect is more valuable than allowing people to keep their self-narratives.

When a physician sees a patient who has a negative illness story, according to Brody, the physician should seek to help the patient have a less negative illness story. If the physician succeeds, then the patient has a higher probability of healing and healing quickly. But is that benefit (a higher probability of healing and healing quickly), worth the cost (losing one's illness story)? For many, it may be. But, for others, having a physician *manipulate* them (the patient) so as to change the patient's story, may be extremely disvalued. If it is more disvalued by the patient than the patient values increasing the probability of getting well quickly, then the physician has *in effect* harmed the patient.

Brody can only disagree with this result (that the patient has been harmed) if he can provide an argument that shows that the patient's preference ordering (keeping his story is more valued than increasing the probability of getting well quickly) is mistaken. No such argument has been given. And, if Engelhardt is correct, no such argument *can* be given in our society.

VIII. MISINTERPRETATION AND CONTINUITY OF CARE

Even if we ignore the moral problems associated with willfully attempting to modify a patient's illness story, Brody's approach is subject to a simple practical problem: patients and physicians may not understand each other's stories. For example, a woman who has broken her hip and tells a story about a long life of hardships may be a person depicting a negative story about a very sick person or she could be telling a positive story about a woman who has succeeded in overcoming many obstacles in her life. While many verbal and non-verbal clues *may* be of help, they need not. In fact, they may cause the physician to misinterpret the patient's story. Furthermore, the physician's personal viewpoint may color the meaning of the patient's story. If a physician has a grandparent who tells a similar story and whose grandparent means by it a very positive account of overcoming obstacles, then the physician may project that grandparent's meaning onto the patient's story when that is not the patient's meaning. The patient may, in the same manner, misinterpret the story the physician is trying to tell.

Even if the physician gets the story right, the physician may not be able to influence (or may influence in an undesirable direction) the story the patient tells herself about her problem. Frequently physician and patient come from different backgrounds. A rich white male physician may have little in common with his poor Hispanic (English speaking) female patient. The physician may not be able to communicate with the patient in a meaningful way. In addition, because of the difference between patient and physician, (and perhaps the patient's prior experience with the health-care profession) the patient may not trust the physician.[5] As a result of misunderstanding and a potential low level of trust (or outright distrust), the physician's attempt to modify the patient's story may go awry.[6]

Here Rosenbaum's views on continuity of care return. A physician who has just met a patient may have great difficulty interpreting that story and determining what to do on the patient's behalf (the patient and physician are moral strangers) and that patient may have difficulty understanding the physician's story about the patient's condition. Continuity of care can help the patient and the physician understand each other. This will allow the physician to understand the patient and to help the physician mold the patient's story in a positive direction. As Rosenbaum notes, the placebo effect is most powerful when it is based on a trusted physician enthusiastically recommending a new therapy.

Thus, when Brody seeks to describe the ideal PPR as one that increases the impact of the placebo effect he must have a PPR in mind that includes continuity of care. This becomes apparent when Brody discusses the shared partnership model of the PPR. According to Brody, the shared partnership model has six components: (1) whole-person focus, (2) physician's knowledge of the patient, (3) caring and empathy, (4) patient trust of physician, (5) care appropriately adapted to conform with the patient's goals, expectations, beliefs, and values, and (6) patient participation and shared decision-making. As can be readily seen, components (1), (2), and (3) are facilitated by continuity of care.[7]

But notice, the argument that continuity of care is a vital component of the PPR is based on the alleged benefits that continuity of care provides. In Brody's case, continuity of care helps the physician and patient to understand each other. This understanding allows the physician to correctly determine if the patient's illness story is negative and, if it is, to change the patient's illness story. As I have shown, doing so is, in Engelhardt's world, morally suspect because it implies a preferred set of values increasing the probability of getting well over keeping control over one's own narratives. Since Brody has provided no argument for such a preference ordering, and Engelhardt has provided strong arguments that such preference orderings are impossible in a liberal secular society, we must reject Brody's argument for a PPR that incorporates continuity of care and care that induces the placebo effect.

IX. GOOD FENCES MAKE GOOD PPRS

Both Rosenbaum and Brody argue for a PPR that links the patient and the physician closely together through a continuous and on-going relationship. Katherine Montgomery in 'A medicine of neighbors' attempts to set some boundaries on how close this relationship should be. In particular, she argues against a PPR based on friendship and argues for a PPR based on neighborliness.

Montgomery believes that a PPR based on friendship brings the physician too close to the patient. This closeness results in a decrease in the physician's ability to understand and treat the patient's problem. This is much like the problem of family members treating other family members. A mother who treats her child, because of the closeness of the relationship, may impose her beliefs about her child (derived from their normal everyday interaction) onto the current situation. When the child expresses discomfort about the current procedure, the mother-physician may fail to pay attention to that complaint because she is used to her child complaining about minor inconveniences in normal life and imposes that image onto the current situation. The mother, because of her deep interest in her child's care, may also pay *too* much attention to the complaints of her child-patient. In either case, the child's care would be better if the mother-physician could be somewhat distant from (or, as Montgomery puts it *disinterested* in) the child.[8] Friendship can, similarly, negatively impact the patient-physician relationship. According to Montgomery, knowing someone too well can interfere with hearing the patient's complaint and providing the most efficacious therapy for that patient.

Montgomery begins her attack on the over-closeness of a medicine of friends by explaining the desire of physicians to have a medicine of friends and why that explanation does not provide warrant for establishing a medicine of friends. A medicine of friends is, first of all, a rejection of the current state of the profession. Physicians feel detached from their patients and desire to spend more time with them in order to truly understand their patient's complaints. Patients also desire a relationship with their physician that allows them to spend more time together so that the patient can feel as if the physician has heard the patient's story. So, the idea of a medicine of friends is a counter-weight to the assembly line approach to medicine that both patients and physicians dislike.

The second reason physicians desire a medicine of friends is in response to the medical decontextualization of the patient. When a patient comes to the physician with a broken arm, the physician focuses on fixing the broken arm.[9] The primary task of the physician is to fix *the problem*, not the patient. The patient *qua* person is reduced to the patient *qua* broken arm. While a decontextualizing mindset is a necessary component of medical practice, it can produce impoverished relationships. The appeal of a medicine of friends is that it helps the physician to recontextualize the patient as a whole person.

Thinking of patients as friends seems to be a fairly good idea and has a great deal of intuitive appeal. So why does Montgomery argue against a friendship model of the PPR? The problem with the friendship model is that it eliminates what she terms 'therapeutic distance' between the patient and the physician.[10] In order for a physician to treat my ills, the physician must focus on my injury or diseased body-part without paying attention to the penumbra of issues that a friend would at that point. She must ignore my other problems at that moment – something a friend should not do.

Montgomery contends that these explanations of the desire for a medicine of friends do not, by themselves, warrant our having a medicine of friends. Decontextualization and short examinations need not result in bad medicine even if they decrease patient and physician satisfaction with the patient-physician encounter. Furthermore, because a medicine based on friendship can impose harms on the patient (by removing therapeutic distance), it should not be adopted. Instead, Montgomery suggests, we should replace the friendship model with a model based on neighborliness. A model of the PPR based on good neighbors contains much of what is good about the friendship model without the problems of losing therapeutic distance. Unlike a friend, a neighbor has an interest in what his neighbor is doing, but is not deeply interested in what the neighbor is doing (while a friend is interested in what the friend is doing regardless of whether the friend has a stake in what the friend is doing). Good neighbors share other features of the ideal PPR: an acceptance of whoever is the neighbor and a willingness to help when crises or problems arise.

X. FRIENDS, NEIGHBORS, AND CHOICES

The central problem with Montgomery's account is that it assumes a view of personal relationships that not all persons share. Montgomery argues that a medicine of friends eliminates or decreases the therapeutic distance necessary for *good* medicine. This implies that there is a universal value that can be attributed to medical outcomes that delineates good medicine from bad medicine, something that Montgomery has not established and that is in direct conflict with Engelhardt's views. On Montgomery's view, a physician who is my friend can act, because of that friendship, in a way that harms me by not treating my illness or injury in the *appropriate* way. For example, the physician who is my friend may give me antibiotics when I do not have a bacterial infection because my friend's goal is to

272   SCOTT DEVITO

soothe my fears (or perhaps my friend's goal is not to lose my friendship). Giving antibiotics to people who do not have bacterial infections is problematic because it increases the number of antibiotic resistant bacteria and in some cases can exacerbate a viral or fungal infection.

But such worries contain an implicit reference to a value system. Perhaps some people value having their (possibly unjustified) fears soothed more than they disvalue the consequences of having their fears soothed. To argue that I ought not have a doctor as a friend because having a friend treat me will increase the odds that certain physical outcomes will occur and those outcomes are bad, implies that a privileged set of values exist and that under that value system the ratio of harm to benefit, in a friendship model of the PPR, is higher than in a more distant relationship. But as Engelhardt has pointed out no such privileged set of values exists in a liberal secular society.

Surely, for some a medicine of friends is bad. If so, let the patient and physician contract to be neighbors instead of friends. Montgomery's work is valuable because she gives us another model of the PPR that those who value therapeutic distance and the consequences of that distance can adopt. But if Engelhardt is correct, then therapeutic distance can not be understood to necessarily provide more benefits than harms for all persons.

XI. CONCLUSION

The arguments of Brody, Montgomery, and Rosenbaum have been shown to come up short under the glare of Engelhardt's moral strangers. Each sought to argue for a particular kind of PPR, but their arguments fall short because they contain, hidden within them, (not argued for) preferred sets of values and value-orderings. Such things are not compatible with the work of Engelhardt and, from that point of view, the views that rely upon them must be rejected.

State University of New York at Buffalo
Buffalo, New York

NOTES

[1] A third possibility is that the patient and physician actually distrust each other. I do not discuss such a situation as its problems are fairly obvious.

[2] Trust also plays a role in the relationship between a medical care organization and the patient. Since my present purpose does not require discussion of trust in this type of relationship, I simply refer the reader to Rosenbaum's paper located within this volume.

[3] Here I assume that the physician does not treat the patient in a manner inconsistent with the patient's wishes or without the patient's consent.

[4] "Tailoring" a conversation need not be considered a sinister act. One can tailor a conversation out of the best of motives and intentions.

[5] Problems of miscommunication are a serious problem in research on alcoholism in minority communities. These communities often feel betrayed by the way they are depicted in this research, thus decreasing trust between the community and the medical-scientific community. For a good discussion of an example of this problem see Foulks, 1989 and responses in the Spring 1989 issue of *American Indian and Native Alaska Mental Health Research 2.*

[6] A lack of trust of the (white) medical establishment is a fairly common problem in minority communities. See, for example, Ballard, *et al.*, 1993 and Roberts, 1996.

[7] A problem for Brody's view, that still remains, is that in many cases of acute patient care a specialist is called in to care for the patient. The elderly woman with the broken hip will be treated by an orthopedic surgeon who (probably) has had no contact with the patient previous to the injury. The trusted primary care physician will not be treating the patient. Thus in acute care situations Brody's shared partnership approach, as well as Rosenbaum's continuity of care approach, fails. Patients and physicians in these situations are moral strangers and little can be done to make them knowledgeable about the other.

[8] Similar problems can arise when patients and physicians know each other for a long time, i.e., when they have a long-term continuous relationship.

[9] Here we assume that no other systemic problems exist (e.g., brittle bone disease) that are related to the broken arm. But even if the problem is part of a wider problem, decontextualization can continue to occur at a higher level. The patient becomes the injured or diseased system whether that system is localized (a broken arm) or wide-spread (brittle bone disease).

[10] Interestingly, Engelhardt believes that there will always be a certain distance between patient and physician (Englehardt, 1996, pp. 291-296).

REFERENCES

Ballard, E.L, Nash, F., Raiford, K., and Harrell, L.E.: 1993, 'Recruitment of black elderly for clinical research studies of dementia: The CERAD experience,' *The Gerontologist* 33, 561-565.

Brody, H.: 2000, 'Can relationships heal – At a reasonable cost?', this volume.

Engelhardt, H.T., Jr.,: 2000, 'Bioethics at the end of the millennium: Fashioning healthcare policy in the absence of a moral consensus,' in this volume.

Englehardt, H.T.: 1996, *The Foundations of Bioethics,* 2nd edition, Oxford University Press, New York.

Foulks, E.F.: 1989, 'Misalliances in the barrow alcohol study,' *American Indian and Native Alaska Mental Health Research* 2, 7-17.

Montgomery, K.: 2000, 'A medicine of neighbors,' in this volume.

Roberts, D.E.: 1996, 'Reconstructing the patient: Starting with women of color,' in Wolf (1996), pp. 116-143.

Rosenbaum, J.R.: 2000, 'Trust, institutions, and the physician-patient relationship: Implications for continuity of care,' in this volume.

Wolf, S.M.: 1996, *Feminism and Bioethics: Beyond Reproduction,* Oxford University Press, New York.

ADRIANNE McEVOY

GENERAL BIBLIOGRAPHY

The following is a compilation of over six hundred references to which the reader can turn for further information on the issues discussed in this volume. While some are brought together from the chapters themselves, there are many which have been added for the reader's convenience. By no means is this considered an exhaustive effort to reference the changes being witnessed in the medical arena. It is, however, a starting point for those who wish to become more familiar with some of the more important facets of economic, genetic, and fiduciary changes in medicine.

I. BIBLIOGRAPHY: THE DILEMMA OF FUNDING HEALTH CARE

Policy makers in the United States are finally admitting something that the American people have known for some time: health care in this country is in a state of financial shambles. In order for medical professionals to continue giving the type of care demanded by their patients, many sacrifices need to be made. Whereas three years ago, "rationing" was a taboo discussion, we are finally coming to terms with the fact that the demand for effective health care far exceeds the supply of therapeutic, diagnostic, and economic capabilities.

The following articles discuss the concerns of patients, their providers, and policy makers as they come to terms with the changing nature of medicine. While medicine will always be considered an art, more and more it is recognized as a business. The professional ethics which guide physicians and other health-care providers are now being influenced by outside forces both social and economical. The problem is not that care has to be managed. The problem is that the care is being mis-managed. While the following is not an exhaustive listing of all texts and articles discussing concerns over health care reform, it should provide the reader with ample opportunity to begin understanding the forces guiding economic changes in health care as well as medicine's reactionary measures.

Stephen Wear, James J. Bono, Gerald Logue and Adrianne McEvoy (eds.), Ethical Issues in Health Care on the Frontiers of the Twenty-First Century, 275–318.
© 2000 *Kluwer Academic Publishers. Printed in Great Britain.*

Cases and Statutes

Arrington v. Group Hospitalization & Medical Services, 806 F.Supp. 287, 290 (D.D.C. 1992).
Barnett v. Kaiser Foundation Health Plan, Inc., 32 F.3d 413 (9th Cir., 1994).
Bucci v. Blue Cross-Blue Shield of Connecticut, 764 F. Supp. 728, 731 (D.Conn. 1991).
Doe v. Group Hospitalization & Medical Services, 3 F.3d 80 (4th Cir. 1993).
Farley v. Benefit Trust Life Insurance. Co., 979 F.2d 653 (8th Cir. 1992).
Fuja v. Benefit Trust Life Insurance. Co., 18 F.3d 1405, 1412 (7th Cir. 1994).
Gee v. Utah State Retirement Bd, 842 P.2d 919, 920-21 (Utah App. 1992).
Goepel v. Mail Handlers Benefit Plan (No. 93-3711, 1993 WL 384498 (D. N.J. 9/24/93)).
Harris v. Blue Cross Blue Shield of Missouri, 995 F.2d 877 (8th Cir. 1993).
Harris v. Mutual of Omaha Companies, 992 F.2d 706, 713 (7th Cir. 1993).
Hinds v. Blue Cross and Blue Shield of Tennessee, No. 3:95-0508, M.D.TN, 12/28/95.
Katskee v. Blue Cross/Blue Shield, 515 NW2d 645, 647 (Neb. 1994).
Loyola University of Chicago v. Humana Insurance Co., 996 F.2d 895 (7th Cir 1993).
Madden v. Kaiser Foundation Hospitals, 552 P.2d 1178 (Cal. 1976).
McGee v. Equicor-Equitable HCA Corp., 953 F.2d 1192 (10th Cir. 1992).
McLeroy v. Blue Cross/Blue Shield of Oregon, Inc., 825 F.Supp. 1064 (N.D. Ga. 1993).
Miller by Miller v. Whitburn, 10 F3d 1315, 1320 (7th Cir. 1993) (citing *Rush v Parham*, 625 F.2d 1150, 1156 (5th Cir. 1980)).
Morris v. Metriyakool, 344 NW 2d 736, 756 (Mich. 1984) (opinion of Justice Ryan).
Muse v. Charter Hospital Winston-Salem Inc., 452 S.E.2d 589 (N.C.App. 1995).
Nesseim v. Mail Handlers Benefits Plan, 995 F.2d 804 (8th Cir., 1993).
Sarchett v. Blue Shield of California, 729 P. 2d 267 (Cal. 1987).
Thomas v. Gulf Health Plan, Inc., 688 F. Supp. 590 (S.D. Ala. 1988).
Wickline v. State of California, 192 Cal. App. 3d 1630 (1987).

Journal Articles and Texts

Aaron, H.J., Schwartz, W.B.: 1984, *The Painful Prescription: Rationing Hospital Care*, Brookings Institution, Washington D.C.
Aaron, H.J., Schwartz, W.B.: 1985, 'Hospital cost control: A bitter pill to swallow,' *Harvard Business Review* 64, 160-167.
Abelson, R.: 1999, 'Ideas & trends: For managed care, free-market shock,' *New York Times*, January 3, 4.
Abraham, E., Wunderink, R., Silverman, H., Perl, T.M., Naraway, S., Levy, H., Bone, R., Wenzel, R.P., Balk, R., Allred, R., Pennington, J.E., and Wherry, J.C.: 1995, 'Efficacy and safety of monoclonal antibody to human tumor necrosis factor in patients with sepsis syndrome,' *Journal of the American Medical Association* 273, 934-941.
Abraham, K.S.: 1981, 'Judge-made law and judge-made insurance: Honoring the reasonable expectations of the insured,' *Virginia Law Review* 67, 1151-1191.
Adler, N.E., Boyce, W.T., Chesney, M.A., Folkman, S., Syme, S.L.: 1993, 'Socioeconomic inequalities in health: No easy solution,' *Journal of the American Medical Association* 269, 3140-3145.
Allen, M.C., Donohue, P.K., Dusman, A.E.: 1993, 'The limit of viability--neonatal

outcome of infants born at 22 to 25 weeks' gestation,' *New England Journal of Medicine* 329, 1597-1601.

American College of Physicians: 1994, 'A national health work force policy,' *Annals of Internal Medicine* 121, 542-546.

American College of Physicians: 1994a, 'Magnetic resonance imaging of the brain and spine: A revised statement,' *Annals of Internal Medicine* 120, 872-875.

American College of Physicians: 1994b, 'The oversight of medical care: A proposal for reform,' *Annals of Internal Medicine* 120, 423-431.

American College of Physicians: 1995, 'Rural primary care,' *Annals of Internal Medicine* 122, 380-90.

American College of Physicians, Philadelphia: 1996, 'The impact of managed care on medical education and physician workforce.'

American College of Physicians, Philadelphia: 1996, 'Universal coverage: Renewing the call to action.'

American College of Physicians: 1996, 'Voluntary purchasing pools: A market model for improving access, quality, and cost in health care,' *Annals of Internal Medicine* 124, 845-853.

American Medical Association, Chicago: 1996, 'Code of Medical Ethics, current opinions with annotations.'

American Nursing Association, Kansas City, Missouri: 1980, 'Nursing: A social policy statement.'

Anders, G.: 1994a, 'More insurers pay for care that's in trials,' *Wall Street Journal* 2/15/94, B-1.

Anders, G.: 1994b, 'Limits on second-eye cataract surgery are lifted by major actuarial firm,' *Wall Street Journal* 12/15/94, B-6.

Anderson, G.F., Hall, M.A., Steinberg, E.P.: 1993, 'Medical technology assessment and practice guidelines: Their day in court,' *American Journal of Public Health* 83, 1635-1639.

Andrulis, D.P.: 1997, 'The urban health penalty: New dimensions and directions in inner-city health care,' *Inner City Health Care*, American College of Physicians, Philadelphia, no. 1.

Angell, M, Kassirer, J.P.: 1996, 'Quality and the medical marketplace -- following elephants,' *New England Journal of Medicine* 335, 883-885.

Anonymous: 1996, 'Insurance reform in a voluntary system: Implications for the sick, the well, and universal health care. American College of Physicians,' *Annals of Internal Medicine* 125(3), 242-249.

Anonymous: 1997, 'Inner-city health care. American College of Physicians,' *Annals of Internal Medicine* 126(6), 485-490.

Antman, K., Schnipper, L.E., Frei, E. III: 1988, 'The crisis in clinical cancer research: Third-party insurance and investigational therapy,' *New England Journal of Medicine* 319, 46-48.

Avorn, J., Chen, M., Hartley, R.: 1982, 'Scientific versus commercial sources of influence on the prescribing behavior of physicians,' *American Journal of Medicine* 73, 4-8.

Ayanian, J.Z., Kohler, B.A., Abe, T., Epstein, A.M.: 1993, 'The relation between health insurance coverage and clinical outcomes among women with breast cancer,' *New England Journal of Medicine* 329, 326-331.

Banta, D. and Gelijns, A.: (1994) 'The future and health care technology: Implications of a system for early identification,' *World Health Statistic Quarterly* 47, 140-148.

Baker, R. (ed.): 1995, *The Codification of Morality: Historical and Philosophical Studies of the Formalization of Western Medical Morality in the Eighteenth and Nineteenth Centuries.* Volume Two: A*nglo-American Medical Ethics and Medical Jurisprudence in the Nineteenth Century*, Kluwer Academic Publishers, Dordrecht, The Netherlands.

Balas, E.A., Kretschmer, R.A.C., Gnann, W., *et al.*: 1998, 'Interpreting cost analyses of clinical interventions,' *Journal of the American Medical Association* 279, 54-57.

Bartlett, J.G.: 1996, 'Protease inhibitors for HIV infection,' *Annals of Internal Medicine* 124, 1086-1088.

Beauchamp, T.L., Childress, J.F.: 1994, *Principles of Biomedical Ethics*, 4th edition, Oxford University Press, New York.

Beauchamp, T.L., McCullough, L.B.: 1984, *Medical Ethics: The Moral Responsibilities of Physicians*, Prentice-Hall, Englewood Cliffs, N.J.

Bennett, C.L., Smith, T.J., George, S.L., Hillner, B.E., Fleishman, S., Niell, H.B.: 1995, 'Free-riding and the prisoner's dilemma: Problems in funding economic analyses of phase III cancer clinic trials,' *Journal of Clinical Oncology* 13, 2457-2463.

Bercu, B.B.: 1996, 'The growing conundrum: Growth hormone treatment of non-growth hormone deficient child,' *Journal of the American Medical Association* 276, 567-568.

Berenson, R.A.: 1997, 'Beyond competition,' *Health Affairs* 16(2), 171-180.

Bergman, R.: 1996, 'Rethinking nursing values,' *Journal of Advanced Nursing* 23, 4.

Bergthold, L.A.: 1995, 'Medical necessity: Do we need it?' *Health Affairs* 14(4), 180-190.

Berk, M.L., Schur, C.L., Cantor, J.C.: 1995, 'Ability to obtain health care: Recent estimates from the Robert Wood Johnson Foundation National Access to Care Survey,' *Health Affairs* 14(3), 139-46.

Berkowitz, R.L.: 1993, 'Should every pregnant woman undergo ultrasonography?' *The New England Journal of Medicine* 329, 874-875.

Berwick, D.M.: 1996, 'Quality comes home,' *Annals of Internal Medicine* 125(10), 839-843.

Bevis, E.O.: 1981, 'Caring: A life force,' in N. Leininger (ed.), *Caring: An Essential Human Need*, Charles B. Slack, Thorofare, New Jersey, pp. 49-59.

Billings, J., Anderson, G.M., Newman, L.S.: 1996, 'Recent findings on preventable hospitalizations,' *Health Affairs* (Fall) 239-249.

Black, H.C.: 1979, *Black's Law Dictionary*, West Publishing Company, Minneapolis, Minnesota.

Blendon, R.J., Brodie, M., Benson, J.: 1995a, 'What happened to Americans' support for the Clinton health plan?' *Health Affairs* 14(2), 7-23.

Blendon, R.J., Brodie, M., Benson, J.: 1995b, 'What should be done now that national health system reform is dead?' *Journal of the American Medical Association* 273, 243-244.

Blendon, R.J., Brodie, M., Benson, J.M., *et al.*: 1998, 'Understanding the managed care backlash,' *Health Affairs* 17(4), 80-94.

Blumberg L.J., Nichols L.M.: 1996, 'First, do no harm: Developing health insurance market reform packages,' *Health Affairs* 15(3), 35-53.

Blumberg, L.J., Liska, D.W.: 1996, 'The uninsured in the United States: A status report,' Urban Institute.

Blumenthal, D.: 1995, 'Health care reform - past and future,' *New England Journal of Medicine* 332, 465-468.

Blumenthal, D., Epstein, A.M.: 1996, 'The role of physicians in the future of quality management,' *New England Journal of Medicine* 335, 1328-1331.

Blumenthal, D.: 1999, 'Health care reform at the close of the 20th century,' *New England Journal of Medicine* 340(24), 1916-1920.

Bodenheimer, T.: 1996, 'The HMO backlash: Righteous or reactionary?' *New England Journal of Medicine* 335, 1601-1604.

Boyles, J.H.: 'U.S. health-care costs,' *Lancet* 347, 694.

Bradshaw, G., Bradshaw, P.: 1995 'The equity debate within the British National Health Services,' *Journal of Nursing Administration* 3, 161-168.

Bridges W.: 1994, 'The end of the job,' *Fortune* 130(6), 46-51.

Brody, B.: 1995, *Ethical Issues in Drug Testing, Approval, and Pricing*, Oxford University Press, New York.

Brook, R.H.: 1997, 'Managed care is not the problem, quality is,' *Journal of the American Medical Association* 278, 1612-1614.

Brook, R.H., McGlynn, E.A., Cleary, P.D.: 1996, 'Measuring quality of care,' *New England Journal of Medicine* 335, 966-970.

Brooten, D. and Naylor, D.: 1995 'Nurses' effect on changing patient outcomes,' *Image* 27, 95-99

Browning, E.S.: 1995, 'Change in health care shakes up the business of drug development,' *Wall Street Journal* 3/28/95, A-1, A-6.

Buchan, W.: 1769, *Domestic Medicine*, Balfour, Auld & Smellie, Edinburgh.

Burner, S.T., Waldo, D.R.: 1995, 'National health expenditure projections, 1994-2005,' *Health Care Finance Review* 16(4), 221-242.

Burnum, J.F.: 1987, 'Medical practice a la mode,' *The New England Journal of Medicine* 317, 1220-1222.

Butler, S.M., Moffit, R.E.: 1995, 'The FEHBP as a model for a new medicare program,' *Health Affairs* 14(4), 47-61

Buto, K.A.: 1994, 'How can Medicare keep pace with cutting-edge technology?' *Health Affairs* 13(3), 137-140.

Calabresi, G., Bobbitt, P.: 1978, *Tragic Choices*, Norton & Co, New York.

Callahan, D., *et al.*: 1996, *The Goals of Medicine: Setting New Priorities*, unpublished manuscript, Hastings Center.

Cantor, J.C., Miles, E.L., Baker, L.C., Barker, D.C.: 1996, 'Physician service to the underserved: Implications for affirmative action in medical education,' *Inquiry* 33, 167-80

Catton, H.: 1998, 'A New Zealand approach,' *Nursing Standard,* 12(32), 22-24.

Center for Studying Health System Change, Washington, D.C.: 1997, 'Charting change: A longitudinal look at the American health system,' Annual Report.

Chalmers, T.C.: 1988, 'Third-party payers and investigational therapy,' *New England Journal of Medicine* 319, 1228.

Chassin, M.R., Brook, R.H., Park, R.E., *et al.*: 1986, 'Variations in the use of medical and surgical services by the Medicare population,' *The New England Journal of Medicine* 314, 285-290.

Chassin, M.R., *et al.*: 1987 'Does inappropriate use explain geographic variations in the use of health care services?' *Journal of the American Medical Association* 258, 2533-2537.

Chervenak, F.A., McCullough, L.B.: 1995, 'The threat to autonomy of the new managed practice of medicine,' *Journal of Clinical Ethics* 6, 320-323.

Cho, M.K., Bero, L.A.: 'The quality of drug studies published in symposium proceedings,' *Annals of Internal Medicine* 124, 485-489.

Clancy, C.M.; Brody, H.: 1995, 'Managed care: Jekyll or Hyde?' *Journal of the American Medical Association,* 273, 338-339.

Clay, J.: 1997, 'The debate over US government's role in health care financing: "Deja vu, all over again",' *International Journal of Public Administration* 20, 1183-1202.

Cleary, P.D., Greenfield, S., Mulley, A.G., *et al.*: 1991, 'Variations in length of stay and outcomes for six medical and surgical conditions in Massachusetts and California,' *Journal of the American Medical Association* 266, 73-79.

Cohen, J. and Stewart, I.: 1994, *The Collapse of Chaos*, Penguin, New York.

Coney, S.: 1996, 'Relentless unraveling of New Zealand's health-care system,' *Lancet* 347, 1825.

Cova, J.L.: 1992, 'A swift response to a "modest" proposal,' *Journal of the National Cancer Institute* 84, 744-745.

Cuttler, L., Silvers, B.J., Singh, J., Marrero, U., Finkelstein, B., Tannin, G., and Neuhauser, D.: 1966, 'Short stature and growth hormone therapy: A national study of physician recommendation patterns,' *Journal of the American Medical Association* 276, 531-537

Davis, K., Schoen, C.: 1998, 'Incremental health insurance coverage: Building on the current system,' in S.H. Altman, U.E. Reinhardt, A.E. Shields, (eds.), *The Future U.S. Health Care System: Who Will Care For the Poor and Uninsured?* Health Administration Press, Chicago, pp. 247-263.

DeBakey, M.: 1998, 'RX for the health care system,' *Wall Street Journal* October 8, A18.

Deeks, S.G., Smith, M., Holodniy, M., and Kahn, J.O.: 1997, 'AHIV-1 protease inhibitors: A review for clinicians,' *Journal of the American Medical Association* 277, 145-53.

Deming, W.E.: 1986, *Out of Crisis*, Massachusetts Institute of Technology, Center for Advanced Engineering Study, Cambridge, Massachusetts.

Department of Commerce: 1995, 'Health insurance coverage -- who had a lapse between 1991 and 1993? Statistical Brief 95-21,' Bureau of the Census, Washington D.C., August.

Detsky, A.S.: 1995, 'Regional variation in medical care,' *New England Journal of Medicine* 333, 589-590.

Deyo, R.A., Psaty, B.M., Simon, G., Wagner, E.H., and Omenn, G.S.: 1997, 'The messenger under attack--intimidation of researchers by special interest groups,' *New England Journal of Medicine* 336, 1176-1180.

Deyo, R.A.: 1994, 'Magnetic resonance imaging of the lumbar spine: Terrific test or tar baby?' *New England Journal of Medicine* 331, 115-116.

Dickens, P.: 1994, *Quality and Excellence in Human Services*, John Wiley, New York.

Donelan, K., Blendon, R.J., Hill, C.A., Hoffman, C., Rowland, D., *et al.*: 1996, 'Whatever happened to the health insurance crisis in the United States? Voices from a national survey,' *Journal of the American Medical Association* 276, 1346-1350.

Downs, F.: 1997, 'Clinical relevance revisited,' Nursing Research 46(3).

Durenberger, D.F., Foote, S.B.: 1994, 'Technology and health reform: A legislative perspective,' *Health Affairs* 13(3), 197-205.

Durham, M.L. 1998, 'Partnerships for research among managed care organizations,' *Health Affairs* 17(1), 111-122.

ECRI: 1995a, 'High-dose chemotherapy with autologous bone marrow transplantation and/or blood cell transplantation for the treatment of metastatic breast cancer,' *Healthy Technology Assessment Information Service: Executive Briefings*, February.

ECRI. 1995b 'Xenograft transplantation: Science, ethics, and public policy,' *Health Technology Assessment News* (September-October), 1-12.

ECRI: 1995c, 'Pallidotomy and thalamotomy for Parkinson's disease,' *Health Technology Assessment News* (June), 1-10.

Eddy, D.M.: 1992, 'Applying cost-effectiveness analysis: The inside story,' *Journal of the American Medical Association* 268, 2575-2582

Eddy, D.M.: 1993, 'Three battles to watch in the 1990s,' *Journal of the American Medical Association* 270, 520-526.

Eddy, D.M.: 1996, 'Benefit language: Criteria that will improve quality while reducing costs,' *Journal of the American Medical Association* 275, 650-657.

Eddy, D.M.: 1997, 'Investigational treatments: How strict should we be?' *Journal of the American Medical Association* 278, 179-185.

Eddy D.M.: 1998, 'Performance measurement: Problems and solutions,' *Health Affairs* 17(4), 7-25.

Editorial: 1994, 'Consumer-first health care,' *Wall Street Journal*, July 21, A-12.

Editorial: 1995, 'Tomorrow's doctoring: Patient, heal thyself,' *The Economist*, February 4, 19-21.

Ellis, J.H., Cohan, R.H., Sonnad, S.S., Cohan, N.S.: 1996, 'Selective use of radiographic low-osmolality contrast media in the 1990s,' *Radiology* 200, 297-311.

Employee Benefits Research Institute: 1995, 'Employee Benefits Research Institute Issue Brief 158: Sources of health insurance and characteristics of the uninsured: Analysis of the March 1994 Current Population Survey,' 1-47.

Employee Benefits Research Institute: 1996, 'Notes: A monthly newsletter,' *EBRI Education and Research Fund*, 17(1), 1-7.

Engelhardt, H.T., Jr: 1995, 'Christian bioethics as non-ecumenical,' *Christian Bioethics* 1 (September), 182-199.

Epstein, A.: 1995, 'Performance reports on quality--prototypes, problems, and prospects,' *New England Journal of Medicine* 333, 57-61.

Epstein, R.S., Sherwood, L.M.: 1996, 'From outcomes research to disease management: A guide for the perplexed,' *Annals of Internal Medicine* 124, 832-837.

Escarce, J.J., Chen, W., Schwartz, S.: 1995, 'Falling cholecystectomy thresholds since the introduction of laparoscopic cholecystectomy,' *Journal of the American Medical Association* 273, 1581-1585.

Etheredge, L., Jones, S.B., Lewin, L.: 1996, 'What is driving health system change?' *Health Affairs* 15(4), 93-104.

Evans, R.G.: 1995, 'Manufacturing consensus, marketing truth: Guidelines for economic evaluation,' *Annals of Internal Medicine* 123, 59-60.

Ewigman, B.G., Crane, J.P., Frigoletto, F.D., LeFevre, M.L., Bain, R.P., McNellis, D., Radius Study Group: 1993, 'Effect of prenatal ultrasound screening on perinatal outcome,' *The New England Journal of Medicine* 329, 821-827.

Faden, R.R., Beauchamp, T.L.: 1986, *A History and Theory of Informed Consent*, Oxford University Press, New York.

Farley, D.O., McGlynn, E.A., Klein, D., RAND Corporation: 1998, 'Assessing quality in managed care: health plan reporting of HEDIS performance measures,' Commonwealth

282   ADRIANNE McEVOY

Fund, New York, September.
Farrell, M.G.: 1997, 'ERISA preemption and regulation of managed health care: The case for managed federalism,' *American Journal of Law & Medicine* 23, 251-289.
Feinstein, A.R., Horwitz, R.I.: 1997, 'Problems in the "evidence" of "evidence-based medicine",' *American Journal of Medicine* 103, 529-535.
Ferguson, J.H., Dubinsky, M., Kirsch, P.J.: 1993, 'Court-ordered reimbursement for unproven medical technology: Circumventing technology assessment,' *Journal of the American Medical Association* 269, 2116-2121.
Fins, J.J.: 1994, 'Prescription for health care reform: A page from the formulary,' *P & T* (August), 750, 753-759.
Fioritti, A., Lo, B., Russo, L., Melega, V.: 1997, 'Reform said or done? The case of Emilia-Romagna within the Italian psychiatric context,' *American Journal of Psychiatry* 154(1), 94-98.
Fisher, E.S., Welch, H.G., Wennberg, J.E.: 1992, 'Prioritizing Oregon's hospital resources: An example based on variations in discretionary medical utilization,' *Journal of the American Medical Association* 267, 1925-1931.
Forman, H.P., McClennan, B.L.: 1994, 'Health services research in radiology: Opportunities and imperatives,' *AJR* 163, 257-261.
Fox, N.: 1993, *Postmodernism, Sociology and Health*, University of Toronto Press, Toronto.
Franks, P., Clancy, C.M., Gold, M.R.: 1993, 'Health insurance and mortality: Evidence from a national cohort,' *Journal of the American Medical Association* 270, 737-741.
Furberg, C.D.: 1995, 'Should dihydropyridines be used as first-line drugs in the treatment of hypertension? The con side,' *Archives of Internal Medicine* 155, 2157-2161.
Gabel, J.R.: 1998, 'On drinking with your competitors after five: Research collaboration in the real world,' *Health Affairs* 17(1), 123-127.
Garber, A.M.: 1994, 'Can technology assessment control health spending?,' *Health Affairs* 13(3), 115-126.
Garber, A.M.: 1992, 'No price too high?' *New England Journal of Medicine* 327, 1676-1678.
Gifford, F.: 1996, 'Outcomes research and practice guidelines: Upstream issues for downstream users,' *Hastings Center Report* 26(2), 38-44.
Ginsburg, P.B., Gabel, J.R.: 1998, 'Tracking health care costs: What's new in 1998,' *Health Affairs* 17(5), 141-146.
Glassman, P.A., Jacobson, P.D., Asch, S.: 1997, 'Medical necessity and defined coverage benefits in the Oregon health plan,' *American Journal of Public Health* 87, 1053-1058.
Glassman, P.A., Model, K.E., Kahan, J.P., Jacobson, P.D., Peabody, J.W.: 1997, 'The role of medical necessity and cost-effectiveness in making medical decisions,' *Annals of Internal Medicine* 126, 152-156.
Glazier, A.K.: 1997, 'Genetic predispositions, prophylactic treatments and private health insurance: Nothing is better than a good pair of genes,' *American Journal of Law & Medicine* 23, 45-68.
Goldberg, R.M.: 1998, 'Why HMOs now love regulation,' *Wall Street Journal* July 17, A14.
Goldsmith, J.; 1994, 'The impact of new technology on health costs,' *Health Affairs* 13(3), 80-81.
Goldsmith, J.C.: 1992, 'The reshaping of health care,' *Healthcare Forum Journal* 34(4),

34-41.

Goldsmith, J.C.: 1993, 'Technology and the end to entitlement,' *Healthcare Forum Journal* 36(5), 16-23.

Goodwin, J.S., Goodwin, J.M.: 1984, 'The tomato effect: Rejection of highly efficacious therapies,' *Journal of the American Medical Association* 251, 2387-2390.

Goold, S.D., Brody, H.: 1995, 'Rationing decisions in managed care settings: An ethical analysis,' in *Health Care Crisis? The Search for Answers,* University Publishing Group.

Gornick, M.E., Eggers, P.W., Reilly, T.W., Mentnech, R.M., Fitterman, L.K., Kucken, L.E., *et al.*: 1996, 'Effects of race and income on mortality and use of services among Medicare beneficiaries,' New England Journal of Medicine 335, 797-798.

Gosfield, A.G.: 1994, 'Clinical practice guidelines and the law: Applications and implications,' in C.B. Callaghan (ed.), *Health Law Handbook*, Thomson Legal Publishing, Inc., Deerfield, Illinois (reprinted in, and with pagination of, NHLA's Legal Issues Related to Clinical Practice Guidelines), pp. 59-95.

Gottsegen, S.W.: 1981, 'A new approach for the interpretation of insurance contracts-- *Great American Insurance Co. v. Tate Construction Co.*,' *Wake Forest Law Review* 17, 140-152.

Grandinetti, D.A.: 1997, 'Add fun and profits to your practice: Do research,' *Medical Economics* 74(25), 67-79.

Gray, C.: 1996, 'Visions of our health care future: Is a parallel private system the answer?' *Canadian Medical Association Journal* 154(7), 1084-1087.

Gray, B.H.: 1992, 'The legislative battle over health services research,' *Health Affairs* 11(4), 38-66.

Gray, D.T., Fyler, D.C., Walker, A.M., Weinstein, M.C., Chalmers, T.C., *et al.*: 1993, 'Clinical outcomes and costs of transcatheter as compared with surgical closure of patent ductus arteriosus,' *New England Journal of Medicine* 329, 1517-1523.

Greco, R.S.: 1998, 'The oasis,' *Lancet*, 351, 1052-1053.

Greenberg, D.S.: 1999, 'Medicare reform hangs in the balance,' *Lancet* 353, 1075.

Greenberg, M.: 1991, 'American cities: Good and bad news about public health,' *Bull N Y Academic Medicine* 67, 17-21.

Greenfield, S., Nelson, E.C., Subkoff, M., *et al.*: 1992, 'Variations in resource utilization among medical specialties and systems of care: Results from the medical outcomes study,' *Journal of the American Medical Association* 267, 1624-1630.

Gregory, J.: 1743, 'Medical notes,' Aberdeen University Library, MS 2206/45.

Gregory, J.: 1765, *A Comparative View of the State and Faculties of Man Compared with those of the Animal World*, J. Dodsley, London.

Gregory, J.: 1770, *Observations on the Duties and Offices of a Physician, and on the Method of Prosecuting Enquiries in Philosophy*, London, W. Strahan and T. Cadell. Reprinted in L.B. McCullough (ed), *John Gregory's Observations and Lectures on the Duties, Offices, and Qualifications of a Physician*, Kluwer Academic Publishers, 1998, pp. 93-159.

Gregory, J.: *Lectures on the Duties and Qualifications of a Physician*, W. Strahan and T. Cadell, London. Reprinted in L.B. McCullough (ed.), *John Gregory's Observations and Lectures on the Duties, Offices, and Qualifications of a Physician*, Kluwer Academic Publishers, 1998, pp. 161-245.

Grimes, D.A.: 1993, 'Technology follies: The uncritical acceptance of medical innovation,' *Journal of the American Medical Association* 269, 3030-3033.

Groves, T.: 1997, 'Primary care: Opportunities and threats. What the changes mean,' *British Medical Journal* 314, 436-438.

Guadagnoli, E., Hauptman, P.J., Avanian, J.Z., Pashos, C.L., McNeil, B.J., Cleary, P.D.: 1995, 'Variation in the use of cardiac procedures after acute myocardial infarction,' *New England Journal of Medicine* 333, 573-578.

Gulick, R.M., Mellors, J.W., Havlir, D., *et al.*: 1997, 'Treatment with indinavir, zidovudine, and lamivudine in adults with human immunodeficiency virus infection and prior antiretroviral therapy,' *New England Journal of Medicine* 337, 734-739.

Hadley, J., Steinberg, E.P., Feder, J.: 1991, 'Comparison of uninsured and privately insured hospital patients: Condition on admission, resource use, and outcome,' *Journal of the American Medical Association* 265, 374-379.

Hadorn, D.C.: 1992, 'Chapter 2: Necessary-care guidelines,' in Hadorn, D.C. (ed.), *Basic Benefits and Clinical Guideline*, Westview Press, Boulder.

Hall, M.A., Anderson, G.F.: 1992, 'Health insurers' assessment of medical necessity,' *University of Pennsylvania Law Review* 140, 1637-1712.

Hall, M.A.: 1993, 'Informed consent to rationing decisions,' *The Milbank Quarterly* 71, 645-668.

Hall, M.A.: 1994, 'Rationing health care at the bedside,' *New York University Law Review* 69 (4-5), 693-780.

Hammer, S.M., Squires, K.E., Hughes, M.D., *et al.*: 1997, 'A controlled trial of two nucleoside analogues plus indinavir in persons with human immunodeficiency virus infection and CD4 cell counts of 200 per cubic millimeter or less,' *New England Journal of Medicine* 337, 725-733.

Hanania, E.G., Kavanagh, J., Hortobagyi, G., Giles, R.E., Champlin, R., Deisseroth, A.B.: 1995, 'Recent advances in the application of gene therapy to human disease,' *The American Journal of Medicine* 99, 537-5521.

Handy, C.: 1990, *The Age of Unreason*, Boston, The Harvard Business School Press.

Handy, C.: 1994, *The Age of Paradox*, Boston, The Harvard Business School Press.

Harrison M.: 1995, 'Implementing reforms in the health care system: Physicians and hospital reforms in four countries,' *The Gertner Institute for Health Policy* (in Hebrew), Tel Aviv.

Havighurst, C.C.: 1995, *Health Care Choices: Private Contracts as Instruments of Health Reform*, The AEI Press, Washington, D.C..

Hays, B., Noris, J., Martin, K., Androwich, I.: 1994, 'Informatics issues for nursing's future,' *Advances in Nursing Science* 16(4), 71-81.

Hayward, R.S.A., Wilson, M.C., Tunis, S.R., Bass, E.B., Guyatt, G., *et al.*: 1995, 'Users' guides to the medical literature: VIII. How to use clinical practice guidelines; A. Are the recommendations valid?' *Journal of the American Medical Association* 274, 570-574.

Hazzard, W.R.: 1997, '2001: An American health care odyssey,' *Annals of Internal Medicine* 126(8), 658-659.

'Healthcare for the poor and uninsured: An uncertain future,' in S.H. Altman, U.E. Reinhardt, and A.E. Shields (eds.), *The Future U.S. Health Care System: Who Will Care For The Poor And Uninsured?* Health Administration Press, Chicago, 1998 pp. 1-22.

Health Rights Hotline, Annual Report: 1998, 'July 1, 1997-June 30, 1998: Consumers in managed care: Problems, solutions and lessons learned from the Health Rights Hotline,'

Sacramento, California.

Health Technology Assessment Information Service: 1995, Executive Briefings, February.

Healy, P.: 1996, 'E-mail: The nurse,' Nursing Standard 10(27), 14.

Hellinger, F.J.: 1998, 'The effect of managed care on quality: A review of recent evidence,' Archives of Internal Medicine 158, 833-841.

Hibbard, J.H., Jewett, J.J., Legnini, M.W., and Tusler, M.: 1997, 'Choosing a health plan: Do large empoyers use the data?' Health Affairs 16(6), 172-180.

Hillman, A.L., Eisenberg, J.M., Puly, M.V., Bloom, B.S., Glick, H., Kinosian, B., and Schwartz, J.S.: 1991, 'Avoiding bias in the conduct and reporting of cost-effectiveness research sponsored by pharmaceutical companies,' New England Journal of Medicine 324, 1362-1365.

Hoffmann, F.: 1749, Medicus Politicus, sive Regulae Prudentiae secundum quas Medicus Juvenis Studia sua et Vitae Rationem Dirigere Debet, in Frederici Hoffmanni, Operum Omnium Physico-Medicorum Supplementum in Duas Partes Distributum, apud Fratres de Tournes, Genevae.

Holahan, J., Winterbottom, C., Rajan, S.: 1995, 'The changing composition of health insurance coverage in the United States,' Urban Institute, Washington, D.C.

Holder, A.R.: 1994, 'Medical insurance payments and patients involved in research,' IRB 16 (1-2), 19-22.

Holoweiko, M.: 1995, 'When an insurer calls your treatment experimental,' Medical Economics 72(17), 171-182.

Horn, S.D., Sharkey, P.D., Tracy, D.M., Horn, C.E., James, B., Goodwin, F.: 1996, 'Intended and unintended consequences of HMO cost-containment strategies: Results from the managed care outcomes project,' The American Journal of Managed Care 2, 253-264.

Hornberger, J., Wrone, E.: 1997, 'When to base clinical policies on observational versus randomized trial data,' Annals of Internal Medicine 127, 697-703.

Hughes, R.G., Davis, T.L., Reynolds, R.C.: 1995, 'Assuring children's health as the basis for health care reform,' Health Affairs 14(2), 158-167.

Hume, D.: 1978, A Treatise of Human Nature, P.H. Nidditch (ed.), 2nd. ed., Oxford, Clarendon Press. (Based on 1739-1740 edition.)

Iezzoni, L.I.: 1997, 'Assessing quality using administrative data,' Annals of Internal Medicine 127, 666-674.

Iglehart, J.K.: 1999, 'The American health care system - expenditures,' New England Journal of Medicine 340, 70-76.

Iglehart, J.K.: 1996, 'The National Committee for Quality Assurance,' New England Journal of Medicine 335, 995-999.

Jacobson, P.D., Rosenquist, C.J.: 1988, 'The introduction of low-osmolar contrast agents in radiology,' Journal of the American Medical Association 260, 1586-1592.

Jensen, M.C., Brant-Zawadzki, M.N., Obuchowski, N., Modic, M.T., Malkasian, D., Ross, J.S.: 1994, 'Magnetic resonance imaging of the lumbar spine in people without back pain,' New England Journal of Medicine 331, 69-73.

Jones, L.: 1995, 'Does prevention save money?' American Medical News 17, 20-22.

Jonsen, A.: 1986, 'Bentham in a box: Technology assessment and health care allocation,' Law, Medicine, and Health Care 14, 172-174

Jonsen, A.R., Siegler, M., Winslade, W.J.: 1992, Clinical Ethics, 3rd ed., McGraw-Hill, New York.

Kahn, C.N.: 1998, 'The AHCPR after the battles,' *Health Affairs* 17(1), 109-110.

Kalb, P.E.: 1990, 'Controlling health care costs by controlling technology: A private contractual approach,' *Yale Law Journal* 99, 1109-1126.

Kassirer, J.P.: 1995, 'Managed care and the morality of the marketplace,' *New England Journal of Medicine* 333, 50-52.

Kassirer, J.P.: 1989, 'Our stubborn quest for diagnostic certainty: A cause of excessive testing,' *New England Journal of Medicine* 320, 1489-1491.

Kassirer, J.P., Angell, M.: 1994, 'The Journal's policy on cost-effectiveness analyses,' *New England Journal of Medicine* 331, 669-670.

Keister, L.W.: 1995, 'With health costs finally moderating, employers' focus turns to quality,' *Managed Care* 10, 20-24.

Kent, C.: 1996, 'Can clinical research thrive (or survive)?' *American Medical News* 3, 44.

Kent, D.L., Haynor, D.R., Longstreth, W.T., Jr., Larson, E.B.: 1994, 'The clinical efficacy of magnetic resonance imaging in neuroimaging,' *Annals of Internal Medicine* 120, 856-871.

Kessler, D.K., Kessler, K.M., Myerburg, R.J.: 1995, 'Ambulatory electrocardiography: A cost per management decision analysis,' *Archives Internal Medicine* 155, 165-69.

Kessler, D.A., Rose, J.L., Temple, R.J., Schapiro, R., Griffin, J.P.: 1994, 'Therapeutic-class wars – drug promotion in a competitive marketplace,' *New England Journal of Medicine* 331, 1350-1353.

Kilborn, P.T.: 1998, 'Voters' anger at HMOs plays as hot political issue,' *New York Times* May 17, 1.

King, R.T., Jr.: 1996, 'How a drug firm paid for university study, then undermined it,' *Wall Street Journal*, April 25, A-1, A-6.

King, S.B., Lembo, N.J., Weintraub, W.S., *et al.*: 1994, 'A randomized trial comparing coronary angioplasty with coronary bypass surgery,' *New England Journal of Medicine* 331, 1044-1050

Kitson, A.: 1996, 'Does nursing have a future?' *British Medical Journal* 313, 1647-1651.

Klakovitch, M.: 1994, 'Connective leadership for the 21st century: A historical perspective and future directions,' *Advances in Nursing Science* 16(4), 42-54.

Kolata, G.: 1995, 'Women rejecting trials for testing a cancer therapy,' *New York Times*, Feb 15, C1.

Kong, S.X., Wertheimer, A.I.: 1998, 'Outcomes research: Collaboration among academic researchers, managed care organizations, and pharmaceutical manufacturers,' *American Journal of Managed Care* 4, 28-34.

Kuttner, R.: 1998, 'The risk-adjustment debate,' *New England Journal of Medicine* 339, 1952-1956.

Kuttner, R.: 1999, 'The American health care system - health insurance coverage,' *New England Journal of Medicine* 340, 163-168.

Lawrence, C.J.: 1975, 'William Buchan: Medicine laid open,' *Medical History* 19, 20-35.

Leape, L.L., Park, R.E., Solomon, D.H., Chassin, M.R., Kosecoff, J., Brook, R.H.: 1989, 'Relation between surgeons' practice volumes and geographic variation in the rate of carotid endarterectomy,' *The New England Journal of Medicine* 321, 653-657.

Leape, L.L., Park, R.E., Solomon, D.H., Chassin, M.R., Kosecoff, J., Brook, R.H.: 1990, 'Does inappropriate use explain small-area variations in the use of health care services?' *Journal of the American Medical Association* 263, 669-672.

Leifer, D.: 1997, 'Health's missing millions,' *Nursing Standard* 11(41), 12.

Leininger, M.: 1981, 'The phenomenon of caring: Important research, questions, and theoretical considerations,' in N. Leininger (ed.), *Caring: An Essential Human Need*, Charles B. Slack, Thorofare, New Jersey.

Levinsky, N.G.: 1984, 'The doctor's master,' *New England Journal of Medicine* 311, 1573-1575.

Levit, K.R., Lazenby, H.C., Braden, B.R.: 1998, 'National health spending trends in 1996,' *Health Affairs* 17(1), 35-51.

Lipson, D.J., De Sa, J.M.: 1996, 'Impact of purchasing strategies on local health care systems,' *Health Affairs* 15(2), 62-76.

Lochhead, M.: 1948, *The Scottish Household in the Eighteenth Century: A Century of Scottish Domestic Life*, The Moray Press, Edinburgh.

Long, S.H., Marquis, M.S.: 1993, 'Gaps in employer coverage: Lack of supply or lack of demand?' *Health Affairs* 12, 282-293.

Long, S.H., Marquis, M.S.: 1994, 'The uninsured "access gap," and the cost of universal coverage,' Health Affairs 13(2), 211-20.

Loveridge, C., Cummings, S.: 1996, *Nursing Management in the New Paradigm*, Aspen, Gaithensburg, Maryland.

Luft, H.S.: 1995, 'Modifying managed competition to address cost and quality,' *Health Affairs* 15(1), 23-38.

Lurie, N., Christianson, J., Finch, M., Moscovice, I.: 1994, 'The effects of capitation on health and functional status of the Medicaid elderly,' *Annals of Internal Medicine* 120, 506-511

Lyotard, J. F.: 1984, *The Postmodern Condition: A Report on Knowledge*, G. Bennington and B. Massumi (trans.), Manchester University Press, Manchester.

Manthey, M.: (1996) 'Impact of health care reform,' *Journal of Nursing Administration* 26(6), 10-12.

Mark, B.: 1995, 'Nursing research policy: The black box of patient outcomes research,' *Image* 27, 42.

Marquis, M.S., Long, S.H.: 1997, 'Federalism and health system reform. Prospects for state action,' *Journal of the American Medical Association* 278(6), 514-517.

Marwick, C.: 1998, '"Bill of rights" for patients sent to Clinton,' *Journal of the American Medical Association* 279(1), 7-8.

Marx, K , Engels, F.: 1960, *The German Ideology*, International Publishers, New York.

Matas, K., Brown, C., Holman, E.: 1996 'Measuring outcomes in nursing centers: Otitis media as a sample case,' *Nurse Practitioner* 21(6), 116-125.

McCord, C., Freeman, H.P.: 1990, 'Excess mortality in Harlem,' *New England Journal of Medicine* 322, 173-177.

McCullough, L.B., Chervenak, F.A.: 1994, *Ethics in Obstetrics and Gynecology*, Oxford University Press, New York.

McCullough, L.B., Chervenak, F.A., Coverdale, J.F., 1996, 'Ethically justified guidelines for defining sexual boundaries between obstetrician-gynecologists and their patients,' *American Journal of Obstetrics and Gynecology* 175, 496-500.

McCullough, L.B.: 1996, 'Reification and synergy in clinical ethics and its adequacy to the managed practice of medicine,' *Journal of Medicine and Philosophy* 21, 1-6.

McCullough, L.B.: 1998, (ed.), *John Gregory's Observations and Lectures on the Duties, Offices, and Qualifications of a Physician*, Kluwer Academic Publishers, Dordrecht, The Netherlands.

McCullough, L.B.: 1998, *John Gregory (1724-1773) and the Invention of Professional Medical Ethics and the Profession of Medicine*, Kluwer Academic Publishers, Dordrecht, The Netherlands.

McDonough, J.E., Hager, C.L., Rosman, B.: 1997, 'Health care reform stages a comeback in Massachusetts,' *New England Journal of Medicine* 336(2), 148-151.

McGivney, W.T.: 1992, 'Proposal for assuring technology competency and leadership in medicine,' *Journal of the National Cancer Institute* 84, 742-744.

McLaugblin, F., Thomas, S., Barter, M.: 1995, 'Changes related to care delivery patterns,' *Journal of Nursing Administration* 25(5), 35-46.

Mechanic, D., Schlesinger, M.: 1996, 'The impact of managed care on patients' trust in medical care and their physicians,' *Journal of the American Medical Association* 275, 1693-1697.

Method of Prosecuting Enquiries in Philosophy, W. Strahan and T. Cadell, London. Reprinted in L.B. McCullough (ed.), *John Gregory's Observations and Lectures on the Duties, Offices, and Qualifications of a Physician*, Kluwer Academic Publishers, 1998, pp. 93-159.

Miles, S.: 1992, 'Medical futility,' *Law, Medicine and Health Care* 20, 310-315.

Miller, R.H., Luft, H.S.: 1994, 'Managed care plan performance since 1980: A literature analysis,' *Journal of the American Medical Association* 271, 1512-9.

Miller, M.G., Miller, L.S., Fireman, B., Black, S.B.: 1994, 'Variation in practice for discretionary admissions,' *Journal of the American Medical Association* 271, 1493-1498.

Milligan, M.A., More, E.S. (eds.): 1994, *The Empathic Practitioner: Empathy, Gender, and Medicine*, Rutgers University Press, New Brunswick, New Jersey, pp. 19-39.

Mohr, W., Mahon, M.: 1996, 'Dirty hands: The underside of marketplace health care,' *Advances in Nursing Science* 19(1), 28-37.

More, E.S.: 1994, '"Empathy" enters the profession of medicine,' in E.S. More, M.A. Milligan (eds.), *The Empathic Practicioner: Empathy, Gender and Medicine*, Rutgers University Press, New Brunswick, New Jersey.

Morreim, E.H.: 1987, 'Cost containment and the standard of medical care,' *California Law Review* 75(5), 1719-1763.

Morreim, E.H.: 1995, *Balancing Act*, Georgetown University Press, Washington, D.C.

Morreim, E.H.: 1997, 'Medicine meets resource limits: Restructuring the legal standard of care,' *University of Pittsburgh Law Journal* 59(1), 1-95.

Morreim, E.H.: 1989, 'Stratified scarcity: Redefining the standard of care,' *Law, Medicine and Health Care* 17, 356-367.

Morreim, E.H.: 1992, 'Rationing and the law,' in M.A. Strosberg, J.M. Wiener, R. Baker, I.A. Fein, (eds.) *Rationing America's Medical Care: The Oregon Plan and Beyond*, Brookings Institution, Washington, D.C., pp. 159-184.

Morreim, E.H.: 1994, 'Of rescue and responsibility: Learning to live with limits,' *Journal of Medicine and Philosophy* 19, 455-470.

Morreim, E.H.: 1995a, 'Diverse and perverse incentives in managed care; bringing the patient into alignment,' *Widener Law Symposium Journal* 1, 89-139.

Morreim, E.H.: 1995b, 'The ethics of incentives in managed care,' *Trends in Health Care, Law and Ethics* 10(1-2), 56-62.

Morreim, E.H.: 1995d, 'Moral justice and legal justice in managed care: The ascent of contributive justice,' *Journal of Law, Medicine, and Ethics* 23, 247-265.

Morreim, E.H.: 1998, 'Revenue streams and clinical discretion,' *Journal of the American Geriatrics Society* 46(3), 331-337.

Morreim, E.H.: 1995c 'Futilitarianism, exoticare, and coerced altruism: The ADA meets its limits,' *Seton Hall Law Review* 25, 101-149.

Mortenson, L.E.: 1989, 'Insurers target chemotherapy payments,' *Wall Street Journal*, May 11.

Murata, S.K.: 1996, 'Here come big changes in your patients' insurance,' *Medical Economics* 73(7), 185-190.

National Public Health and Hospital Institute: 1995, *Urban Social Health*,Washington, D.C.

Nelson, A.F., Quiter, E.S., Solberg, L.I.: 1998, 'The state of research within managed care plans: 1997 survey,' *Health Affairs* 17(1),128-138.

Neumann, P.J., Zinner, D.E., and Paltiel, A.D.: 1996, 'The FDA and regulation of cost-effectiveness claims,' *Health Affairs* 15(3), 54-71.

New York Times: 1996, 'Word for Word/H.M.O Contracts,' *New York Times*, September 22, Week in Review, 7 (National Edition).

NIH Technology Assessment Panel on Gaucher Disease: 1996, 'Gaucher disease: Current issues in diagnosis and treatment,' *Journal of the American Medical Association* 275, 548-553

O'Brien, C.L.: 1996, 'Direct contracting: Potential legal and regulatory barriers,' *Minnesota Medicine* 79, 21-25.

O'Connell, L: 1994, 'Ethicists and health care reform: An indecent proposal?' *Journal of Medicine and Philosophy* 19 (5), 419-424.

Office of Technology Assessment Washington, D.C.: 1992, 'Does health insurance make a difference?' background paper.

Ogden, J.: 1998, 'Tired of waiting,' *Nursing Standard* 12(23), 14.

Pappas, G., Queen, S., Hadden, W., Fisher, G.: 1993, 'The increasing disparity in mortality between socioeconomic groups in the United States, 1960 and 1986,' *New England Journal of Medicine* 329, 103-109

Pearson, S.D., Goulart-Fisher, D., Lee, T.H.: 1995, 'Critical pathways as a strategy for improving care: Problems and potential,' *Annals of Internal Medicine* 123, 941-948.

Pellegrino, E.D.: 1993, 'The metamorphosis of medical ethics: A 30-year retrospective,' *Journal of the American Medical Association* 269, 1158-1163.

Pellegrino, E.D., Caplan, A., Goold, S.D.: 1998, 'Doctors and ethics, morals and manuals,' *Annals of Internal Medicine* 128(7), 569-571.

Pelligrino, E., Thomasma, D.C.: 1988, *For the Patient's Good*, Oxford University Press, New York.

Percival, T.: 1803, *Medical Ethics, or a Code of Institutes and Precepts, Adapted to the Professional Conduct of Physicians and Surgeons*, Printed by J. Russell, for J. Johnson, St. Paul's Church Yard & R. Bickerstaff, Strand, London.

Peters, W.P., Rogers, M.C.: 'Variation in approval by insurance companies of coverage for autologous bone marrow transplantation for breast cancer,' *New England Journal of Medicine* 330, 473-477.

Pew Health Profession Commission: 1995, 'Critical challenges: Revitalizing the health care professions for the twenty-first century,' UCSF Center for the Health Professions, San Francisco.

Pilote, L., Califf, R.M., Sapp, S., Miller, D.P., Mark, D.B., Weaver, D., Gore, J.M.,

Armstrong, P.W., Ohman, M., Topol, E.J., for the GUSTO-1 Investigators: 1995, 'Regional variation across the United States in the management of acute myocardial infarction,' *New England Journal of Medicine* 333, 565-572.

Porter, D., Porter, R.: 1989, *Patient's Progress: Doctors and Doctoring in Eighteenth-Century England*, Stanford University Press, Stanford, CA.

Porter, R.: 1987, 'A touch of danger: The man-midwife as sexual predator,' in G.S. Rousseau and R. Porter (eds.), *Sexual Underworlds of the Enlightenment*, Manchester University Press, Manchester, pp. 206-232.

Power, E.J.: 1995, 'Identifying health technologies that work,' *Journal of the American Medical Association* 274, 205.

Raak, A., Mur-Veeman, I.: 1996, 'Home care policy in the Netherlands reforming legalization to facilitate the provision of multi-discipline home care,' *Health Policy* 36, 37-51.

Ray, W.A.: 1997, 'Policy and program analysis using administrative data banks,' *Annals of Internal Medicine* 127, 712-718.

Reiser, S.J.: 1994, 'Criteria for standard versus experimental therapy,' *Health Affairs* 13(3), 127-136.

Rennie, D.: 1997, 'Thyroid storm,' *Journal of the American Medical Association* 277, 1238-1243.

Report to the Board of Trustees, AMA: 1995, 'Direct contracting with employers: A strategy to increase physician involvement in the current health care market--an update to B of T Report 27 (A-95),' *B of T Report I-95*.

Rice, T., Pourat, N., Levan, R., *et al.*: 1998, 'Policy report: Trends in job-based health insurance coverage,' UCLA Center for Health Policy Research, Los Angeles.

Riesch, K.: 1990 'A review of the state of the art of research on nursing centers,' *Perspectives in Nursing* 4, 91-104.

Risse, G.: 1986, *Hospital Life in Enlightenment Scotland: Care and Teaching at the Royal Infirmary of Edinburgh*, Cambridge University Press, Cambridge.

Robinson, J.C.: 1995, 'Health care purchasing and market changes in California,' *Health Affairs* 14(4), 117-130.

Rodwin, M.: 1993, *Medicine, Money, and Morals: Physicians' Conflicts of Interest*, Oxford University Press, New York.

Romig, C.L.: 1997, 'Health care reform--the consolation prize,' *AORN Journal* 65(5), 974-977.

Roper, W.L., Cutler, C.M.: 1998, 'Health plan accountability and reporting: Issues and challenges,' Health Affairs 17(2), 152-155.

Rose, F.: 1990, 'A new age for business,' *Fortune* 122(9), 156-164.

Rosenbaum, S., Frankford, D.M., Moore, B., Borzi, P.: 1999, 'Who should determine when health care is medically necessary?' *New England Journal of Medicine* 340(3), 229-232.

Rovne, J.: 1998, 'U.S. President's Commission ends in discord,' *Lancet* 351, 890

Rowland, D., Feder, J., Keenan, P.S.: 1998, 'The problem of the uninsured - it's real and getting worse: Uninsured in America: The causes and consequences,' in S.H. Altman, U.E. Reinhardt and A.E. Shields, (eds.), *The Future U.S. Health Care System: Who Will Care for the Poor and Uninsured?* Health Administration Press, Chicago, pp. 25-44.

Rowland, D., Lyons, B., Salganicoff, A., Long, P.: 1994, 'Profile of the uninsured in America,' *Health Affairs* 13(2), 283-287.

Rowland, D.: 1993, 'Kaiser Commission on the future of Medicaid. Subcommittee on Select Revenue Measures' Committee on Ways and Means, U.S. House of Representatives, June 29.

Rutter, T.: 1998, 'Health features prominently in the State of the Union address,' *British Medical Journal* 316(7129), 415.

Schauffler, H.H., and Rodriguez, T.: 1996, 'Exercising purchasing power for preventive care,' *Health Affairs* 15(1), 73-85.

Scheiber, G.J., Poullier, J.P., Greenwald, L.M.: 1994, 'Health system performance in OCED countries, 1980-1992,' *Health Affairs* 1(Fall), 100-112.

Schneiderman, L.: 1988, *The Psychology of Social Change*, Human Sciences Press, New York.

Schroeder, S.A.: 1996, 'The medically uninsured--will they always be with us?' *New England Journal of Medicine* 334(17). 1130-1133.

Scott, R.A., Aiken, L.H., Mechanic, D., Moravcsik, J.: 1995, 'Organizational aspects of caring,' *Milbank Quarterly* 73, 77-95.

Scovern, H.: 1988, 'A physician's experiences in a for-profit staff-model HMO,' *New England Journal of Medicine* 319, 787-790.

Selden, T.H., Banthin, J.S., Cohen, J.W.: 1999, 'Waiting in the wings: Eligibility and enrollment in the State Children's Health Insurance Program,' *Health Affairs*18(2), 126-133.

Shambaugh, G.E., Jr.: 1996, 'US health-care costs,' *Lancet* 347, 694.

Shani, M.: 1995, 'Health care reform in Israel,' in J. Shemer, M. Vienonen (eds.), *Reforming Health Care System*, Geffen, Jerusalem.

Shelton, D.L.: 1996, 'Drugs offer hope--at a price,' *American Medical News* 3, 22.

Shewry, S., Hunt, S., Ramey, J., Bertko, J.: 1996, 'Risk adjustment: The missing piece of market competition,' *Health Affairs* 15(1), 171-181.

Short, P.F., Banthin, J.S.: 1995, 'New estimates of the underinsured younger then 65 years,' *Journal of the American Medical Association* 274, 1302-1306.

Short, P.F., Klerman, J.A.: 1998, *Targeting long-and short-term gaps in health insurance. Improving health care coverage and affordability series*, Commonwealth Fund, New York.

Slomski, A.J.: 1996, 'Here they come: Price-conscious patients,' *Medical Economics* 73(8), 40-46.

Smith, S., Freeland, M., Heffler, S., McKusick, D.: 1998, 'The next ten years of health spending: What does the future hold?' *Health Affairs* 17(5), 128-124

SoRelle, R.: 1999, 'Patient protection: Health reform on the line,' *Circulation* 99(13), 1651-1652.

Soumerai, S.B., Ross-Degnan, D., Fortess, E.E., Abelson J.: 1993, 'A critical analysis of studies of state drug reimbursement policies: Research in need of discipline,' *Milbank Quarterly* 71(2), 217-252.

Spitzer, A.: 1998, 'Nursing in the health care system of the postmodern world: Crossroads, paradoxes and complexity,' *Journal of Advanced Nursing* 28(1), 164-171.

Starr, P.: 1982, *The Social Transformation of American Medicine*, Basic Books, New York.

Steinberg, E.P., Tunis, S., Shapiro, D.: 1995, 'Insurance coverage for experimental technologies,' *Health Affairs*14(4), 143-158.

Stelfox, H.T., Chua, G., O'Rourke, K., Detsky, A.S.: 1998, 'Conflict of interest in the debate over calcium-channel antagonists,' *New England Journal of Medicine* 338, 101-

106.

Sticbler, J.: 1994, 'System development and integration in health care,' *Journal of Nursing Administration* 24(10), 48-53.

Strosberg, M.A., Wiener, J.M., Baker, R. (eds): 1992, *Rationing America's Medical Care: The Oregon Plan and Beyond*, Brookings Institution, Washington, D.C.

Swanson, M.: 1990, 'Providing care in the NICU: Sometimes an act of love,' *Advances in Nursing Science* 13, 60-73.

Swanson, M.: 1991, 'Empirical development of a middle range theory of caring,' *Nursing Research* 40, 161-166.

Swartz, K.: 1998, 'All uninsured are not the same,' in S.H. Altman, U.E. Reinhardt, A.E. Shields (eds.), *The Future U.S. Health Care System: Who Will Care for the Poor and Uninsured?* Health Administration Press, Chicago, pp. 45-66.

Swartz, K.: 1994, 'Dynamics of people without health insurance: Don't let the numbers fool you,' *Journal of the American Medical Association* 271, 64-66.

Tabbush, V., Swanson, G.: 1996, 'Changing paradigms in medical payment,' *Archives of Internal Medicine* 156(4), 357-360.

Tannock, I.F.: 1987, 'Treating the patient, not just the cancer,' *New England Journal of Medicine* 317, 1534-35.

Task Force on Principles for Economic Analysis of Health Care Technology: 1995, 'Economic analysis of health care technology: A report on principles,' *Annals of Internal Medicine* 122, 61-70.

ten Have, J.A.M.J.: 1995, 'Medical technology assessment and ethics: Ambivalent relations,' *Hastings Center Report* 25(5), 13-19.

Terry, K.: 1996, 'Can functional-status surveys improve your care?' *Medical Economics* 73(14), 126-144.

The President's Advisory Commission on Consumer Protection and Quality in the Health Care Industry: 1998, 'Quality first: better health care for all Americans: Final report to the President of the United States,' Government Printing Office, Washington, D.C.

Thomasma, D.C., Muraskas, J., Marshall, P.A., Myers, T., Tomich, P., O'Neill, J.A.: 1996, 'The ethics of caring for conjoined twins: The Lakeberg twins,' *Hastings Center Report* 26(4), 4-12.

Thompson, D.F.: 1993, 'Understanding financial conflicts of interest,' *New England Journal of Medicine* 329, 573-576.

Thorpe, K.E., Shields, A.E., Gold, H., Altman, S., Shactman, D.: 1995, 'Anticipating the number of uninsured Americans and the demand for uncompensated care: The combined impact of proposed Medicaid reductions and erosion of employer-sponsored insurance,' paper presented to the Council on the Economic Impact of Health Care Reform, Washington, D.C., November 8.

Toffler, A.: 1990, *Power Shift*, Morrow, New York.

Tong, R.: 1993, *Feminine and Feminist Ethics*, Wadsworth Publishing Company, Belmont, California.

Ulman, H.L.: 1990, *The Minutes of the Aberdeen Philosophical Society*, Aberdeen University Press, Aberdeen, Scotland.

United States General Accounting Office: 1996, 'Health insurance: Coverage of autologous bone marrow transplantation for breast cancer,' April, GAO/HEHS-96-83.

Urban Institute: 1996, 'Potential effects of congressional welfare reform legislation on family incomes,' Washington, D.C.

Van Dusen, L.: 1997, 'This business called medicine,' *Canadian Medical Association Journal* 157(12), 1724-1725.

Vaughan, B.: 1998, 'Developing nursing practice [editorial],' *Journal of Clinical Nursing* 7(3), 199-200.

Ventura, M.J.: 1998, 'Can these nurses make a difference?' *RN* 61(7), 4749.

Volunteer Trustees Foundation for Research and Education: 1995, 'State attorneys general's authority to police the sale and conversion of not-for-profit hospitals and HMOs,' Washington, D.C.

Wakefield, M.K.: 1996, 'Federal health initiatives: A status report,' *Dermatology Nursing* 8(3), 199-200.

Waldrop, M.: 1992, *Complexity*, Simon and Schuster, New York.

Waters, A., Doult, B.: 1998, 'The big issue,' *Nursing Standard* 12(23), 12.

Watson, J.: 1981, 'Some issues related to a science of caring,' in N. Leininger (ed.), *Caring: An Essential Human Need*, Charles B. Slack, Thorofare, New Jersey, pp. 59-69.

Wear, S.: 1993, *Informed Consent: Patient Autonomy and Physician Beneficence within Clinical Medicine*, Kluwer Academic Publishers, Dordrecht, The Netherlands.

Weis, D., Schank, M.: 1991, 'Professional values and empowerment: A role for continuing education,' *Journal of Continuing Education in Nursing* 22(2), 50-53.

Weissman, J.S., Epstein, A.M.: 1994, *Falling Through the Safety Net*, Johns Hopkins University Press, Baltimore.

Weissman, J.S., Gastonis, C., Epstein, A.M.: 1992, 'Rates of avoidable hospitalization by insurance status in Massachusetts and Maryland,' *Journal of the American Medical Association* 268, 2388-2394.

Welch, W.P., Miller, M.E., Welch, H.G., Fisher, E.S., Wennberg, J.E.: 1993, 'Geographic variation in expenditures for physicians' services in the United States,' *The New England Journal of Medicine* 328, 621-627.

Wells, K.B., Sturm, R.: 1995, 'Care for depression in a changing environment,' *Health Affairs* 14(3), 78-89.

Wennberg, J.E., Freeman, J.L., Culp, W.J.: 1987, 'Are hospital services rationed in New Haven or overutilized Boston?' *The Lancet* 1, 1185-1188.

Wennberg, J.E.: 1986, 'Which rate is right?' *The New England Journal of Medicine* 314, 310-311.

Wennberg, J.E.: 1990, 'Outcomes research, cost containment, and the fear of rationing,' *The New England Journal of Medicine* 323, 1202-1204.

Wennberg, J.E.: 1991, 'Unwanted variations in the rule of practice,' *Journal of the American Medical Association* 265, 1306-1307.

Wetzell, S.: 1996, 'Consumer clout,' *Minnesota Medicine* 79(2), 15-19.

White House Domestic Policy Council: 1993, *The President's Health Security Plan*, Times Books, New York.

Whitehead, M., Gustafsson, RA., Diderichsen, F.: 1997, 'Why is Sweden rethinking its NHS style reforms?' *British Medical Journal* 315(7113), 935-939.

Whyte, H.E., Fitzhardinge, P.M., Shennen, A.T., et al.: 1993, 'Extreme immaturity: Outcome of 568 pregnancies of 23-26 weeks gestation,' *Obstetrics and Gyenecology* 82, 1-7.

Whytt, R.: 1765, *Observations on the Nature, Causes and Cure of those Disorders Which Have Been Commonly Called Nervous, Hypochondriac or Hysteric . . .* 2nd. ed., corrected, Balfour, Edinburgh.

Winslow R.: 1998, 'Health-care inflation revives in Minneapolis despite cost-cutting,' *Wall Street Journal*, May 19, A1.

Winslow, R.: 1996, 'Study questions safety and cost of heart device,' *Wall Street Journal*, September 18, B-1, B-12.

Winterbottom, C.: 1993, *Trends in Health Insurance Coverage: 1988-1991*, Urban Institute, Washington, D.C.

Winterbottom, C., Liska, D.W., Obermaier, K.M.: 1995, *State-level Data Book on Health Care Access and Financing*, 2nd ed., Urban Institute, Washington, D.C..

Wolf, G., Boland, S., Aukerman, M.: 1994, 'A transformational model for the practice of professional nursing. Part 1, the model,' *Journal of Nursing Administration* 24(4), 51-57.

Wong, J.: 1997, 'Health care finance in the US: Past, present, and future,' *International Journal of Public Administration* 20, 1297-1315.

Wong, E.T., Lincoln, T.L.: 1983, 'Ready! Fire! . . . Aim!' *Journal of the American Medical Association* 250, 2510-2513.

II. BIBLIOGRAPHY: THE HUMAN GENOME PROJECT

The Human Genome Project (HGP) is a nine year old international effort to map and sequence the entire human genome. At present time, the HGP includes scientists from the United States, the United Kingdom, Canada, France, Germany and Japan all working to develop a biological "periodic table" which will make it possible not only to identify signatures from each building block, but in time develop treatment options based on our newfound genetic knowledge (Lander, 1996). The HGP has successfully completed all its major goals up through 1998 and has set forth a plan which calls for the completion of the project through an emphasis on DNA sequencing by 2003, 2 years ahead of the original schedule.

As advancements are being made daily, the following articles are by no means a comprehensive listing of all of the journal contributions to this difficult, and for many, ethically problematic project. Advances in genetics will clearly influence clinical decision making in terms of patients' relationships with their providers, their friends and families, and inevitably their finances. The discussions of the revolution in medicine resulting from the HGP include the translation of genetic discoveries into meaningful medical diagnostics and therapeutics as well as the ethical and social problems that come hand in hand with the unraveling of the human genetic code.

GENERAL BIBLIOGRAPHY 295

Journal Articles and Texts

Adams, J.: 1814, *A Treatise on the Supposed Hereditary Properties of Diseases,* J. Callow Publishers, London.

Altimore, M.: 1982, 'The social construction of a scientific controversy,' *Science, Technology and Human Values* (Fall), 24-31.

American Society of Human Genetics, American College of Medical Genetics.: 1995, 'Points to consider: Ethical, legal and psychosocial implications of genetic testing in children and adolescents,' *American Journal of Human Genetics* 57, 1233-1241.

Andrews, L. *et al.*: 1994, *Assessing Genetic Risks,* National Academy Press, Washington, D.C..

Angier, N.: 1993, 'Scientists isolate gene that causes cancer of colon,' *New York Times,* December 3, A1.

Annas, G.: 1994, 'Rules for "gene banks:" Protecting privacy in the genetics age,' in T. Murphy and M. Lappe, (eds.), *Justice and the Human Genome Project,* Berkeley, University of California Press, pp. 75-91.

ASHG Report: 1995, 'Report from the ASHG Information and Education Committee: Medical school core curriculum in genetics,' *American Journal of Human Genetics* 56, 535-553.

Baird, P.A.: 1990, 'Genetics and health care: A paradigm shift,' *Perspectives in Biology and Medicine* 33, 203-213.

Bauer, J.: 1942, *Constitution and Disease,* Grune and Stratton, New York.

Beasley, M.G., Costello, P.M., and Smith, I.: 1994, 'Intellectual status of young adults with phenylketonuria (PKU),' in J.-P. Farriaux and J.-L. Dhondt (eds.), *New Horizons in Neonatal Screening,* Amsterdam and New York, Elsevier Science, pp. 109-110.

Benjamin, C.M., Adam, S., Wiggins, S., Theilmann, J.L., Copley, T.T., Bloch, M., *et al.*: 1994, 'Proceed with care: Direct predictive testing for Huntington Disease,' *American Journal of Human Genetics* 55, 606-617.

Beutler, E.: 1996, 'The cost of treating Gaucher's disease,' *Nature Medicine* 2(5), 523.

Billings, P., Kohn, M., De Cuevas, M., Beckwith J., Alper, J.S., and Natowicz, M.R.: 1992, 'Discrimination as a consequence of genetic screening,' *American Journal of Human Genetics* 50, 476-482.

Bloche, M.G.: 1996, 'Clinical counseling and the problem of autonomy-negating influence,' in R.R. Faden and N.E. Kass, (eds.), *HIV, AIDS & Childbearing: Public Policy, Private Lives,* Oxford University Press, New York, pp. 257-319.

Bosch, X.: 1998, 'Geneticists discuss ethics of human genome project,' *Lancet* 352, 1448.

Brandt J.: 1989, 'Presymptomatic diagnosis of delayed onset diseases with linked DNA markers: The experience of HD,' *Journal of the American Medical Association* 261, 3108-3114.

Brandt, A.: 1987, *No Magic Bullet: A Social History of Venereal Disease in the United States Since 1880,* Oxford University Press, New York.

Brenton, D.P. and Lilburn, M.: 1996, 'Maternal phenylketonuria: A study from the United Kingdom,' *European Journal of Pediatrics* 155 [Suppl 1], S177-S180.

Brock D.: 1992, 'The Human Genome Project and human identity,' *Houston Law Review* 29, 19-21.

Brownlee, S., Silbemer, J.: 1991, 'The age of genes,' *US News and World Report,* November 4, 60.

Brzeszinsky, Z.: 1995, 'The new dimension of human rights,' Morgenthau Memorial Lecture.

Burgard, P. *et al.*: 1996, 'Intellectual development of the patients of the German collaborative study of children treated for Phenylketonuria,' *European Journal of Pediatrics* 155 [Suppl 1], S33-S38.

Burke, W., Thomson, E., Khoury, M.J., *et al.*: 1998, 'Hereditary hemochromatosis: Gene discovery and its implications for population-based screening,' *Journal of the American Medical Association* 280, 172-178.

Cambien, F.: 1992, 'Deletion polymorphism in the gene for angiotensin converting enzyme – a potent risk factor for myocardial infarction,' *Nature* 359, 5641-644.

Caskey, C.T.: 1993, 'Presymptomatic diagnosis: A first step toward genetic health care,' *Science* 262, 48-49.

Charo, A., Rothenberg, K.: 1994, '"The Good Mother:" The limits of reproductive accountability and genetic choice,' in K. Rothenberg and E. Thomson (eds.), *Women and Prenatal Testing: Facing the Challenges of Genetic Technology,* Ohio State University Press, Columbus, Ohio, pp. 105-131.

Churchill, F.: 1976, 'Rudolf Virchow and the pathologist's criteria for the inheritance of acquired characteristics,' *Journal of the History of Medicine* 31, 117-148.

Ciocco, A.: 1932, 'The historical background of the modern study of constitution,' *Bulletin of the History of Medicine* 4, 23-28.

Clarke, A.: 1990, 'Genetics, ethics, and audit,' *The Lancet* 335, 1145-1147.

Clayton, E.W.: 1997, 'Legal and ethical commentary: The dangers of reading duty too broadly,' *Journal of Law, Medicine & Ethics* 25, 19-21.

Clydesdale, F.: 1989, 'Present and future food science and technology in industrialized countries,' *Food Technology* (September), 134-146.

Collingwood, R. G.: 1974, 'Three senses of the word "cause",' in T. Beauchamp (ed.), *Philosophical Problems of Causation,* Dickenson Publishing Co., Encino, California, pp. 118-126.

Collins, F., Galas, D.: 1993, 'A new five-year plan for the US Human Genome Project,' *Science* 262, 43-46.

Collins, F.S., Guyer, M.S., Charkravarti, A.: 1997, 'Variations on a theme: Cataloging human DNA sequence variation,' *Science* 278, 1580-1581.

Collins, F.S., Patrinos, A., Jordan, E., Chakravarti, A., Gesteland, R., Walters, I.: 1998, 'New goals for the U.S. Human Genome Project: 1998-2003,' *Science* 282, 682-689.

Collins, F.S.: 1992, 'Positional cloning: Let's not call it reverse anymore,' *Nature Genetics* 1, 3-6.

Collins, F.S.: 1995, 'Positional cloning: From perditional to traditional,' *Nature Genetics* 9, 347-350.

Compliance manual, Section 902: 1995, Equal Employment Opportunity Commission, 1995.

Comry, J., (ed.): 1906, *Black's Medical Dictionary 1st Edition,* A. and C. Black, Ltd., London.

Cook, R.: 1989, *Mutation,* Putnam, New York.

Cook-Deegan, R.: 1994, *The Gene Wars: Science, Politics and the Human Genome,* W.W. Norton, New York.

Castle,W.E., *et al.*: 1912, *Heredity and Eugenics,* University of Chicago Press, Chicago

Culver, K.W.: 1990, 'The splice of life: Gene therapy comes of age,' *The Sciences* 1(7),

18-24.

Culver, K.W.: 1994, *Gene Therapy: A Handbook for Physicians*, Mary Ann Liebert, New York.

Davenport, C.B.: 1912, 'The inheritance of physical and mental traits of man and their application to eugenics,' in W.E. Coulter, *et al.*, (eds.) *Heredity and Eugenics*, University of Chicago Press, Chicago, pp. 269-288.

Davis, B.: 1992, 'Germ-line gene therapy: Evolutionary and moral considerations,' *Human Gene Therapy* 3, 361-365.

Deloukas, P., Schuler, G.D., Gyapay, G., *et al.*: 1998, 'A physical map of 30,000 human genes,' Science 282, 744-746.

Deloukas, P., Schuler, G.D., Gyapay, G., Beasley, E.M., Soderlund, C., Rodriguez-Tome, P., Hui, L., Matise, T.C., McKusick, K.B., Beckmann, J.S., Bentolila, S., Bihoreau, M., Birren, B.B., Browne, J., Butler, A., Castle, A.B., Chiannilkulchai, N., Clee, C., Day, P.J., Dehejia, A., Dibling, T., Drouot, N., Duprat, S., Fizames, C., Bentley, D.R., *et al.*: 1998, 'A physical map of 30,000 human genes,' *Science* 282, 744-746.

Department of Health and Human Services, Department of Energy: 1990, *Understanding Our Genetic Inheritance: The U.S. Human Genome Project: The first five years: FY 1991-1995*, Government Printing Office, Washington, D.C. (NIH publication no. 90-1590.)

DeRisi, J., Penland, L., Brown, P.O., *et al.*: 1996, 'Use of a cDNA microarray to analyse gene expression patterns in human cancer,' *Nature Genetics* 14, 457-60.

Dewitt, P.E.: 1994, 'The genetic revolution,' *Time*, January 17, 46-47.

Dewitt, P.E.: 1989, 'The perils of treading on heredity,' *Time*, March 20, 70.

Diamond, A.: 1994, 'Phenylalanine levels of 6-10 mg/dl may not be as benign as once thought,' *Acta Paediatrica* Supplement 407, 89-91.

Dobzhansky, T.: 1962, *Mankind Evolving,* Yale University Press, New Haven.

Douglas, M.: 1975, 'Deciphering a meal,' in *Implicit Meanings: Essays in Anthropology*, Routledge and Kegan Paul, London, pp. 249-275.

Draper, E.: 1992, *Risky Business: Genetic Testing and Exclusionary Practices in the Hazardous Workplace,* Cambridge University Press, New York.

Duster, T.: 1989, *Backdoor to Eugenics*, Routledge Publishing Co., New York.

Edelson, P.J.: 1994 'Lessons from the history of genetic screening in the US: Policy, past, present, and future,' unpublished ms.

Edlin, J.G.: 1987, 'Inappropriate use of genetic terminology in medical research: A public health issue,' *Perspectives in Biology and Medicine* 31, 47-56.

Engelhardt, H.T.: 1984, 'Clinical problems and the concept of disease,' in L. Nordenfelt and B. Lindahl (eds.), *Health, Disease and Causal Explanation in Medicine*, D. Reidel Publishers, Boston, pp. 27-41.

Epstein, C.J.: 1997, '1996 ASHG Presidential Address. Toward the 21st Century,' *American Journal of Human Genetics* 60, 1-9.

Faden, R.: 1994, 'Reproductive genetic testing, prevention and the ethics of mothering,' in E. Thomson and K. Rothenberg (eds.), *Women and Prenatal Testing: Facing the Challenges of Genetic Technology,* Ohio State University Press, Columbus, Ohio, pp. 88-98.

Fasten, N.: 1935, *Principles of Genetics and Eugenics*, Ginn and Co., NewYork.

Faust, D., Libon, D., and Pueschel, S.: 1986-87, 'Neuropsychological functioning in treated phenylketonuria,' *International Journal of Psychological Medicine* 16, 169-177.

Feder, J.N., Gnirke, A., Thomas, W., *et al.*: 1996, 'A novel MHC class I-like gene is mutated in patients with hereditary haemochromatosis,' *Nature Genetics* 13, 399-408.

Fisch, R.O. *et al.*: 1997, 'Phenylketonuria: Current dietary treatment practices in the United States and Canada,' *Journal of the American College of Nutrition* 16, 147-151.

Fox-Keller, E.: 1991, 'Genetics, reductionism and normative uses of biological information,' *Southern California Law Review* 65, 285-291.

Friedman, T.: 1990, 'The Human Genome Project – Some implications of extensive "reverse genetic" medicine,' *American Journal of Human Genetics* 46, 407-414.

Friedman, T.: 1996, 'Human gene therapy: An immature genie, but one that is certainly out of the bottle,' *Nature Medicine* 2(2), February, 145.

Froguel, P., Zouali, H., Vionnet, N., *et al.*: 1993, 'Familial hyperglycemia due to mutations in glucokinase: Definition of a subtype of diabetes mellitus,' *New England Journal of Medicine* 328, 697-702.

Garrison, F.: 1929, *An Introduction to the History of Medicine*, W.B. Saunders, Philadelphia.

Gasser, T., Muller-Myhsok, B., Wszolek, Z.K., *et al.*: 1998, 'A susceptibility locus for Parkinson's disease maps to chromosome 2p13,' *Nature Genetics* 18, 262-265.

Geller, G. and Holtzman, N.A.: 1995, 'A qualitative assessment of primary care physicians' perceptions about the ethical and social implications of offering genetic testing,' *Qualitative Health Research* 5, 97-116.

Gert, B., Berger, E., Cahill, G. *et al.*: 1996, *Morality and the New Genetics,* Jones and Bartlett Publishing Co., Boston.

Giardiello, F.M., Brensinger, J.D., Petersen, G.M.: 1997, 'The use and interpretation of commercial APC gene testing for familial adenomatous polyposis,' *New England Journal of Medicine* 336, 823-827.

Gilbert, W.: 1992, 'A vision of the Grail,' in D. Kevles and L. Hood (eds.), *The Code of Codes : Scientific and Social Issues in the Human Genome Project*, Harvard University Press, Boston, 83-98.

Goldenberg, S.: 1990, *New York Times,* Cartoon, September 16.

Golub, E.S.: 1997, *The Limits of Medicine: How Science Shapes our Hope for the Cure*, University of Chicago Press, Chicago.

Green, R.M.: 1997, 'Parental autonomy and the obligation not to harm one's child genetically,' *Journal of Law, Medicine & Ethics* 25, 5-15.

Guyer, M. and Collins F.C.: 1993, 'The Human Genome Project and the future of medicine,' *American Journal of Diseases of Children* 147, 1145-1152.

Hacia, J.G., Brody, L.C., Chee, M.S., Fodor, S.P., Collins, F.S.: 1996, 'Detection of heterozygous mutations in BRCA1 using high density oligonucleotide arrays and two-colour fluorescence analysis,' *Nature Genetics* 14, 441-447.

Hacia, J.G., Makalowski, W., Edgemon, K., *et al.*: 1998, 'Evolutionary sequence comparisons using high-density oligonucleotide arrays,' *Nature Genetics* 18, 155-158.

Haller, J.: 1981, *American Medicine in Transition: 1840-1910*, University of Illinois Press, Chicago.

Harding, A. E.: 1992, 'Growing old: The most common mitochondrial disease of all?' *Nature Genetics* 2, 51-252.

Hayes, A., Costa, T., Scriver, C.R., and Childs, B.: 1985, 'The effect of mendelian disease on human health II: Response to treatment,' *American Journal of Medical Genetics* 2, 243-255.

Health Insurance Portability and Accountability Act: 1996, HR 3103.

Hesslow, G.: 1984, 'What is a genetic disease?' in L. Nordenfelt and B. Lindahl (eds.), *Health, Disease and Causal Explanation in Medicine*, D. Reidel Publishers, Boston, pp. 183-193.

Health and Human Services Press Office: 1998, 'HHS forms genetic testing advisory board [press release],' Washington D.C., August 7. Available at: http://www.hhs.gov/news/press/1998.html.

Hoffman, C., Rice, D. and Sung, Hai-Yen: 1996, 'Persons with chronic conditions: Their prevalence and costs,' *Journal of the American Medical Association* 276, 1473-1479.

Holtzman, N.A., Watson, M.S., (eds.): 1998, *Promoting safe and effective genetic testing in the United States: final report of the Task Force on Genetic Testing*, Johns Hopkins University Press, Baltimore.

Holtzman, N.: 1989, *Proceed with Caution: Predicting Genetic Risks in the Recombinant DNA Era*, Johns Hopkins University Press, Baltimore, Maryland.

Hood, L.: 1988, 'Biotechnology and medicine of the future,' *Journal of the American Medical Association* 259, 1837-1844.

Hood, L.: 1992, 'Biology and medicine in the twenty-first century,' in D.J. Kevles and L. Hood, (eds.), *The Code of Codes: Scientific and Social Issues in the Human Genome Project*, Harvard University Press, Cambridge.

Hubbard, R. and Wald, E.: 1993, *Exploding the Gene Myth: How Genetic Information is Produced and Manipulated by Scientists, Physicians, Employers, Insurance Companies, Educators, and Law Enforcers,* Beacon Press, Boston.

Hubbard, R; Wald, I: 1993, *Exploding the Gene Myth*, Colophon Books, Boston.

Hudson, K.L., Rothenberg, K.H., Andrews, L.B., Kahn, M.J., Collins, F.S.: 1995, 'Genetic discrimination and health insurance: An urgent need for reform,' *Science* 270, 391-393

Hudson, R.: 1987, *Disease and Its Control: The Shaping of Modern Thought*, Praeger Press, New York.

Hull, R.: 'On getting "genetic" out of "genetic disease",' in J. Davis (ed.), *Contemporary Issues in Biomedical Ethics*, Humana Press, Clifton, New Jersey, pp. 71-87.

Jasper, J. and Nelkin, D.: 1994, *The Animal Rights Crusade*, The Free Press, New York.

Jones, A.C., Yamamura, Y., Almasy, L., *et al.*: 1998, 'Autosomal recessive juvenile parkinsonism maps to 6q25.2-q27 in four ethnic groups: Detailed genetic mapping of the linked region,' *American Journal of Human Genetics* 63, 80-87.

Juengst, E.: 1993, 'Causation and the conceptual scheme of medical knowledge,' in C. Delkeskamp-Hayes and M.A.G. Cutter (eds.), *Science, Technology and the Art of Medicine*, Kluwer, Dordrecht, pp. 127-152.

Juengst, E.: 1995, 'The ethics of prediction: Genetic risk and the physician-patient relationship,' *Genome Science and Technology* 1, 21-36.

Kahn, P.: 1996, 'Coming to grips with genes and risk,' *Science* (October 25), p. 497.

Kalverboer, A.F.: 1994, 'Social behavior and task orientation in early-treated PKU,' *Acta Paediatrica* Supplement 407, 104-105.

Karanjawala, Z.E., Collins, F.S.: 1998, 'Genetics in the context of medical practice,' *Journal of the American Medical Association* 280, 1533-1534.

Kass, E.H.: 1971, 'Infectious disease and social change,' *Journal of Infectious Diseases* 123, 110-114.

Kenen, R. H., Schmidt, R. M.: 1978, 'Stigmatization of carrier status: Social implications of heterozygote screening,' *American Journal of Public Health* 49, 116-120.

Kevles, D.J. 1985, *In the Name of Eugenics: Genetics and the Uses of Human Heredity.* Alfred A. Knopf, New York.

Kevles, D.J., Hood, L., (eds.): 1992, *The Code of Codes: Scientific and Social Issues in the Human Genome Project,* Harvard University Press, Cambridge.

Kirkman, H.N.: 1982, 'Projections of a rebound in frequency of mental retardation from phenylketonuria,' *Applied Research in Mental Retardation* 3, 319-28.

Koch, R. *et al.*: 1993, 'The North American Collaborative Study of Maternal Phenylketonuria. Status Report 1993,' *American Journal of Diseases of Children* 147, 1224-1230.

Koch, R. *et al.*: 1994, 'The International Collaborative Study of Maternal Phenylketonuria: Status Report 1994,' *Acta Paediatrica Supplement* 407, 111-119.

Koch, R., Yusin, M., Fishler, K.: 1985, 'Successful adjustment to society in young adults with phenylketonuria.' *Journal of Inherited Metabolic Disorders* 8, 209-211.

Kolata, G.: 1997, 'Tests to assess risks for cancer raising questions,' *New York Times,* March 27, A9.

Kononen, J., Bubendorf, L., Kallioniemi, A., *et al.*: 1998, 'Tissue microarrays for high-throughput molecular profiling of tumor specimens,' *Nature Medicine* 4, 844-847.

Koshland, D.: 1992 'Elephants, montrosities and the law,' *Science* 225, 777.

Krimsky, S.: 1982, *Genetic Alchemy,* Massachusetts Institute of Technology Press, Cambridge, MA.

Kroll, P.: 1990, 'The gene healers: Curing inherited diseases,' *The Plain Truth: A Magazine of Understanding* 55(8), 8.

Kuivenhoven, J.A., Jukema, J.W., Zwinderman, A.H., *et al.*: 'The role of a common variant of the cholesteryl ester transfer protein gene in the progression of coronary atherosclerosis,' *New England Journal of Medicine* 338, 86-93.

Lander, E.S.: 1996, 'The new genomics: Global views of biology,' *Science* 274, 536-539.

Laughlin, H.H.: 1914, 'Report of the Committee to Study and to Report on the Best Practical Means of Cutting Off the Defective Germ-Plasm in the American Population. I. The scope of the committee's work,' Bulletin No. 10A, Eugenics Record Office, Cold Spring Harbor, New York.

Leach, F.S., Nicolaides, N.C., Papadopoulos, N., *et al.*: 1993, 'Mutations of a mutS homolog in hereditary nonpolyposis colorectal cancer,' *Cell* 75, 1215-1225.

Leroy, E., Boyer, R., Auburger, G., *et al.*: 1998, 'The ubiquitin pathway in Parkinson's disease,' *Nature* 395, 451-452.

Levy, H.L., Ghavami, M.: 1996, 'Maternal Phenylketonuria: A metabolic teratogen,' *Teratology* 53, 176-184.

Levy, H.L.: 1991, 'Nutritional therapy in inborn errors of metabolism,' in R.J. Desnick (ed.), *Treatment of Genetic Diseases,* Churchill and Livingstone, New York.

Lewontin, R.C.: 1992, 'The dream of the human genome,' *The New York Review of Books* 39; reprinted in R.C. Lewontin, *The Doctrine of DNA,* Penguin Books, New York, pp. 60-83.

Lewontin, R.C.: 1993, *The Doctrine of DNA,* Penguin Books, New York.

Lin, J.H.: 1998, 'Divining and altering the future: Implications from the Human Genome Project,' *Journal of the American Medical Association* 280, 1532.

Mabry, C.C.: 1991, 'Status report on Phenylketonuria treatment,' *American Journal of Diseases of Childhood* 145, 33.

Magnus, D.: 1996, 'Gene therapy and the concept of genetic disease,' in *Ethics and*

Genetics: A Global Conversation, G. McGee (ed.), HTTP://www.med.upenn.edu/ bioethic/genetics.html.

Malkin, D, Li F. P., Strong L.C., *et al.*: 1990, 'Germ-line p53 mutations in a familial syndrome of breast cancer, sarcomas and other neoplasms,' *Science* 250, 1233-1238.

Mann, G.: 1964, 'The concept of predisposition,' *Journal of Environmental Health* 8, 840-845.

Mann, J.M.: 1997, 'Medicine and public health, ethics and human rights,' *Hastings Center Report* 27 (May-June), 6-13.

Markel, H.: 1992, 'The stigma of disease: The implications of genetic screening,' *American Journal of Medicine* 93, 209-215.

Marteau, T., Richards, M., (eds.): 1996, *The Troubled Helix*, Cambridge University Press, Cambridge.

Marteau, T.M., Drake, H.: 1995, 'Attributions for disability: The influence of genetic screening,' *Social Science and Medicine* 40, 1127-1132.

Mathieu, D.: 1996, *Preventing Prenatal Harm: Should the State Intervene?* Georgetown University Press, Washington, D.C.

Mayeux, R., Ottoman, R., Maestre, G., *et al.*: 1995, 'Synergistic effects of traumatic head injury and apolipoprotein E4 in patients with Alzheimer's disease,' *Neurology* 45, 555-557.

Mayr, E.: 1963, *Animal Species and Evolution*, Harvard University Press, Boston.

Mazzocco, M.M. *et al.*: 1994, 'Cognitive development among children with early-treated phenylketonuria,' *Developmental Neuropsychology* 10, 133-151.

McGee, G.: 1997, *The Perfect Baby: A Pragmatic Approach to Genetics*, Rowman and Littlefield, Lanham, MD.

McKeown, T.: 1976, *The Role of Medicine: Dream, Mirage or Nemesis*, Nuffield Provincial Hospitals Trust, London.

McKinlay, J.B., McKinlay, S.M.: 1977, 'The questionable contribution of medical measures to the decline in mortality in the United States in the twentieth century,' *Health and Society* (Summer), 405-428.

Medical Research Council: 1993, 'Phenylketonuria due to phenylalanine hydroxylase deficiency: An unfolding story,' *British Medical Journal* 306 (January 9), 115-119.

Miki, Y., Swensen, J., Shattuck-Eidens, D., *et al.*: 1994, 'A strong candidate for the breast and ovarian cancer susceptibility gene BRCA1,' *Science* 226, 66-71.

Millner, B.N.: 1993, 'Insurance coverage of special foods needed in the treatment of phenylketonuria,' *Public Health Reports* 108 (Jan.-Feb.), 60-65.

Milunsky, A.M., Annas, G.J. (eds.): 1980, *Genetics and the Law II,* Plenum Press, New York.

Mitka, M.: 1998, 'Genetics research already touching your practice,' *American Medical News* April 6, News section, 3.

Montagu, A.: 1959, *Human Heredity*, World Publishing, Cleveland.

Montgomerey, G.: 1990, 'Ultimate medicine,' *Discover* 11, March, 60.

Motulsky, A.G.: 1980, 'Governmental responsibilities in genetic diseases,' in A.M. Milunsky, G.J. Annas, (eds.), *Genetics and the Law II*, Plenum Press, New York, pp. 237-245.

Murray J.C., Buetow, K.H., Weber, J.L., *et al.*: 1994, 'A comprehensive human linkage map with centimorgan density,' *Science* 265, 2049-2054.

Murray, R.: 1974, 'Genetic disease and human health: A clinical perspective,' *The*

Hastings Center Report, 4-7.

Murray, T.H.: 1996. *The Worth of a Child,* University of California Press, Berkeley.

National Action Plan on Breast Cancer: 1995, 'One family's experience,' at the Workshop on Genetic Discrimination, sponsered by the National Action Plan and the ELSI Working Group of the NIH Human Genome Project, Bethesda, Maryland, July 11.

National Center for Genome Resources: 1996, *National Survey of Public and Stakeholders' Attitudes and Awareness of Genetic Issues,* Schulman, Ronca, and Bucuvalas, Washington, D.C..

National Human Genome Research Institute Web site. Available at: http://www.nhgri.nih.gov.

National Human Genome Research Institute: 1998, 'Genome project leaders announce intent to finish sequencing the human genome two years early [press release],' Bethesda, Maryland, September 12. Available at: http://www.nhgri.nih.gov/NEWS.

National Institutes of Health: 1997, 'Task force on genetic testing: Promoting safe and effective genetic testing in the United States, final report,' Bethesda, Maryland. Available at: http://www.nhgri.nih.gov/ELSI/TFGT_final/.

National Research Council Committee on Mapping and Sequencing the Human Genome: 1988, *Mapping and Sequencing the Human Genome,* National Academy Press, Washington, D.C.

Nelkin, D., Lindee, S.: 1995, *The DNA Mystique: The Gene as a Cultural Icon,* W.H. Freeman, New York.

Nelkin, D., Tancredi, L.: 1989, *Dangerous Diagnostics: The Social Power of Biological Information,* Basic Books, New York.

Nolan, K., Swenson, S.: 1988, 'New tools, new dilemmas: Genetic frontiers,' *Hastings Center Report* 40-46.

O'Flynn, M.E.: 1992, 'Newborn screening for Phenylketonuria: Thirty years of progress,' *Current Problems in Pediatrics* 22 (April), 159-165.

Office of Technology Assessment, U.S. Congress: 1988, 'Newborn screening for congenital disorders,' in *Healthy Children: Investing in the Future,* USGPO, Washington, D.C., pp. 93-116.

Palomaki, G.E.: 1994, 'Population-based prenatal screening for the fragile X syndrome,' *Journal of Medical Screening* 1, 65-72.

Papadopoulos N., Nicolaides, N.C., Wei, Y.F., *et al.*: 1994, 'Mutation of a mutL homolog in hereditary colon cancer,' *Science* 263, 1625-1629.

Patients' Bill of Rights Act. S.2330, 1998.

Paul, D. :1984, 'Eugenics and the left,' *Journal of the History of Ideas* 45, 567-590.

Paul, D.B., Edelson, P.J.: 1998, 'The struggle over metabolic screening,' in S. de Chadarevian and H. Kamminga (eds.), *Molecularising Biology and Medicine: New Practices and Alliances, 1930s-1970s,* Harwood Academic Publishers, Reading, pp. 203-220.

Paul, D.B., Spencer, H.G.: 1995, 'The hidden science of eugenics,' *Nature* 374, 302-304.

Paul, D.B.: 1992, 'Eugenic anxieties, social realities, and political choices,' *Social Research* 59 (Fall), 663-683.

Paul, D.B.: 1997, 'From eugenics to medical geneticsm,' *Journal of Policy History* 9, 96-116.

Paul, D.B.: 1998, 'Competing agendas, converging stories: The case of PKU,' in *The Politics of Heredity: Essays on Eugenics, Biomedicine, and the Nature-Nurture Debate,*

SUNY Press, Albany, and M. Fortun and E. Mendelsohn, (eds.), *The Practices of Human Genetics*, Kluwer, Dordrecht (in press).

Pennington, B.F., *et al.*: 1985, 'Neuropsychological deficits in early-treated phenylketonuric children,' *American Journal of Mental Deficiency* 5, 467-474.

Penrose, L.S.: 1946, 'Phenylketonuria: A problem in eugenics,' *Lancet* (June 29), 949-953.

Poirier, J., Delisle, M-C, Quirion, R., *et al.*: 1995, 'Apolipoprotein E4 allele as a predictor of cholinergic deficits and treatment outcome in Alzheimer disease,' *Procedures of the National Academy of Science U.S.A.* 92, 12260-12264.

Polymeropoulos, M.H., Lavedan, C., Leroy, E., *et al.*: 1997, 'Mutation in the alpha-synuclein gene identified in families with Parkinson's disease,' *Science* 276, 2045-2047.

Powers, M.: 1996, 'The moral right to have children,' in R.R. Faden and N.E. Kass, (eds.), *HIV, AIDS & Childbearing: Public Policy, Private Lives,* Oxford University Press, New York, pp. 320-344

Preventive Services Task Force: 1996, *Guide to Clinical Preventive Services*, 2nd ed., Williams and Wilkins, Baltimore.

Proctor, R.: 1995, *Cancer Wars*, Basic Books, New York.

Rapp, R.: 1988, 'Chromosomes and communication,' *Medical Anthropology Quarterly,* 143-157.

Rather, L.J.: 1959, 'Towards a philosophical study of the idea of disease,' in C. Brooks and P. Cranefield (eds.), *The Historical Development of Physiological Thought*, Hafner Publishing Co., New York, pp. 351-375.

Reed, S.: 1964, *Parenthood and Heredity*, John Wiley, New York.

Reilly, P.: 1991, *The Surgical Solution: A History of Involuntary Sterilization in the U.S*, Baltimore, Maryland, Johns Hopkins University Press.

Richmond, P.A.: 1954, 'American attitudes towards the germ theory of disease, 1860-1880,' *Journal of the History of Medicine* 9, 428-454.

Ris, M.D. *et al.*: 1994, 'Early-treated Phenylketonuria: Adult neuropsychologic outcome,' *Journal of Pediatrics* 124, pp. 388-92.

Roblin, R.: 1979, 'Human genetic therapy: Outlook and apprehensions,' in G. Chacko (ed.), *Health Handbook* , Amsterdam, North Holland Publishing Co, 104-114.

Rogers, M: 1977, *Biohazzard,* Knopf, New York.

Rosenberg, C.: 1976, *No Other Gods: On Science and American Social Thought*, John Hopkins University Press, Baltimore, Maryland

Rosenberg. S.S.: 1962, 'A new life for Karen,' *Family Weekly* (December 16), 6-7.

Rosenfeld, A.: 1992, 'The medical story of the century,' *Longevity*, May, 42-53.

Rothenberg, K., Fuller, B., Rothstein, M., *et al.*: 1997, 'Genetic information and the workplace: Legislative approaches and policy challenges,' *Science* 275, 1755-1757.

Rothstein, M.: 1992, 'Genetic discrimination in employment and the Americans with Disabilities Act,' *Houston Law Review* 29, 23-85.

Roussea, F., *et al.*: 1991, 'Direct diagnosis by DNA analysis of the Fragile X syndrome of mental retardation," *New England Journal of Medicine* 325, 1673-1681.

Savill, J.: 1997, 'Science, medicine, and the future. Prospecting for gold in the human genome,' *British Medical Journal* 314, 43-45.

Saxton M.: 1988, 'Prenatal screening and discriminatory attitudes about disability,' in E. Baruch, A. D'Adamo, J. Seager (eds.), *Embryos, Ethics and Women's Rights*, Haworth Press, New York.

304   ADRIANNE McEVOY

Scanlon, C., Fibison, W.: 1995, 'Managing genetic information: Implications for nursing practice,' *American Nurses Publishing*, 1-50.

Schafer, A.J., Hawkins, J.R.: 1998, 'DNA variation and the future of human genetics,' *Nature Biotechnology* 16, 33-39.

Schmidt, H.: 1996, 'Intelligence and professional career in young adults treated early for phenylketonuria,' *European Journal of Pediatrics* 155 [Suppl 1], S97-S100.

Schuler, G.D., Boguski, M.S., Stewart, E.A., Stein, L.D., Gyapay, G., Rice, K., White, R.E., Rodriguez-Tome, P., Aggarwal, A., Bajorek, E., Bentolila, S., Birren, B.B., Butler A., Castle, A.B., Chiannilkulchai, N., Chu, A., Clee, C., Cowles, S., Day, P.J., Dibling, T., Drouot, N., Dunham, I., Duprat, S., East, C., Hudson, T.J., *et al.*: 1996, 'A gene map of the human genome,' *Science* 274, 540-546.

Smith, I., Beasley, M.G., Ades, A.E.: 1990, 'Intelligence and quality of dietary treatment in phenylketonuria,' *Archives of Disease in Childhood* 65, 472-478.

Smith, I., Beasley, M.G., Ades, A.E.: 1991, 'Effects on intelligence of relaxing the low phenylalanine diet in phenylketonuria,' *Archives of Disease in Childhood* 66, 311-316.

Smith, I.: 1994, 'Treatment of phenylalanine hydroxylase deficiency,' *Acta Paediatrica Supplement* 407, 60-65.

Stanbury, J., *et al.*: 1983, 'Inborn errors of metabolism in the 1980's,' in J. Stanbury, *et al.*, (eds)., *The Metabolic Basis of Inherited Diseases*, McGraw Hill, Inc., New York.

Stephenson, F.: 1888, 'Temperament and diathesis in disease,' *Medical Record* 34, 362.

Stewart, M.: 1991, *Prodigy*, Harper Collins, New York.

Stokinger, H.D., Scheel, L.D.: 1973, 'Hypersusceptibility and genetic problems in occupational medicine: A consensus report,' *Journal of Occupational Medicine* 15, 564-573.

Strohman, R.C.: 1993, 'Ancient genomes, wise bodies, unhealthy people: Limits of a genetic paradigm in biology and medicine,' *Perspectives in Biology and Medicine* 37, 112-145.

Tauber, A.I. and Sarkar, S.: 1992, 'The Human Genome Project: Has blind reductionism gone too far?' *Perspectives in Biology and Medicine* 35, 220-235.

Treacy, E., Childs, B., Scriver, C.R.: 1995, 'Response to treatment in hereditary metabolic disease: 1993 survey and 10-Year comparison,' *American Journal of Human Genetics* 56, 359-367.

Ueda K, Fukushima H, Masliah E, *et al.*: 1993, 'Molecular cloning of cDNA encoding an unrecognized component of amyloid in Alzheimer disease,' *Procedures of the National Acadamy of Science U.S.A.* 90, 11282-11286.

US Department of Health and Human Services, US Department of Energy: 1990, 'Understanding our genetic inheritance: The US Genome Project: The first five years. FY 1991-1995 (DOE/ER-0452P),' National Technical Information Service, Springfield, Virginia.

van Dijck, J.: 1998, *ImagEnation: Popular Images of Genetics*, New York University Press, New York.

Venter, J.C., Adams, M.D., Sutton, G.G., Kerlavage, A.R., Smith, H.O., Hunkapiller, M.: 1998, 'Shotgun sequencing of the human genome,' *Science* 280, 1540-1542.

Vivigen Genetic Reppository brochure: 1991, Sante Fe, New Mexico, June 15.

Wachbroit, R.: 1994, 'Distinquishing genetic disease and genetic susceptibility,' *American Journal of Medical Genetics* 53, 236-240.

Waisbren, S.E. and Levy, H.L.: 1991, 'Agoraphobia in Phenylketonuria,' *Journal of*

Inherited Metabolic Disorders 3, 149-153.

Waisbren, S.E. *et al.*: 1994, 'Review of neuropsychological functioning in treated phenylketonuria: An information processing approach,' *Acta Paediatrica Supplement* 407, 98-103.

Wang, D.G., Fan, J-B., Siao, C-J., *et al.*: 1998, 'Large-scale identification, mapping and genotyping of single-nucleotide polymorphisms in the human genome,' *Science* 280, 1077-1082.

Watson, J., Cook-Deegan, R.: 1990, 'The Human Genome Project and international health,' *Journal of the American Medical Association* 263, 3322-3324.

Weglage, J. *et al.*: 1993, 'School performance and intellectual outcome in adolescents with Phenylketonuria,' *Acta Paediatrica* 82, 582-586.

Weglage, J. *et al.*: 1995, 'Neurological findings in early treated phenylketonuria,' *Acta Paediatrica* 84, 411-415.

Weglage, J. *et al.*: 1996a, 'Deficits in selective and sustained attention processes in early treated children with Phenylketonuria - result of impaired frontal lobe functions,' *European Journal of Pediatrics* 155, 200-204.

Weglage, J. *et al.*: 1996b, 'Psychosocial aspects in Phenylketonuria,' *European Journal of Pediatrics* 155 [Suppl 1] S101-S104.

Welsh, M.C. *et al.*: 1990, 'Neuropsychology of early treated Phenylketonuria: Specific executive function deficits,' *Child Development* 61, 1697-1713.

Wertz, D.C.: 1997a, 'Society and the not-so-new genetics: What are we afraid of? Some Future Predictions from a social scientist,' *Journal of Contemporary Health Law and Policy* 13, 299-346.

Wertz, D.C.: 1997b, 'Data distributed at the Workshop on Eugenic Thought and Practice: A Reappraisal towards the End of the Twentieth Century,' (May 26-29), Van Leer Institute, Jerusalem. (Published in part as Wertz, D.C.: 1998, 'Eugenics is alive and well: A survey of genetics professionals around the world,' *Science in Context* 11, 493-510.)

Wexler, N.: 1992, 'Clairvoyance and caution,' in D.J. Kelves and L. Hood (eds.) *The Code of Codes*, Harvard Univerity Press, Cambridge.

Wingerson, L.: 1982, 'Searching for depression gene,' *Discover* February 4, 60-64.

Winzeler, E.A., Richards, D.R., Conway, A.R., *et al.*: 1998, 'Direct allelic variation scanning of the yeast genome,' *Science* 281, 1194-1197.

Wooster, R., Bignell, G., Lancaster, J., *et al.*: 1995, 'Identification of the breast cancer susceptibility gene BRCA2,' *Nature* 378, 789-792. [Erratum, 1996, *Nature* 379, 749.]

Yamagata, K., Oda, N., Kaisaki, P.J., *et al.*: 1996, 'Mutations in the hepatocyte nuclear factor-1(alpha) gene in maturity-onset diabetes of the young (MODY3),' *Nature* 384, 455-458.

Yount, R.: 1994, 'Pursuit of excellence' in D. Knight (ed.), *The Clarion Awards*, Doubleday, New York.

III. BIBLIOGRAPHY: THE PHYSICIAN-PATIENT RELATIONSHIP

It hardly needs to be pointed out that there has been an exponential increase in knowledge and technology witnessed in the medical field these last fifty years. Hand in hand with this technological revolution has been a revolution in the relationship between the patient and her medical care provider. The combination of medical advancements coupled with the increasing control patients have over their own health care decisions has radically changed the physician-patient relationship. A field which was once dominated by paternalism is now focused on the relatively new concept of patient autonomy. The dynamic nature of the provider-patient relationship can also be dramatically affected by changes in economic tides.

The following discussions deal with the nature of the patient-provider relationship and all the ethical dilemmas inherent in such interactions. Many basic principles, including trust, honesty, benevolence, non-maleficence, autonomy and communication are offered as behavioral guides for a difficult yet necessary relationship between medical professionals and their patients. Once again, the following is not an all-inclusive listing but it will give a start to anyone interested in pursuing the dynamics of this difficult, yet rewarding relationship.

Journal Articles and Texts

American Medical Association Council on Ethical and Judicial Affairs: 1994, 'Ethical issues in health care system reform: The provision of adequate health care,' *Journal of the American Medical Association* 272, 1056-1062.

American Medical Association: 1997, 'American Medical Association Council on Ethical and Judicial Affairs Code of Medical Ethics Current Opinions With Annotations, 1996-1997,' Chicago, Illinois.

American Medical Association: 1996, 'Physician opinion on health care issues-1996,' Chicago, Illinois.

American Psychiatric Association: 1994, *Diagnostic and Statistical Manual of Mental Disorders*, 4th ed., American Psychiatric Association, Washington, D.C.

Andersen, R., Anderson, O.W., Smedby. B.: 1968, 'Perception of and response to symptoms of illness in Sweden and the United States,' *Medical Care* 6, 18-30.

Aristotle: 1985, *Nicomachean Ethics*, T. Irwin (trans.), Hackett, Indianapolis, VIII 3, 1156a20 et seq., pp. 211ff.

Ashraf, H.: 1999, 'Cross-culture communication needed,' *Lancet* 353, 910.

Auden, W.H.: 1991, 'The art of healing: In memoriam David Protetch, MD,' in R. Reynolds, J. Stone (eds.), *On Doctoring: Stories, Poems, Essays,* Simon and Schuster, New York, pp. 168-70.

Baker, L.C., Cantor, J.C.: 1993, 'Physician satisfaction under managed care,' *Health*

Affairs 12, 258-270.

Balint, J., Shelton, W.: 1996, 'Regaining the initiative: Forging a new model of the patient physician relationship,' *Journal of the American Medical Association* 275, 887-891.

Balint, M.: 1957, *The Doctor, His Patient, and the Illness*, International Universities Press, New York, N.Y.

Ballard, E.L, Nash, F., Raiford, K., and Harrell, L.E.: 1993 'Recruitment of black elderly for clinical research studies of dementia: The CERAD experience,' *The Gerontologist* 33, 561-565.

Banks, M.H., Beresford, S.A., Morrell, D.C., Waller, J.J., Watkins, C.J.: 1975, 'Factors influencing demand for primary medical care in women aged 20-44 years: A preliminary report,' *International Journal of Epidemiology* 4, 189-195.

Barsky, A.J. III.: 1979, 'Patients who amplify bodily sensations,' *Annals of Internal Medicine* 91, 63-70.

Barsky, A.J., Klerman, G.L.: 1983, 'Overview: Hypochondriasis, bodily complaints, and somatic styles,' *American Journal of Psychiatry* 140, 273-283.

Barsky, A.J.: 1988, 'The paradox of health,' *New England Journal of Medicine* 318, 414-418.

Bass, M.J., Buck, C., Turner, L., *et al.*: 1986, 'The physician's actions and the outcome of illness in family practice,' *Journal of Family Practice* 23, 43-47.

Bass, M.J., McWhinney, I.R., Dempsey, J.B., *et al.*: 1986, 'Predictors of outcome in headache patients presenting to family physicians – a one year prospective study,' *Headache Journal* 26, 285-294.

Becker, M.H., Drachman, R.H., Kirscht, J.P.: 1974, 'A field experiment to evaluate various outcomes of continuity of physician care,' *American Journal of Public Health* 64, 1062-1070.

Bellah, R., Madsen, R., Sullivan, W.M., Swidler, A. and Tipton, S.M.: 1985, *Habits of the Heart: Individualism and Commitment in American Life,* University of California Press, Berkeley, p. 115-123.

Berger, J.: 1967, *A Fortunate Man*, Holt, New York.

Berwick, D.M.: 1994, 'Eleven worthy aims for clinical leadership of health system reform,' *Journal of the American Medical Association* 272, 797-802.

BeSaw, L.: 1996, 'A matter of concern,' *Texas Medicine* 92, 22-27.

Bindman, A.B., Grumbach, K., Osmond, D., Vranizan, K., Stewart, A.L.: 1996, 'Primary care and the receipt of preventive service,' *Journal of General Internal Medicine* 11, 269-276.

Blackhall, L.J., Murphy, S.T., Frank, G., Michel, V., Azen, S.: 1995, 'Ethnicity and attitudes toward patient autonomy,' *Journal of the American Medical Association* 274, 820-825.

Bloche, M.G.: 1999, 'Clinical loyalties and the social purposes of medicine,' *Journal of the American Medical Association* 281, 268-274.

Blumberg, M.Z.: 1999, 'Ethics and managed care can coexist with a free market,' *Archives of Internal Medicine* 159(12), 1375-1376.

Blumenthal, D.: 1995, 'Effects of market reforms on doctors and their patients,' *Health Affairs* 15, 170-184.

Bobo, L., Womeodu, R.J., Knox, A.L. Jr.: 1991, 'Principles of intercultural medicine in an internal medicine program,' *American Journal of Medical Science* 302, 244-248.

Boelen, C.: 1992 'Medical education reform: The need for global action,' *Academic*

Medicine 67, 745-749.

Booth, W.C.: 1988, *The Company We Keep: An Ethics of Fiction*, University of California Press, Berkeley.

Borenstein, D.B.: 1996, 'Does managed care permit appropriate use of psychotherapy?' *Psychiatric Services* 47, 971-974.

Borkan, J., Reis, S., Hermoni, D., Biderman, A.: 1995, 'Talking about the pain: A patient-centered study of low back pain in primary care,' *Social Science Medicine* 40, 977-988.

Borkan, J.M., Quirk, M., Sullivan, M.: 1991, 'Finding meaning after the fall: Injury narratives from elderly hip fracture patients,' *Social Science Medicine* 33, 947-957.

Botelho, R.J., McDaniel, S.H., Jones, J.E.: 1990, 'A family systems approach to a Balint-style group: A report on a CME demonstration project for primary care physicians,' *Family Medicine* 22, 293-295

Botelho, R.J.: 'A negotiation model for the doctor-patient relationship,' *Family Practice* 9, 210-208.

Branch, W.T, and Suchman, A.: 1990, 'Meaningful experiences in medicine,' *American Journal of Medicine* 88, 56-59.

Brody, H.: 1980, *Placebos and the Philosophy of Medicine: Clinical, Conceptual and Ethical Issues*, University of Chicago Press, Chicago.

Brody, H.: 1985, 'Placebo effect: An examination of Gurnbaum's definition,' in L. White, B. Tursky, and G. E. Schwartz (eds.), *Placebo: Theory, Research, and Mechanisms*, Guilford, New York.

Brody, H.: 1986, 'The placebo response. Part 1. Exploring the myths. Part 2. Use in clinical practice,' *Drug Therapy* 16 (7), 106-131.

Brody, H.: 1987, *Stories of Sickness*, Yale University Press, New Haven, Connecticut.

Brody, H.: 1994, 'My story is broken, can you help me fix it? Medical ethics and the joint construction of narrative,' *Literary Medicine* 13, 79-92.

Brody, H., Goold, S.D. (in preparation): 'Jekyll, Hyde, and Gresham: The future of managed care in an unregulated market.'

Brown, P., Levinson, S.C.: 1978, *Politeness: Some Universals in Language Usage*, Cambridge University Press, Cambridge.

Broyard, A.: 1992, *Intoxicated by My Illness*, Clarkson Potter, New York.

Bruner, J.: 1986, *Actual Minds, Possible Worlds*, Harvard University Press, Cambridge, MA.

Bulger, R.J.: 1990, 'The demise of the placebo effect in the practice of scientific medicine — a natural progression or an undesirable aberration?' *Transactions of the American Clinical and Climatological Association* 102, 285-293.

Bullough, B., Bullough, V.L.: 1972, *Poverty, Ethnic Identity, and Health Care*, Appleton-Century-Crofts, New York.

Byrne, P.S., Long, B.E.L.: 1976, *Doctors Talking to Patients*, HMSO, London.

Callahan, D.: 1984, 'Autonomy: A moral good, not a moral obsession,' *Hastings Center Report* 14 (5), 40-42.

Callahan, S.: 1988, 'The role of emotion in ethical decision-making,' *Hastings Center Report* 8:3, 9-14.

Calman, K.C., Royston, G.H.D.: 1997, 'Risk language and dialectics,' *British Medical Journal* 315, 939-942.

Cameron, L., Leventhal, E.A., Leventhal, H.: 1995, 'Seeking medical care in response to symptoms and life stress,' *Psychosomatic Medicine* 57, 37-47.

Campbell, T.L.: 1986, 'The family's impact on health: A critical review and annotated bibliography,' *Family Systems Medicine* 1986(4), 135-328.

Cannon, R.O., III: 1995, 'The sensitive heart: A syndrome of abnormal cardiac pain perception,' *Journal of the American Medical Association* 273, 883-887.

Carrillo, J.E., Green, A.R., Betancourt, J.R.: 1999, 'Cross-cultural primary care: A patient-based approach,' *Annals of Internal Medicine* 130(10), 829-834.

Cassel, E.J.: 1982, 'The nature of suffering and the goals of medicine,' *New England Journal of Medicine* 306, 639-645.

Cassell, E.J.: 1991, *The Nature of Suffering and the Goals of Medicine*, Oxford University Press, New York.

Chao, J.: 1988, 'Continuity of care: Incorporating patient perceptions,' *Family Medicine* 20, 333-337.

Charles, C., Gafni, A., Whelan, T.: 1997, 'Shared decision-making in the medical encounter: What does it mean? (Or it takes at least two to tango),' *Social Science Medicine* 44, 681-92.

Childress, J.F., Siegler, M.: 1984, 'Metaphors and models of doctor-patient relationships: Their implications for autonomy,' *Theoretical Medicine* 5 17-30.

Chugh, U., Dillman, E., Kurtz, S.M., Lockyer, J., Parboosingh, J.: 1993, 'Multicultural issues in medical curriculum: implications for Canadian physicians,' *Medical Teacher* 15, 83-91.

Ciompi, L.: 1991, 'Affects as central organising and integrating factors: A new psychosocial/biological model of the psyche,' *British Journal of Psychiatry* 159, 97-105.

Connelly, J.: 1998, 'Emotions, ethics, and decisions in primary care,' *Journal of Clinical Ethics* 9(3), 225-234.

Coulthard, R.M., Ashby, M.C.: 1976, 'A linguistic description of doctor-patient interviews,' in M. Wadsworth, D. Robinson (eds.), *Studies in Everyday Medical Life*, Robinson, London.

Coupland, J., Robinson, J., Coupland, N.: 1994, 'Frame negotiation in doctor-elderly patient consultations,' *Discourse and Society* 5, 89-124.

Cousins, N. 1979, *Anatomy of an Illness as Perceived by the Patient: Reflections on Healing and Regeneration*, Norton, New York.

Cykert, S., Hansen, C., Layson, R., Joines, J.: 1997, 'Primary care physicians and capitated reimbursement,' *Journal of General Internal Medicine* 12, 192-194.

Damasio, A.R.: 1994, *Descartes' Error: Emotion, Reason, and the Human Brain*, GP Putnam's Sons, New York.

De La Cancela, V., Guarnaccia, P.J., Carrillo, J.E.: 1986, 'Psychosocial distress among Latinos: A critical analysis of ataques de nervios,' *Humanity and Society* 10, 431-447.

Demers, R.Y., Altamore, R., Mustin, H., Kleinman, A., Leonardi, D.: 1980, 'An exploration of the dimensions of illness behavior,' *Journal of Family Practice* 11, 1085-1092.

Dietrich, A.J., Marton, K.I.: 1982, 'Does continuous care from a physician make a difference?' *Journal of Family Practice* 15, 929-937.

Dougherty, C.J.: 1992, 'Ethical values at stake in health care reform,' *Journal of the American Medical Association* 268, 2409-2412.

Drew, P., Heritage, J., (eds.): 1992, *Analyzing Talk at Work*, Cambridge University Press, Cambridge.

Drossman, D.A., McKee, D.C., Sandler, R.S., et al.: 1988, 'Psychosocial factors in the irritable bowel syndrome: A multivariate study of patients and nonpatients with irritable bowel syndrome,' Gastroenterology 95(3), 701-708.

Eisenberg, L.: 1977, 'Disease and illness. Distinctions between professional and popular ideas of sickness,' Culture, Medicine, and Psychiatry 1, 9-23.

Eisenbruch, M.: 1989, 'Medical education for a multicultural society,' Medical Journal of Australia 151, 574-576, 579-580.

Ellwood, P.M., Lundberg, G.D.: 1996, 'Managed care: A work in progress,' Journal of the American Medical Association 276, 1083-1086.

Elwyn, G., Gwyn, R.: 1999, 'Narrative based medicine: Stories we hear and stories we tell: Analysing talk in clinical practice,' British Medical Association 318, 186-188.

Emanuel, E.J., Dubler, N.N.: 1995, 'Preserving the physician-patient relationship in the era of managed care,' Journal of the American Medical Association 1273, 323-329.

Emanuel, E.J., Emanuel, L.L.: 1992, 'Four models of the physician-patient relationship,' Journal of the American Medical Association 267, 2221-26.

Emanuel, E.J., Emanuel, L.L.: 1996, 'What is accountability in health care?' Annals of Internal Medicine 124, 229-239.

Engel, G.L.: 1977, 'The need for a new medical model: A challenge for biomedicine,' Science 196, 129-136.

Engelhardt, H.T., Jr.: 1996, The Foundations of Bioethics, 2nd edition, Oxford University Press, New York.

Entralgo, P. L.: 1969, Doctor and Patient, Frances Partridge (trans.), McGraw-Hill, New York.

Epstein, R.M., Campbell, T.L., Cohen-Cole, S.A., McWhinney, I.R., Smilkstein, G.: 1993, 'Perspectives on patient-doctor communication,' Journal of Family Practice 37, 377-388.

Epstein, R.M.: 1995, 'Communication between primary care physicians and consultants,' Archives of Family Medicine 4, 403-409.

Epstein, R.M., Quill, T.E., McWhinney, I.R.: 1999, 'Somatization reconsidered: Incorporating the patient's experience of illness,' Archives of Internal Medicine 159(3), 215-222

Erde, E.L., Jones, A.H.: 1983, 'Diminished capacity, friendship and medical paternalism: Two case studies from fiction,' Theoretical Medicine 4, 303-22.

Fabrega, H. Jr.: 1991, 'Somatization in cultural and historical perspective,' in L.J. Kirmayer, J.M. Robbins (eds.), Current Concepts of Somatization: Research and Clinical Perspectives, American Psychiatric Press Inc, Washington, DC, pp. 181-199.

Feinstein, J.S.: 1993, 'The relationship between socioeconomic status and health: A review of the literature,' Milbank Quarterly 71, 279-322.

Feldman, D.S., Novack, D.H., Gracely, E.: 1998, 'Effects of managed care on physician-patient relationships, quality of care, and the ethical practice of medicine: A physician survey,' Archives of Internal Medicine 158(15), 1626-1632.

Fisher, W.R.: 1984, 'Narration as a human communication paradigm: The case of public moral argument,' Communication Monographs 51, 1-22.

Ford, C.V.: 1983, The Somatizing Disorders: Illness as a Way of Life, Elsevier Biomedical Press, New York.

Foucault, M.: 1975, The Birth of the Clinic: An Archeology of Medical Perception, A.M. Sheridan Smith (trans.), Vintage, New York.

Foulks, E.F.: 1989, 'Misalliances in the barrow alcohol study,' *American Indian and Native Alaska Mental Health Research* 2, 7-17.

Fox, R.: 1959, *Experiment Perilous: Physicians and Patients Facing the Unknown*, Free Press, Glencoe, Illinois.

Fox, E.: 1997, 'Predominance of the curative model of medical care: A residual problem,' *Journal of the American Medical Association* 278, 761-763.

Frank, A.: 1994, 'Reclaiming an orphan genre: First-person narratives of illness,' *Literature and Medicine* 13, 1-21.

Frank, A.W.: 1995, *The Wounded Storyteller: Body, Illness, and Ethics*, University of Chicago Press, Chicago.

Franks, P., Clancy, C.M.: 1992, 'Gatekeeping revisited: Protecting patients from overtreatment,' *New England Journal of Medicine* 327, 424-429.

Franks, P., Clancy, C.M., Nutting, P.A.: 1992, 'Gatekeeping revisited-- protecting patients from overtreatment,' *New England Journal of Medicine* 327, 424-429.

Freeling, P., Gask, L.: 1998, 'Sticks and stones: Changing terminology is no substitute for good consultation skills,' *British Medical Journal* 317, 1028-1029.

Fried, C.: 1976, 'The lawyer as friend: The moral foundations of the lawyer-client relation,' *Yale Law Review* 85, 1060-89.

Frost, R.: 1915, 'Mending wall,' *North of Boston*, Holt, New York.

Geertz, C.: 1973, *The Interpretation of Cultures*, Basic Books, New York.

Goldstein, E., Bobo, L., Womeodu, R., Kaufman, L., Nathan, M., Palmer, D., et al.: 1996, 'Intercultural medicine,' in N.M. Jensen, J.A. Van Kirk (eds.), *A Curriculum for Internal Medicine Residency: The University of Wisconsin Program*, American College of Physicians, Philadelphia.

Greenfield, S., Kaplan, S., Ware, J.E.: 1985, 'Expanding patient involvement in care: Effects on patient outcomes,' *Annals of Internal Medicine* 102, 520-528.

Greenfield, S., Rogers, W., Mangotich, M., Carney, M.F., Tarlov, A.R.: 1995, 'Outcomes of patients with hypertension and non-insulin-dependent diabetes mellitus treated by different systems and specialists: Results from the Medical Outcomes Study,' *Journal of the American Medical Association* 274, 1436-1444.

Griffith, J.L., Griffith, M.E.: 1994, *The Body Speaks: Therapeutic Dialogues for Mind-Body Problems*, Basic Books, New York, NY.

Grunbaum, A.: 1985, 'Explication and implications of the placebo concept,' in L. White, B. Tursky, G.E. Schwartz (eds.), *Placebo: Theory, Research, and Mechanisms*, Guilford, New York, pp. 9-36.

Halm, E.A., Causino, N., Blumenthal, D.: 1997, 'Is gatekeeping better than traditional care? A survey of physicians' attitudes,' *Journal of the American Medical Association* 278, 1677-1681.

Harwood, A., (ed.): 1981, *Ethnicity and Medical Care*, Harvard University Press, Cambridge, MA

Harrington, A. (ed.): 1997, *The Placebo Effect: An Interdisciplinary Exploration*, Harvard University Press, Cambridge, MA.

Hautman, M.A.: 1979, 'Folk health & illness beliefs,' *Nurse Practitioner* 4, 24-34.

Hawkins, A.H.: 1993, *Reconstructing Illness: Studies in Pathography*, Purdue University Press, West Lafayette, Indiana.

Helman, C.G.: 1994, *Culture, Health and Illness: An Introduction for Health Professionals*, 3rd edition, Butterworth-Heinemann, Boston.

312   ADRIANNE McEVOY

Helman, C.G.: 1981, 'Disease versus illness in general practice,' *Journal of the Royal College of General Practice* 31, 548-552.
Hewson, M.G., Kindy, P.J., Van Kirk, J., Gennis, V.A., Day, R.P.: 1996, 'Strategies for managing uncertainty and complexity,' *Journal of General Internal Medicine* 11, 481-485.
Hillman, J.: 1983, *Healing Fiction*, Spring Publications, Woodstock, Connecticut.
Hunter, K.M.: 1991, *Doctors' Stories: The Narrative Structure of Medical Knowledge*, Princeton University Press, Princeton.
Illingworth, P.M.L.: 1988, 'The friendship model of physician/patient relationship and patient autonomy,' *Bioethics* 2, 22-36.
Ingelfinger, F.: 1980, 'Arrogance,' *New England Journal of Medicine* 303, 1507-1511.
Institute of Medicine: 1994, *Defining Primary Care: An Interim Report*, National Academy Press, Washington, DC.
Inui, T.S., Carter, W.B.: 1985, 'Problems and prospects for health services research on provider-patient communication,' *Medical Care* 23, 521-538.
James, D.N.: 1989, 'The friendship model: A reply to Illingworth,' *Bioethics* 3, 42-146.
Kao, A.C., Green, D.C., Zaslavsky, A.M., Koplan, J.P., Cleary, P.D.: 1998, 'The relationship between method of physician payment and patient trust,' *Journal of the American Medical Association* 280(19), 1708-1714.
Kaplan, C., Lipkin, M. Jr, Gordon, G.H.: 1988, 'Somatization in primary care: Patients with unexplained and vexing medical complaints,' *Journal of General Internal Medicine* 3, 177-190.
Kaplan, S.H., Gandek, B., Greenfield, S., Rogers, W., Ware, J.E.: 1995, 'Patient and visit characteristics related to physicians' participatory decision-making style: Results from the Medical Outcomes Study,' *Medical Care* 33, 1176-1781.
Kaplan, S.H., Greenfield, S., Ware, J.E.: (1989): 'Assessing the effects of physician-patient interactions on the outcomes of chronic disease,' *Medical Care* 27, S110-S127.
Kaptchuk, T.J.: 1983, *The Web That Has No Weaver: Understanding Chinese Medicine*, Congdon & Weed, New York.
Katon, W, Kleinman, A.: 1980, 'Doctor-patient negotiation and other social science strategies in patient care,' in L. Eisenberg, A. Kleinman, (eds.) *The Relevance of Social Science for Medicine*, Reidel, Boston.
Kennell, J.H., Klaus, M.H., McGrath, S., et al.: 1991, 'Continuous emotional support during labor in a U.S. hospital,' *Journal of the American Medical Association* 265, 2197-2201.
Kerr, E.A., Hays, R.D., Mittman, B.S., Siu, A.L., Leake, B., Brook, R.H.: 1997, 'Primary care physicians' satisfaction with quality of care in California capitated medical groups,' *Journal of the American Medical Association* 278, 308-312.
Kirmayer, L.J., Robbins, J.M., Dworkind, M., Yaffe, M.J.: 1993, 'Somatization and the recognition of depression and anxiety in primary care,' *American Journal of Psychiatry* 150, 734-741.
Kirmayer, L.J., Robbins, J.M.: 1991, 'Three forms of somatization in primary care: Prevalence, co-occurrence, and sociodemographic characteristics,' *Journal of Nervous Mental Disorders* 179, 647-655.
Kirmayer, L.J.: 1988, 'Mind and body as metaphors: Hidden values in biomedicine,' in M. Lock, D.R. Gordon (eds.) *Biomedicine Examined*, Kluwer Academic Publishers; Boston, pp. 57-93.

Kleinman, A., Eisenberg, L., Good, B.: 1978, 'Culture, illness, and care: Clinical lessons from anthropologic and cross-cultural research,' *Annals of Internal Medicine* 88, 251-258.

Kleinman, A., Kleinman, J.: 1991, 'Suffering and its professional transformation: Toward an ethnography of interpersonal experience,' *Culture, Medicine, and Psychiatry* 15, 275-301.

Kleinman, A.: 1980, *Patients and Healers in the Context of Culture: An Exploration of the Borderland between Anthropology, Medicine, and Psychiatry*, University of California Press, Berkeley.

Kleinman, A.: 1989, *Illness Narratives: Suffering, Healing and the Human Condition*, Basic Books, New York.

Knickman, Jr, Hughes, R.G., Taylor, H., Binns, K., Lyons, M.P.: 1996, 'Tracking consumers' reactions to the changing health care system,' *Health Affairs* 15, 21-32.

Koopman, C., Eisenthal, S., Stoeckle, J.D.: 1984, 'Ethnicity in the reported pain, emotional distress and requests of medical outpatients,' *Social Science Medicine* 18, 487-490.

Kroenke, K., Mangelsdorff, A.D.: 1989, 'Common symptoms in ambulatory care: Incidence, evaluation, therapy, and outcome,' *American Journal of Medicine* 86, 262-266.

Laine, C., Davidoff, F.: 1996, 'Patient-centered medicine: A professional evolution,' *Journal of the American Medical Association* 275,152-156.

Launer, J.: 1999, 'Narrative based medicine: A narrative approach to mental health in general practice,' *British Medical Journal* 318, 117-119.

Leopold, N., Cooper, J., Clancy, C.: 1996, 'Sustained partnership in primary care,' *Journal of Family Practice* 42, 129-137.

Levenstein, J.H.: 1984, 'The patient-centred general practice consultation,' *South African Family Practice* 5, 276-82.

Lipowski, Z.J.: 1988, 'Somatization: The concept and its clinical application,' *American Journal of Psychiatry* 145, 1358-1368.

Lo, B., Quill, T., Tulsky, J.: 1999, 'Discussing palliative care with patients. American College of Physicians-American Society of Internal Medicine End-of-Life Care Consensus Panel,' *Annals of Internal Medicine* 130, 744-749.

Low, S.M.: 1984, 'The cultural basis of health, illness and disease,' *Social Work Health Care* 9, 13-23.

Lum, C.K., Korenman, S.G.: 1994, 'Cultural-sensitivity training in U.S. medical schools,' *Academic Medicine* 69, 239-241.

Lurie, N., Slater, J., McGovern, P., Ekstrum, J., Quam, L., Margolis, K.: 'Preventive care for women: Does the sex of the physician matter?' *New England Journal of Medicine* 329, 478-482.

Lysaught, M.T.: 1992, 'Who is my neighbor?' *Second Opinion* 18 (October), 59-67.

Malterud, K.: 1994, 'Key questions – a strategy for modifying clinical communication. Transforming tacit skills into a clinical method,' *Scandinavian Journal of Primary Health Care* 12, 121-127.

Manson, A.: 1995, 'Why primary care physicians should not be restrictive gatekeepers,' *Journal of General Internal Medicine* 10, 145-146.

Marvel, M.K., Epstein, R.M., Flowers, K., Beckman, H.B.: 1999, 'Soliciting the patient's agenda: Have we improved?' *Journal of the American Medical Association* 281, 283-287.

May, W.: 1983, *The Physician's Covenant: Images of the Healer in Medical Ethics*, Westminster, Philadelphia.

McArthur, J.H., Moore, F.D.: 1997, 'The two cultures and the health care revolution,' *Journal of the American Medical Association* 277, 985-989.

McCullough, L.B.: 1994, 'Should we create a health care system in the United States?' *Journal of Medicine and Philosophy* 19, 483-490.

McWhinney, I.R.: 1972, 'Beyond diagnosis: An approach to the integration of behavioral science and clinical medicine,' *New England Journal of Medicine* 287, 384-387.

Mechanic, D.: 1995, 'Sociological dimensions of illness behavior,' *Social Science Medicine* 41, 1207-1216.

Miles, S.H.; Koepp, R.: 1995, 'Comments on the AMA report "Ethical Issues in Managed Care",' *Journal of Clinical Ethics* 6, 306-311.

Mishler, E.: 1984, *The Discourse of Medicine: Dialectics of Medical Interviews*, Ablex, Norwood, New Jersey.

Montgomery Hunter, K.: 1991, *Doctors' Stories: The Narrative Structure of Medical Knowledge*, Princeton University Press, Princeton.

Montgomery Hunter, K.: 1996, '"Don't think zebras:" Uncertainty, interpretation, and the place of paradox in clinical education,' *Theoretical Medicine* 5, 1-17.

Morse, D.S., Suchman, A.L., Frankel, R.M.: 1997, 'The meaning of symptoms in 10 women with somatization disorder and a history of childhood abuse,' *Archives of Family Medicine* 6, 468-476.

Muecke, M.A.: 1983, 'Caring for Southeast Asian refugee patients in the USA,' *American Journal of Public Health* 73, 431-438.

Novack, D.: 1995, 'Therapeutic aspects of the clinical encounter,' in M. Lipkin Jr, S.M. Putnam, A. Lazare (eds.), *The Medical Interview: Clinical Care, Education, and Research*, Springer-Verlag, New York, pp. 32-49.

Novack, D.H., Suchman, A.L., Clark, W., Epstein, R.M., Najberg, E., Kaplan, C.: 1997, 'Calibrating the physician: Personal awareness and effective patient care,' *Journal of the American Medical Association* 278, 502-509.

Novack, D.H.: 1987, 'Therapeutic aspects of the clinical encounter,' *Journal of General Internal Medicine* 2, 346-355.

Othmer, E.: 1994, *The Clinical Interview Using DSM-IV*, American Psychiatric Press, Washington, DC.

Pachter, L.M.: 1994, 'Culture and clinical care. Folk illness beliefs and behaviors and their implications for health care delivery,' *Journal of the American Medical Association* 271, 690-694.

Pantilat, S.Z., Alpers, A., Wachter, R.M.: 1999, 'A new doctor in the house: Ethical issues in hospitalist systems,' *Journal of the American Medical Association* 282(2), 171-174.

Peabody, F.W. 1927, 'The care of the patient,' *Journal of the American Medical Association* 88, 877-882.

Pellegrino, E.: 1979, *Humanism and the Physician*, University of Tennessee, Knoxville.

Pellegrino, E.D.: 1994, 'First Annual Nicholas J. Pisacano Lecture: Words can hurt you: Some reflections on the metaphors of managed care,' *Journal of the American Board of Family Practitioners* 7, 505-510.

Pellegrino, E.D. and Thomasma, D.C.: 1993, *The Virtues in Medical Practice*, Oxford University Press, New York.

Pendleton, D., Schofield, T., Tate, P., Havelock, P.: 1984, *The consultation: An Approach to Learning and Teaching*, Oxford University Press, Oxford

Pennebaker, J.W.: 1982, *The Psychology of Physical Symptom*, Springer-Verlag, New York.

Potter, J., Wetherell, M.: 1987, *Discourse and Social Psychology*, Sage, London.

Poulton, J., Rylance, G.W., Johnson, M.R.: 1986, 'Medical teaching of the cultural aspects of ethnic minorities: Does it exist?' *Medical Education* 20, 492-497.

Povar, G., Moreno, J.: 1988, 'Hippocrates and the health maintenance organization: A discussion of ethical issues,' *Annals of Internal Medicine* 109, 419-424.

Price, R.: 1994, *A Whole New Life: An Illness and a Healing*, Atheneum, New York.

Proceedings of the National Conference on Cultural Competence and Women's Health Curricula in Medical Education: 1995, U.S. Department of Health and Human Services, Washington, D.C.

Quesada, G.M.: 1976, 'Language and communication barriers for health delivery to a minority group,' *Social Science Medicine* 10, 323-327.

Quill, T.E.: 1987, 'Somatization disorder: One of medicine's blind spots,' *Journal of the American Medical Association* 254, 3075-3079.

Quill, T.E.: 1995, 'Barriers to effective communication,' in M. Lipkin Jr, S.M. Putnam, A. Lazare (eds.), *The Medical Interview: Clinical Care, Education, and Research*, Springer-Verlag, New York, pp. 110-21.

Quill, T.E., Brody, H.: 1996, 'Physician recommendations and patient autonomy: Finding a balance between physician power and patient choice,' *Annals of Internal Medicine* 125, 763-769.

Quill, T.E., Cassel, C.K.: 1995, 'Nonabandonment: A central obligation for physicians,' *Annals of Internal Medicine* 122, 368-74.

Reagan, M.D.: 1987, 'Physicians as gatekeepers,' *New England Journal of Medicine* 317, 1731-1734

Rabinow, P.: 1977, *Reflections on Fieldwork in Morocco*, Berkeley, University of California Press.

Redelmeier, D.A., Molin, J.P., Tibshirani, R.J.: 1995, 'A randomised trial of compassionate care for the homeless in an emergency department,' *Lancet* 345, 1131-1134.

Retchin, S.M., Brown, R.S., Yeh, S.J., Chu, D., Moreno, L.: 1997, 'Outcomes of stroke patients under the Medicare fee for service and managed care,' *Journal of the American Medical Association* 278, 119-124.

Rhodes, R.: 1995, 'Love thy patient: Justice, caring, and the doctor-patient relationship,' *Cambridge Quarterly of Healthcare and Bioethics* 4, 434-447.

Rivo, M.L., Satcher, D.: 1993, 'Improving access to health care through physician workforce reform. Directions for the 21st century,' *Journal of the American Medical Association* 270, 1074-1078.

Robbins, J.M., Kirmayer, L.J., Cathebras, P., Yaffe, M.J., Dworkind, M.: 'Physician characteristics and the recognition of depression and anxiety in primary care,' *Medical Care* 32, 795-812.

Roberts, D.E.: 1996, 'Reconstructing the patient: Starting with women of color,' in S. Wolf (ed.), *Feminism and Bioethics: Beyond Reproduction*, pp. 116-143.

Rosenbaum, E.E.: 1988, *A Taste of My Own Medicine: When the Doctor Is the Patient*, New York, Random House.

Rosenberg, C.E.: 1987, *The Care of Strangers: The Rise of America's Hospital System*,

Basic Books, New York.

Roter, D., Lipkin, M. Jr., Korsgaard, A.: 1991, 'Sex differences in patients' and physicians' communication during primary care medical visits,' *Medical Care* 29, 1083-1093.

Roter, D.L., Hall, J.A., Kern, D.E., Barker, L.R., Cole, K.A., Roca, R.P.: 1995, 'Improving physicians' interviewing skills and reducing patients' emotional distress. A randomized clinical trial,' *Archives of Internal Medicine* 155, 1877-1884.

Rubenstein, H.L., O'Connor, B.B., Nieman, L.Z., Gracely, E.J.: 1992, 'Introducing students to the role of folk and popular health belief-systems in patient care,' *Academic Medicine* 67, 566-568.

Rudebeck, C.E.: 1992, 'General practice and the dialogue of clinical practice: On symptoms, symptom presentations, and bodily empathy,' *Scandinavian Journal Primary Health Care*, supplement 1, 1-87.

Safran, D.G., Tarlov, A.R., Rogers, W.H.: 1994, 'Primary care performance in fee-for-service and prepaid health care systems,' *Journal of the American Medical Association* 271, 1579-1586.

Salkovskis, P.M.: 1992, 'Psychological treatment of noncardiac chest pain: The cognitive approach,' *American Journal of Medicine* 92(5A), 1145-1215.

Schlesinger, M., Dorwart, R.A., Epstein, S.S.: 1996, 'Managed care constraints on psychiatrists' hospital practices,' *American Journal of Psychiatry* 153, 256-260.

Seaburn, D.B.: 1995, 'Language, silence and somatic fixation,' in, S.H. McDaniel (ed.), *Counseling Families With Chronic Illness*, American Counseling Association, New York, pp. 49-66.

Seligman, M.E.: 1995, 'The effectiveness of psychotherapy: The consumer reports study,' *American Psychology* 50, 965-974.

Sharpe, M., Peveler, R., Mayou, R.: 1992, 'The psychological treatment of patients with functional somatic symptoms: A practical guide,' *Journal of Psychosomatic Research* 36, 515-529.

Silverman, D.: 1987, *Communication and Medical Practice: Social Relations and the Clinic*, Sage, Bristol.

Silverman, D.: 1997, *Discourses of Counselling*, Sage, London.

Slonim, M.B.: 1991, *Children, Culture, and Ethnicity: Evaluating and Understanding the Impact*, Garland, New York.

Smith, R.C.; Hoppe, R.B.: 1991, 'The patient's story: Integrating the patient- and physician-centered approaches to interviewing,' *Annals of Internal Medicine* 115, 470-477.

Smith, W.E.: 1948, 'Country doctor,' *Life Magazine*.

Sontag, S.: 1989, *AIDS and Its Metaphors*, Collins Publishers, Toronto, Ontario.

Sorlie, P.D., Backlund, E., Keller, J.B.: 1995, 'US mortality by economic, demographic, and social characteristics: The National Longitudinal Mortality Study,' *American Journal of Public Health* 85, 949-956.

Sosa, R., Kennell, J.H., Robertson, S., *et al.*: 1980, 'The effect of a supportive companion on perinatal problems, length of labor and mother-infant interaction,' *New England Journal of Medicine* 303, 597-600.

Sox, H.C.: 1999, 'The hospitalist model: Perspectives of the patient, the internist, and internal medicine,' *Annals of Internal Medicine* 130(4 Pt 2), 368-372,

Speckens, A.E., van Hemert, A.M., Spinhoven, P., Hawton, K.E., Bolk, J.H., Rooijmans, H.G.: 1995, 'Cognitive behavioural therapy for medically unexplained physical symptoms: A randomised controlled trial,' *British Medical Journal* 311, 1328-1332.

Spector, R.E.: 1992, *Cultural Diversity in Health and Illness*, 3rd edition, Appleton & Lange, Norwalk, Connecticut.

Spiro, H., McCrea Curnen, M.G., Peschel, E., St. James, D., (eds.): 1993, *Empathy and the Practice of Medicine: Beyond Pills and the Scalpel*, Yale University Press, New Haven, Connecticut.

Stanford, A.F. and King, N.M.P.: 1992, 'Patient stories, doctor stories, and true stories: A cautionary reading,' *Literature and Medicine* 11 185-11199.

Starfield, B., Wray, C., Hess, K., Gross, R., Birk, P.S.; D'Lugoff, B.C.: 1981, 'The influence of patient-practitioner agreement on outcome of care,' *American Journal of Public Health* 71, 127-132.

Starfield, B: 1992, *Primary Care: Concept, Education, and Policy*, Oxford University Press, New York.

Stewart, M., Brown, J.B., Donner, A., McWhinney, I.R., Oates, J., Weston, W.W.: 1997, 'Final report: The impact of patient-centered care on patient outcomes in family practice,' Grant 1046, University of Western Ontario, London, Ontario.

Stewart, M., Brown, J.B., Weston, W.W., McWhinney, I.R., McWilliam, C.L., Freeman, T.R.: 1995, *Patient-Centered Medicine: Transforming the Clinical Method*, Sage Publications, Thousand Oaks, California.

Stoeckle, J.D., Barsky, A.J.: 1981, 'Attributions: Uses of social science knowledge in the "doctoring" of primary care,' in L. Eisenberg, A. Kleinman (eds.) *The Relevance of Social Science for Medicine*, D. Reidel Publishing Co, Dordrecht, the Netherlands, pp. 223-240.

Stone, J.: 1981, 'He makes a house call,' in *All This Rain*, Louisiana State University Press, pp. 4-5.

Taussig, M.T.: 1980, 'Reification and the consciousness of the patient,' *Social Science and Medicine* 143, p. 12.

The SUPPORT Principal Investigators: 1995, 'A controlled trial to improve care for seriously ill hospitalized patients: The study to understand prognoses and preferences for outcomes and risks of treatment (SUPPORT),' *Journal of the American Medical Association* 274, 1591-1598.

Todd, K.H., Samaroo, N., Hoffman, J.R.: 1993, 'Ethnicity as a risk factor for inadequate emergency department analgesia,' *Journal of the American Medical Association* 269, 1537-1539.

Tolstoy, L.: 1960, 'The Death of Ivan Ilych,' in *The Death of Ivan Ilych and Other Stories*, A. Maude (trans.), Signet, New York.

Tuckett, D., Boulton, M., Olson, I., Williams, A.: 1985, *Meetings Between Experts: An Approach to Sharing Ideas in Medical Consultations*, Tavistock Publications, London.

Varela, F.J., Thompson, E., Rosch, E.: 1991, *The Embodied Mind: Cognitive Science and Human Experience*, MIT Press, Cambridge.

Veatch, R.M.: 1972, 'Models for ethical medicine in a revolutionary age,' *Hastings Center Report* 2, 5-7.

Ware, J.E., Bayliss, M.S., Rogers, W.H., Kosinski, M., Tarlov, A.R.: 1996, 'Differences in 4-year health outcomes for elderly and poor, chronically ill patients treated in HMO and fee-for-service systems,' *Journal of the American Medical Association* 276, 1039-1047.

Ware, J.E., Rogers, W.H., Ross Davies, A., *et al.*: 1986, 'Comparison of health outcomes at a health maintenance organization with those of fee for service care,' *Lancet* 1, 1017-

1022.

Ware, N.C., Kleinman, A.: 1992, 'Culture and somatic experience: The social course of illness in neurasthenia and chronic fatigue syndrome,' *Psychosomatic Medicine* 54, 546-560.

Wasson, J.H., Sauvigne, A.E., Mogielnicki, R.P., Frey, W.G., Sox, C.H., Gaudette, C., Rockwell, A.: 1984, 'Continuity of outpatient medical care in elderly men: A randomized trial,' *Journal of the American Medical Association* 252, 2413-2417.

Weiss, L.J., Blustein, J.: 1996, 'Faithful patients: The effect of long-term physician-patient relationships on the costs and use of health care by older Americans,' *American Journal of Public Health* 86, 1742-1747.

Weissburg, D.J.: 1993, 'Managed care organizations and confidential patient information: The need for a uniform standard,' *Journal of Health Care Finance* 21, 42-46.

Welch, M.: 1998, 'Required curricula in diversity and cross-cultural medicine: The time is now,' *Journal of the American Medical Womens Association* 53(Suppl), 121-123.

Welch, W.P., Miller, M.E., Welch, H.G., Fisher, E.S., Wennberg, J.E.: 1993, 'Geographic variations in expenditures for physicians' services in the United States,' *New England Journal of Medicine* 328, 621-627.

White, K.L., Williams, T.F., Greenberg, B.G.: 1961, 'The ecology of medical care,' *New England Journal of Medicine* 265, 885-892.

White, L., Tursky, B., Schwartz, G.E. (eds.): 1985, *Placebo: Theory, Research, and Mechanisms*, Guilford Press, New York.

Wodak, R.: 1996, *Disorders of Discourse*, Longman, London.

Wolf, S.M.: 1996, *Feminism and Bioethics: Beyond Reproduction*, Oxford University Press, New York.

Wood, M.L.: 1991, 'Naming the illness: The power of words,' *Family Medicine* 23, 534-538.

Yelin, E.H., Criswell, L.A., Feigenbaum, P.G.: 1996, 'Health care utilization and outcomes among persons with rheumatoid arthritis in fee-for-service and prepaid group practice settings,' *Journal of the American Medical Association* 276, 1048-1053.

Zborowski, M.: 1952, 'Cultural components in response to pain,' *Journal of Social Issues* 8, 16-30.

Zola, I.K.: 1966, 'Culture and symptoms-an analysis of patients' presenting complaints,' *American Sociological Review* 31, 615-630.

Zoloth-Dorfman, L., Rubin, S.: 1995, 'The patient as commodity: Managed care and the question of ethics,' *Journal of Clinical Ethics* 6, 339-357.

NOTES ON CONTRIBUTORS

James J. Bono, Ph.D., is Associate Professor of History and Medicine at the University at Buffalo. He is a Past President of the Society for Literature and Science and an editor of the award-winning journal *Configurations*. He is the author of *The Word of God and the Languages of Man: Interpreting Nature in Early Modern Science and Medicine, Ficino to Descartes* (1995), has been a member of the Institute for Advanced Study at Princeton, and an Eccles Fellow at the Tanner Humanities Center (University of Utah). With the aid of a NSF grant for 1999-2000, he will complete a new book, *Figuring Science: Metaphor, Narrative, and the Cultural Location of Scientific Revolutions* to be published by Stanford University Press.

Howard Brody, MD, Ph.D., is Professor of Family Practice and Philosophy, and Director of the Center for Ethics and Humanities in the Life Sciences, Michigan State University, East Lansing. He is author of *Stories of Sickness* (1987) and *The Healer's Power* (1992).

Scott DeVito, Ph.D., has taught in the Department of History and Philosophy of Science at the University of Pittsburgh and at the Department of Philosophy at Bowling Green State University. He is currently a Visiting Assistant Professor at the University at Buffalo where he teaches Biomedical Ethics. He has published in *The British Journal for the Philosophy of Science, ISIS, The Journal of Medicine and Philosophy, Philosophy of Science,* and *Public Affairs Quarterly.*

H. Tristram Engelhardt, Jr., Ph.D., M.D. is Professor, Department of Medicine, as well as Department of Community Medicine and Obstetrics and Gynecology, and Member, Center for Medical Ethics and Health Policy, Baylor College of Medicine; also Professor, Department of Philosophy, Rice University; and Adjunct Research Fellow, Institute of Religion.

Eric T. Juengst, Ph.D., is Associate Professor of Biomedical Ethics at the Case Western Reserve University School of Medicine. He currently sits

Stephen Wear, James J. Bono, Gerald Logue and Adrianne McEvoy (eds.), Ethical Issues in Health Care on the Frontiers of the Twenty-First Century, 319–322.
© 2000 *Kluwer Academic Publishers. Printed in Great Britain.*

on the Recombinant DNA Advisory Committee for the National Institutes of Health, the National Ethics Committee for the March of Dimes, and the DNA Advisory Board for the Federal Bureau of Investigation. From 1990 to 1994 he was the first Chief of the Ethical, Legal and Social Implications Branch of the National Center for Human Genome Research at the N.I.H.

Gerald Logue, M.D., is Professor, and Vice-Chairman for Education of the Department of Medicine and the University at Buffalo. He is also Co-Director of the Center for Clinical Ethics and Humanities in Health Care and Chief of the Division of Hematology at the University at Buffalo. He formerly was the Chief of Staff at the Buffalo Department of Veterans Affairs Medical Center and served on the National VA Headquarters Bioethics Advisory Committee.

Laurence B. McCullough, Ph.D., is Professor of Medicine and Medical Ethics and Faculty Associate of the Huffington Center on Aging, Baylor College of Medicine, Houston, Texas. He is author of *John Gregory and the Invention of Professional Medical Ethics and the Profession of Medicine* (Kluwer, 1998), editor *of John Gregory's Writings on Medical Ethics and Philosophy of Medicine* (Kluwer, 1998), and co-editor *of Surgical Ethics* (Oxford, 1998).

Adrianne McEvoy, M.A., is a doctoral student in the University at Buffalo Philosophy Department. Her concentration is biomedical ethics, especially the physician-patient relationship and areas of end-of-life decision making. She is a research associate for the Center for Clinical Ethics and Humanities in Health Care at the University of Buffalo.

Kathryn Montgomery, Ph.D., is Professor of Medicine and of Medical Ethics and Humanities at Northwestern University Medical School and Director of its Medical Ethics and Humanities Program. She is the author of *Doctors' Stories: The Narrative Structure of Medical Knowledge* (1991) and the forthcoming *Is Medicine a Science? Rationality in an Uncertain Practice.*

Jonathan Moreno, Ph.D., is Emily Davie and Joseph S. Kornfeld Professor of Biomedical Ethics and Director of the Center for Biomedical Ethics, University of Virginia, Charlottesville, Virginia. He was senior staff to the president's Advisory Committee on Human Radiation Experiments and is a consultant to the National Bioethics Advisory Commission. His books include *Deciding Together: Bioethics and Moral Consensus* (Oxford 1995), and *Undue Risk: Secret State Experiments With Humans from the Second World War to the Gulf War and Beyond* (W.H. Freeman 1999).

E. Haavi Morreim, Ph.D., is a Professor in the College of Medicine, University of Tennessee, in the Department of Human Values and Ethics, and has a joint appointment as Professor in the Department of Preventive Medicine, Division of Health Services and Policy Research. Her writings on the ethical and legal implications of medicine's changing economics have appeared in journals of law, medicine, and ethics. Her book, *Balancing Act: The New Medical Ethics of Medicine's New Economics*, first appeared in 1991 and was republished in paperback by Georgetown University Press in 1995.

John Naughton, M.D., is Professor of Medicine, Physiology, Rehabilitation Medicine and Social and Preventive Medicine and Former Dean, School of Medicine and Biomedical Sciences, and former Vice President for Clinical Affairs at the University at Buffalo.

Dorothy Nelkin, Ph.D., holds a University Professorship at New York University, teaching in the Department of Sociology and the School of Law. She is a member of the National Academy of Sciences' Institute of Medicine, and a Fellow and former director of the American Association for the Advancement of Science. She has been a Guggenheim Fellow, and a former president of the Society for the Social Studies of Sciences.

Diane B. Paul, Ph.D., is Professor of Political Science and Co-director of the Program in Science, Technology, and Public Policy at the University of Massachusetts at Boston. She has held fellowships at MIT, the Wissenschaftskolleg zu Berlin, and the Humanities Research Institute of the University of California. Her most recent books are *Controlling Human Heredity: 1865 to the Present* (1995) and *The Politics of Heredity: Essays on Eugenics, Biomedicine, and the Nature-Nurture Debate* (1998).

Julie Rothstein Rosenbaum, M.D., is the Chief Resident in the Department of Medicine at New York University-Downtown Hospital. A graduate of Yale University School of Medicine, Dr. Rosenbaum completed her residency in Internal Medicine at the New York Hospital-Cornell Medical Center. She majored in Biomedical Ethics at Brown University and also worked as a research assistant at the Hastings Center.

Stephen Wear, Ph.D. is Associate Professor, Department of Medicine with adjunct appointments in Gynecology-Obstetrics and Philosophy and is Co-Director of the Center for Clinical Ethics and Humanities in Health Care at the University at Buffalo. The second edition of his book *Informed Consent: Patient Autonomy and Clinician Beneficence within Health Care* was recently published by Georgetown University Press.

INDEX

abortion 2, 3, 11, 31
  as treatment 175, 181-182, 184
AIDs/HIV 73, 139, 181-183, 209
American Medical Association 40,
  53, 174, 222
assisted suicide 3, 11, 31
autonomy 5, 36, 83, 184, 209, 221,
  253, 306

Beauchamp, Tom 5-7, 20
beneficence 5, 58-60, 306
Brody, Howard 199, 259, 266-269

cancer 66, 73, 75, 78, 129, 131,
  134, 137, 158, 176, 191, 195
BRCA1 or BRCA2 2, 159, 160,
  176
casuistry 5, 7, 8, 21
Catholicism 7, 25, 31
Childress, James 5-7, 20
Clinton administration 21, 27
confidentiality 45, 136, 162, 215
conflicts of interests 36-37, 39, 51,
  53-69, 222, 225, 236
continuity of care 199, 221-223,
  234-237, 241-242, 244, 249-
  250, 252, 259, 262-266, 268-
  269
contract law regarding medical
  allocation 84-92
cost-consciousness 12, 22, 27-28,
  65-72, 87-93, 173, 194, 241,
  242-245, 259, 266-267
Cook-Deegan, Robert 127-128, 147

cystic fibrosis 141, 160-161, 193-
  194

disclosure 52-54
disease 42, 48-49, 58, 60, 66, 117
  120, 127-150, 167, 171-186,
  191-196, 261
diversity 8-13, 19, 22, 25-33, 50,
  181
  cultural 8-13, 25-33, 50
  economic 22, 29-32
  moral 8-9, 11-13, 22, 25, 32 33,
  216, 260

Emanuel, Linda and Ezekiel 207,
  297
Engelhardt, H. Tristram 1-13, 19
  24, 181, 259-261, 263, 266,
  268-269, 271-272
Enlightenment project 2, 4, 19
eugenics 118-119, 122, 136, 138,
  141, 165, 171-173, 180-183,
  185
euthanasia 2-3, 31

false consciousness 22, 26-29
Food and Drug Administration 65
  66, 68, 72
fee-for-service 27, 36, 54-55, 59,
  66, 243, 252

gag orders 22, 51-54
genetics 12, 116-122, 127-150,
  155-168, 171-186, 275, 294

323

Philosophy and Medicine

1. H. Tristram Engelhardt, Jr. and S.F. Spicker (eds.): *Evaluation and Explanation in the Biomedical Sciences.* 1975   ISBN 90-277-0553-4
2. S.F. Spicker and H. Tristram Engelhardt, Jr. (eds.): *Philosophical Dimensions of the Neuro-Medical Sciences.* 1976   ISBN 90-277-0672-7
3. S.F. Spicker and H. Tristram Engelhardt, Jr. (eds.): *Philosophical Medical Ethics.* Its Nature and Significance. 1977   ISBN 90-277-0772-3
4. H. Tristram Engelhardt, Jr. and S.F. Spicker (eds.): *Mental Health.* Philosophical Perspectives. 1978   ISBN 90-277-0828-2
5. B.A. Brody and H. Tristram Engelhardt, Jr. (eds.): *Mental Illness.* Law and Public Policy. 1980   ISBN 90-277-1057-0
6. H. Tristram Engelhardt, Jr., S.F. Spicker and B. Towers (eds.): *Clinical Judgment.* A Critical Appraisal. 1979   ISBN 90-277-0952-1
7. S.F. Spicker (ed.): *Organism, Medicine, and Metaphysics.* Essays in Honor of Hans Jonas on His 75th Birthday. 1978   ISBN 90-277-0823-1
8. E.E. Shelp (ed.): *Justice and Health Care.* 1981
                    ISBN 90-277-1207-7; Pb 90-277-1251-4
9. S.F. Spicker, J.M. Healey, Jr. and H. Tristram Engelhardt, Jr. (eds.): *The Law-Medicine Relation.* A Philosophical Exploration. 1981   ISBN 90-277-1217-4
10. W.B. Bondeson, H. Tristram Engelhardt, Jr., S.F. Spicker and J.M. White, Jr. (eds.): *New Knowledge in the Biomedical Sciences.* Some Moral Implications of Its Acquisition, Possession, and Use. 1982   ISBN 90-277-1319-7
11. E.E. Shelp (ed.): *Beneficence and Health Care.* 1982   ISBN 90-277-1377-4
12. G.J. Agich (ed.): *Responsibility in Health Care.* 1982   ISBN 90-277-1417-7
13. W.B. Bondeson, H. Tristram Engelhardt, Jr., S.F. Spicker and D.H. Winship: *Abortion and the Status of the Fetus.* 2nd printing, 1984   ISBN 90-277-1493-2
14. E.E. Shelp (ed.): *The Clinical Encounter.* The Moral Fabric of the Patient-Physician Relationship. 1983   ISBN 90-277-1593-9
15. L. Kopelman and J.C. Moskop (eds.): *Ethics and Mental Retardation.* 1984
                    ISBN 90-277-1630-7
16. L. Nordenfelt and B.I.B. Lindahl (eds.): *Health, Disease, and Causal Explanations in Medicine.* 1984   ISBN 90-277-1660-9
17. E.E. Shelp (ed.): *Virtue and Medicine.* Explorations in the Character of Medicine. 1985   ISBN 90-277-1808-3
18. P. Carrick: *Medical Ethics in Antiquity.* Philosophical Perspectives on Abortion and Euthanasia. 1985   ISBN 90-277-1825-3; Pb 90-277-1915-2
19. J.C. Moskop and L. Kopelman (eds.): *Ethics and Critical Care Medicine.* 1985
                    ISBN 90-277-1820-2
20. E.E. Shelp (ed.): *Theology and Bioethics.* Exploring the Foundations and Frontiers. 1985   ISBN 90-277-1857-1

Philosophy and Medicine

21. G.J. Agich and C.E. Begley (eds.): *The Price of Health.* 1986
    ISBN 90-277-2285-4
22. E.E. Shelp (ed.): *Sexuality and Medicine.* Vol. I: Conceptual Roots. 1987
    ISBN 90-277-2290-0; Pb 90-277-2386-9
23. E.E. Shelp (ed.): *Sexuality and Medicine.* Vol. II: Ethical Viewpoints in Transition.
    1987   ISBN 1-55608-013-1; Pb 1-55608-016-6
24. R.C. McMillan, H. Tristram Engelhardt, Jr., and S.F. Spicker (eds.): *Euthanasia
    and the Newborn.* Conflicts Regarding Saving Lives. 1987
    ISBN 90-277-2299-4; Pb 1-55608-039-5
25. S.F. Spicker, S.R. Ingman and I.R. Lawson (eds.): *Ethical Dimensions of Geriatric
    Care.* Value Conflicts for the 21th Century. 1987   ISBN 1-55608-027-1
26. L. Nordenfelt: *On the Nature of Health.* An Action-Theoretic Approach. 2nd,
    rev. ed. 1995   SBN 0-7923-3369-1; Pb 0-7923-3470-1
27. S.F. Spicker, W.B. Bondeson and H. Tristram Engelhardt, Jr. (eds.): *The Contra-
    ceptive Ethos.* Reproductive Rights and Responsibilities. 1987
    ISBN 1-55608-035-2
28. S.F. Spicker, I. Alon, A. de Vries and H. Tristram Engelhardt, Jr. (eds.): *The Use
    of Human Beings in Research.* With Special Reference to Clinical Trials. 1988
    ISBN 1-55608-043-3
29. N.M.P. King, L.R. Churchill and A.W. Cross (eds.): *The Physician as Captain of
    the Ship.* A Critical Reappraisal. 1988   ISBN 1-55608-044-1
30. H.-M. Sass and R.U. Massey (eds.): *Health Care Systems.* Moral Conflicts in
    European and American Public Policy. 1988   ISBN 1-55608-045-X
31. R.M. Zaner (ed.): *Death: Beyond Whole-Brain Criteria.* 1988
    ISBN 1-55608-053-0
32. B.A. Brody (ed.): *Moral Theory and Moral Judgments in Medical Ethics.* 1988
    ISBN 1-55608-060-3
33. L.M. Kopelman and J.C. Moskop (eds.): *Children and Health Care.* Moral and
    Social Issues. 1989   ISBN 1-55608-078-6
34. E.D. Pellegrino, J.P. Langan and J. Collins Harvey (eds.): *Catholic Perspectives
    on Medical Morals.* Foundational Issues. 1989   ISBN 1-55608-083-2
35. B.A. Brody (ed.): *Suicide and Euthanasia.* Historical and Contemporary Themes.
    1989   ISBN 0-7923-0106-4
36. H.A.M.J. ten Have, G.K. Kimsma and S.F. Spicker (eds.): *The Growth of Medical
    Knowledge.* 1990   ISBN 0-7923-0736-4
37. I. Löwy (ed.): *The Polish School of Philosophy of Medicine.* From Tytus
    Chałubiński (1820–1889) to Ludwik Fleck (1896–1961). 1990
    ISBN 0-7923-0958-8
38. T.J. Bole III and W.B. Bondeson: *Rights to Health Care.* 1991
    ISBN 0-7923-1137-X

Philosophy and Medicine

Philosophy and Medicine

55. E. Agius and S. Busuttil (eds.): *Germ-Line Intervention and our Responsibilities to Future Generations.* 1998   ISBN 0-7923-4828-1
56. L.B. McCullough: *John Gregory and the Invention of Professional Medical Ethics and the Professional Medical Ethics and the Profession of Medicine.* 1998
                                                                ISBN 0-7923-4917-2
57. L.B. McCullough: *John Gregory's Writing on Medical Ethics and Philosophy of Medicine.* 1998 [CiME-1]   ISBN 0-7923-5000-6
58. H.A.M.J. ten Have and H.-M. Sass (eds.): *Consensus Formation in Healthcare Ethics.* 1998 [ESiP-2]   ISBN 0-7923-4944-X
59. H.A.M.J. ten Have and J.V.M. Welie (eds.): *Ownership of the Human Body.* Philosophical Considerations on the Use of the Human Body and its Parts in Healthcare. 1998 [ESiP-3]   ISBN 0-7923-5150-9
60. M.J. Cherry (ed.): *Persons and Their Bodies.* Rights, Responsibilities, Relationships. 1999   ISBN 0-7923-5701-9
61. R. Fan (ed.): *Confucian Bioethics.* 1999 [APSiB-1]   ISBN 0-7923-5853-8
62. L.M. Kopelman (ed.): *Building Bioethics.* Conversations with Clouser and Friends on Medical Ethics. 1999   ISBN 0-7923-5853-8
63. W.E. Stempsey: *Disease and Diagnosis.* 2000   PB ISBN 0-7923-6322-1
64. H.T. Engelhardt (ed.): *The Philosophy of Medicine.* Framing the Field. 2000
                                                                ISBN 0-7923-6223-3
65. S. Wear, J.J. Bono, G. Logue and A. McEvoy (eds.): *Ethical Issues in Health Care on the Frontiers of the Twenty-First Century.* 2000   ISBN 0-7923-6277-2

KLUWER ACADEMIC PUBLISHERS – DORDRECHT / BOSTON / LONDON